MECHANICS OF ELASTIC AND INELASTIC SOLIDS
Editors: S. Nemat-Nasser and G. Æ. Oravas

G. M. L. Gladwell, Contact problems in the classical theory of elasticity. 1980. ISBN 90-286-0440-4.

G. Wempner, Mechanics of solids with applications to thin bodies. 1982. ISBN 90-286-0880-X.

T. Mura, Micromechanics of defects in solids. Second, revised edition, 1987. ISBN 90-247-3343-X (HB), and ISBN 90-247-3256-5 (PB).

R. G. Payton, Elastic wave propagation in transversely isotropic media. 1983. ISBN 90-247-2843-6.

S. Nemat-Nasser, H. Abé, S. Hirakawa, Hydrawlic fracturing and geothermal energy. 1983. ISBN 90-247-2855-X.

S. Nemat-Nasser, R. J. Asaro, G. A. Hegemier (eds.), Theoretical foundation for large-scale computations for non-linear material behavior. 1984. ISBN 90-247-3092-9.

N. Cristescu, Rock rheology. 1989. ISBN 90-247-3660-9.

Rock Rheology

N. Cristescu

Department of Mechanics
University of Bucharest
Romania

Kluwer Academic Publishers
Dordrecht / Boston / London

Library of Congress Cataloging in Publication Data CIP

Cristescu, N.
Rock rheology.

(Mechanics of elastic and inelastic solids ; 7)
Bibliography: p.
Includes indexes.
1. Rock deformation. 2. Rocks--Creep.
3. Rheology. I. Title. II. Series.
QE604.C75 1988 622'.2 87-34781
ISBN 90-247-3660-9

Published by Kluwer Academic Publishers,
P.O. Box 17, 3300 AA Dordrecht, The Netherlands.

Kluwer Academic Publishers incorporates the publishing programmes of
D. Reidel, Martinus Nijhoff, Dr W. Junk, and MTP Press.

Sold and distributed in the U.S.A. and Canada
by Kluwer Academic Publishers,
101 Philip Drive, Norwell, MA 02061, U.S.A.

In all other countries, sold and distributed
by Kluwer Academic Publishers Group,
P.O. Box 322, 3300 AH Dordrecht, The Netherlands.

Printed in The Netherlands

Contents

Preface

The present book is the result of work carried out over a period of about ten years by the author and his co-workers in order to describe more accurately the slow irreversible deformation in time of the rocks surrounding underground openings. To begin with, our efforts were directed toward a better understanding of the mechanical behaviour of rocks and to the formulation of more precise mathematical models for their dominant mechanical properties, mainly irreversible dilatancy and/or compressibility during creep. Subsequent efforts were focused on finding improved solutions to important mining and oil engineering problems, such as, for instance, the creep of rocks around wells and tunnels, short-term failure which may occur around an underground opening, damage and failure which take place after long-time intervals, the tunnel support analysis incorporating rock creep, etc.

The book is the result of a great number of questions posed either by mining engineers or by the author himself, and of the corresponding answers (unfortunately often only partial answers). This dialogue must certainly be continued in order to improve the models and to formulate models for other kinds of rocks, or, ultimately to obtain solutions for other important engineering problems. It is hoped that the book will also contribute to a better description, by means of mathematical models, of the mechanical behaviour of rocks. The examples given in the book are illustrative only, pending more accurate and complete experimental data and improved experimental techniques; the aim has been to stress suggested procedures rather than specific models.

For some years the present monograph has been used as a textbook for a course on rock mechanics taught by the author: that is why a few exercises have been added at the end of each chapter. These are meant to help the understanding of the main text. The author's objective is to teach the students to think for themselves, rather than to accumulate a certain body of knowledge. Thus they are encouraged to perform diagnostic tests on rocks, to select the dominant mechanical properties among those shown, and then to use these data in order to formulate for themselves a constitutive equation appropriate for the kind of rock considered and for the practical problem they are subsequently expected to solve, and finally to learn to use simple and clear logic when tackling a mining engineering problem. In all cases the procedures and models given in the book are only a starting point which may suggest the finding of more appropriate procedures and models for the specific rock or problem under consideration.

The references given at the end of the book are in no way exhaustive; they are based mainly on the papers and books kindly sent to the author by corresponding authors, editors, or publishers. The author expresses his gratitude to all of them.

The author is indebted to Prof. S. Nemat-Nasser, the editor of the series, for editing the present text. He is also grateful to his colleagues I. Suliciu and D. Massier for solving various related problems; these are included in the present text and in most cases are published exclusively here. Discussions and extended collaboration with the mining engineer D. Fotă have been stimulating. He and E. Medveş have cooperated in the analysis of rock—support interaction (Chapter 13). Besides these two mining engineers, many other scientists and engineers from the Geomechanics Laboratory in Bucharest have also collaborated, mainly in the collection of experimental data. As ever, Cornelia Cristescu participated in all stages of the preparation of the manuscript, in the hard task of analysing the experimental data, and in helping with numerical computations.

Bucharest N. CRISTESCU

1

Introduction

The present volume is devoted to topics in rock mechanics, mainly to rock rheology. Generally, rock mechanics is considered here to be a branch of the mechanics of continua, dealing with the study of the response of rocks when they are subjected to various kinds of loadings and possibly simultaneously to the influence of other factors such as temperature, time, humidity, chemical agents, etc. It is far from easy to give a comprehensive definition of the concept of what rock is; in what follows this concept is considered to be a primary one, known by everyone from their personal everyday experience. Rock mechanics has many important applications, mainly in mining and petroleum engineering, geophysics, seismology, civil engineering, geothermal energy extraction, radioactive waste isolation, etc.

Rocks are a mixture of various solid mineral or organic grains bound together by binding agents (of various kinds). Typical for rocks is the presence of a great number of pores and microcracks which during loading may close, open, or multiply. Any mathematical model formulated for most rocks must take into account the mechanisms of microcracks opening and/or closing, of slip along microcracks, and of the growth of microcracks.

The pores existing in rocks may sometimes be filled, more or less, with a liquid (water, crude oil, etc.) and/or with a gas (air, natural gas, etc.). Thus the deformation of the rock is also certainly influenced by the properties (sometimes changing) and the motion of these filling substances.

It is obvious that rocks are generally non-homogeneous and anisotropic. That is why the properties established by laboratory tests must be checked again by *in situ* measurements. Sometimes the non-homogeneity or anisotropy of rocks can be disregarded, but at other times it cannot. In the present monograph isotropic properties will generally be considered, since it is thought that as a first approximation such an approach is reasonable in many cases (Panet (1969)).

There are various criteria of rock classification. Most are of great practical importance, but of less significance from the point of view followed here. We must certainly consider hard and soft rocks, competent rocks (those in which an underground opening does not need a support to stand) and incompetent ones.

These concepts are obviously relative, since at a great depth, for instance, all rocks are incompetent anyway.

The history of rock mechanics seems to start with the paper by Coulomb (1773), in which he established a failure criterion for rocks. In an improved version, this criterion is widely used even today. Besides this failure condition, or some other similar one formulated in terms of stresses, a linear elastic constitutive equation has further been accepted for rocks. Thus a complete set of equations, i.e. a mathematical model, has become available for the solution of various engineering problems. Many such solved problems can be found, for instance, in the books by Obert and Duvall (1967), Jumikis (1979), Baklashov and Kartoziya (1975), Galperin and Shafarenko (1977), Jaeger and Cook (1979), Goodman (1980), Hoek and Brown (1980), Coates (1981), Bieniawski (1984), and others. This classical point of view, i.e. the assumption of a linearly elastic constitutive equation with which a failure condition formulated in terms of stresses is associated, will not be adopted in the present monograph. For classical solutions and approaches (from linear elasticity) we refer the reader to one of the excellent books mentioned above. For convenience, however, some results are mentioned briefly in Chapter 5. Sometimes, when the classical solution cannot be found in a complete form, or when more specific details are necessary, additional information will be given in subsequent chapters.

In the present monograph, the main objectives are to describe the creep of rocks around underground openings, typical properties of rocks such as dilatancy or compressibility, failure during creep, etc. For this purpose inelastic constitutive equations must be formulated, and these are much more involved than Hooke's law. One possible procedure is to try to adapt to rocks some of the models that already exist for metals. This procedure may be tricky, however, since there are fundamental differences between the mechanism of deformation and damage of metals and that of rocks.

For instance, for metals the so-called elastic 'constants' are really constants; although this statement is not absolutely true even for metals, in many cases it can be considered reasonable and nowadays it is a quite generally accepted assumption. With rocks, it is not very clear how these elastic 'constants' might be precisely defined. There are several methods of determining them (see Chapters 2 and 6). Thus the values of the elastic 'constants', which are not really 'constant', depend on the method used to determine them and even on the worker who is carrying out the experiment. An exception (again within certain limits) makes the dynamic methods of determining these constants (see Chapter 6) of significance for seismology and for the point of view adopted in the present monograph. As will be seen later, the elastic parameters are instrumental not only for the definition of elastic properties of rocks but also for an accurate definition of their irreversible deformation.

Furthermore, for metals, at room temperature and for stresses below the yield stress, the stress—strain curves are not influenced at all by a change of the rate of deformation; above the yield stress, these curves are influenced by the rate of deformation if it is changed by several orders of magnitude. With rocks, however, it is quite different: the stress—strain curves are highly loading-rate-dependent, starting from the smallest applied stresses. On the other hand, rocks creep even for quite small stresses, and irreversible deformations become evident also for quite small applied stresses. It follows, therefore, that in general the yield stress for rocks must be considered to be practically zero for all stress levels of significance in mining or petroleum engineering problems. Thus rock mechanics becomes *rock rheology*.

Another significant difference between rock mechanics and metal mechanics consists in the way in which the reference configuration is chosen in order to describe the deformation of the two kinds of materials. For metals, as a rule, one chooses as reference configuration the stress-free, strain-free configuration: for rocks, such a configuration does not exist. For rocks the actual *in situ* configuration (before any excavation) is the natural configuration with respect to which any further deformation must be related. The rock specimens we receive in the laboratory are in the so-called relaxed state: a stress-free but not a strain-free state. The genuine natural reference configuration for rock-like materials would rather be the one in which all microcracks are closed (when the rock is subjected to a certain hydrostatic pressure σ_0) since this configuration would correspond to the minimal existing damage in the rock structure (see Chapter 9).

While, for metals, inelasticity can be explained in principle by means of the mechanics and physics of dislocation nucleation and propagation, for most rocks it is mainly the mechanisms of closure and/or opening of microcracks (and sometimes of pores) and their multiplication which explain most of the major inelastic properties; dilatancy, compressibility, creep, failure, etc. Dislocation mechanisms also have a role in individual crystal deformation.

The rock rheology approach used in the present book is constructive in the following sense: first the experimental data needed for the formulation of a constitutive equation are presented, then the procedure required for explicit formulation of such a constitutive equation is described in detail, and finally it is shown how the formulated elastic/viscoplastic constitutive equation can be used to solve some specific problems.

In Chapters 2 and 3 the diagnostic tests necessary to recognize the main mechanical properties of the rock under consideration are presented. Chapter 4 presents typical properties which most rocks possess, e.g. dilatancy and compressibility of volume. It is shown how mathematical models can be formulated to describe such phenomena and how these models are to be included in a general formulation of the constitutive equation.

Further, some classical results which are used in the next few chapters are mentioned: the formulation of some problems of linear elasticity with special emphasis on plane strain elasticity in polar coordinates and on the deformation of a thick-walled tube subjected to internal and external pressures. At the end of Chapter 5 the linear viscoelastic model is introduced, with thermodynamic restrictions and the inequalities to be satisfied by the constitutive parameters, etc. This constitutive equation, which will actually be used for rocks, can describe volumetric creep producing compressibility only. Because of its simplicity many solutions of practical problems can be found by using such a constitutive equation, although it is almost impossible to determine the constitutive coefficients which would satisfy all the required restrictions.

Since the elastic parameters play a special role in the formulation of inelastic constitutive equations, their determination and possible dependence on the stress state are discussed in Chapter 6.

After the introductory chapters described above, the next two chapters are devoted to the formulation of constitutive equations. The experimental data used for this purpose and the procedure to be followed are shown. Detailed examples are given. These constitutive equations can describe either dilatancy or compressibility, depending on the properties of the particular rock and on the stress state. Both elastic/plastic (Chapter 7) and elastic/viscoplastic (Chapter 8) constitutive equations are given. It is important to stress that for the formulation of the constitutive equation only a few general assumptions are postulated. The main characteristics of the constitutive equations come from the experimental data. Chapter 9 is devoted to the formulation of a model to describe damage and failure of rocks, using an energy criterion included in the constitutive equation.

The final four chapters are devoted to various applications. Chapter 10 considers the stress states existing *in situ* before excavation, and how this primary state is introduced as an initial datum when various constitutive equations (elastic, elastic/plastic, viscoelastic, elastic/viscoplastic) are used. Further, creep dilatancy/compressibility of rocks around vertical shafts and oil wells is examined in Chapter 11, considering the rock to be either viscoelastic or elastic/viscoplastic. Possible damage or failure is also taken into account. Similar problems, this time for horizontal tunnels, are considered in Chapter 12. The last chapter is devoted to rock—support interaction and to a theory of tunnel lining, with special emphasis on yieldable supports.

Appendix 1 is an extension of Chapter 5, showing how the growth of a crack existing in an elastic body depends on the orientation of the applied stresses. Appendix 2 gives additional details of the numerical methods used to obtain some of the solutions discussed in Chapter 11 and 12.

Units

Stress: $1\ \mathrm{N\,m^{-2}} = 1\ \mathrm{Pa} = 1.020 \times 10^{-5}\ \mathrm{kgf\,cm^{-2}} = 9.869 \times 10^{-6}\ \mathrm{atm}$
$= 1 \times 10^{-5}\ \mathrm{bar} = 1.450 \times 10^{-4}\ \mathrm{psi} = 1\ \mathrm{J\,m^{-3}}$

Viscosity: $1\ \mathrm{poise} = 1\ \mathrm{P} = 10^{-1}\ \mathrm{N\,s\,m^{-2}}$

2

Mechanical properties of rocks

2.1. Diagnostic tests

The mechanical properties of rocks are studied in order to understand the way in which they may deform and fail, the final objective being to predict their behaviour under known loading conditions.

The mechanical properties of rocks are quite diverse, and the scatter of experimental data is generally much more significant than for other kinds of materials.

It is important to recall that laboratory experiments reproduce only partially the conditions under which rocks deform *in situ*. The dimensions of specimens used in the laboratory are quite small compared to those involved *in situ*. Thus the influence of faults, joints, bedding, and other planes of weakness cannot generally be studied in the laboratory. Even the drilling of specimens disturbs the rock from its original state *in situ*. Larger specimens may be more likely to represent more realistically the properties existing *in situ*. For a discussion of size effects see Obert and Duvall (1967), Pratt *et al.* (1972) and Brace (1981).

Several diagnostic tests are carried out in the laboratory in order to establish those mechanical properties which are dominant in certain circumstances and which have to be included in the data necessary for the formulation of a mathematical model of rock deformation, i.e. of a constitutive equation. Some diagnostic tests common in most laboratories are mentioned below.

(*a*) *Standard uniaxial compression tests* generally last only a few minutes. The deformation and failure of the rock specimen under uniaxial compressive stress are observed. Thus we can get some tentative information concerning the elastic and viscoplastic properties of the rock, the dilatancy or compressibility of the volume, the damage and failure, etc.

(*b*) '*Triaxial*' *tests* provide the same information. This time, besides the axial loading, a lateral hydrostatic (confining) pressure is also applied to the rock specimen. Thus all properties mentioned in (*a*) are obtained for various superimposed hydrostatic pressures. In 'true' triaxial tests all three principal stresses can be varied independently and the corresponding strains can be measured.

(*c*) In *creep tests* a constant or nearly constant uniaxial or triaxial load is applied for a long period of time (weeks, months, or even years), while the strain variation in time is recorded.

(*d*) Experiments of *hydrostatic* (or nearly hydrostatic) *volume compressibility* are designed to reveal the laws of volume compressibility. These are done at high pressures.

(*e*) *Dynamic tests* record: the velocities of various waves propagating in rock, the attenuation of these waves during deformation, the reflection from free surfaces or cracks, the dynamic tensile strength, etc. The dynamic tests reveal the dynamic values of the elastic constants (see Chapter 6).

Rocks are *triphasic* bodies, i.e. they are composed of a solid skeleton (of mass M_s and volume V_s) and of pores and microcracks which may possibly be filled partially with a *gas* — in most cases air — (of volume V_a) and/or with liquids — water, oil, etc. — (of mass M_w and volume V_w). What distinguishes rock from other continuous bodies is the presence of *pores* and/or *microcracks* which result in specific mechanical properties. The *porosity* n is defined as

$$n = \frac{V_v}{V} = \frac{\text{volume of voids}}{\text{total volume}} \tag{2.1.1}$$

sometimes given as a percentage.

The *humidity* or *water content*, i.e. the ratio between the weight of the water contained in the voids and the weight of the dry rock itself, plays a significant role in the mechanical response of the rock. The influence of humidity on the mechanical properties of rocks has been discussed by various authors (see for instance Dreyer (1973, p. 99), and Vutukuri *et al.* (1974, pp. 50—60)).

For further details concerning the way in which rock specimens are to be tested in the laboratory, specimen sizes, etc., see for instance, Stamatiu (1962), Obert and Duvall (1967), Jumikis (1979), Coates (1981), ISRM (1979, pp. 135—156), Vutukuri *et al.* (1974, Chapters 1 and 2).

The main conclusions resulting from typical diagnostic tests are presented briefly below, without technical details. The aspects which are significant in constructing a mathematical model for a rock-like material will be emphasized.

2.2. Unconfined uniaxial compressive tests

These are the simplest and most commonly used tests performed on cylindrical specimens of diameter d and height ℓ (where $\ell > d$, always).

The cylindrical specimen is subjected to an axial force F only. If A is the area of the cross-section, and if during the test this area does not vary too much, then the axial stress is obtained from

$$\sigma = \frac{F}{A}\,. \tag{2.2.1}$$

This concept of stress is therefore used in the sense of Piola—Kirchhoff. If the area of the cross-section is varied significantly during the compression test, then

F divided by the actual cross-sectional area is the stress in the sense of Cauchy. With hard rocks this area does not vary too much. By convention, *compressive stresses are assumed to be positive*. The lateral surface of the specimen remains stress-free.

During the tests both longitudinal strain $\varepsilon_1 = (\ell_0 - \ell)/\ell_0$ and diametric strain $\varepsilon_2 = (d_0 - d)/d_0$ are recorded. Here ℓ_0 and ℓ are the initial and current lengths of the specimen, and d_0 and d the initial and current diameters. Since $\ell_0 > \ell$, it follows that $\varepsilon_1 > 0$, and since $d_0 < d$, $\varepsilon_2 < 0$.

The previous assumption, that during compression testing a single axial stress is involved, is certainly too simple. In fact, due to the friction between the machine platens and the end surface of the specimen, a significant friction force occurs: formula (2.2.1) would be correct if this friction were missing. This consideration has been discussed by various authors (see, for instance, Al-Chalabi and Huang (1974) and Gustkiewicz (1975), who consider also a possible bending of the specimen). In the discussion which follows the simplified uniaxial stress-state will be accepted.

For most rocks the *loading rate* (or strain rate) may significantly influence the compression results. In most laboratories, either the loading rate is disregarded or a standard constant loading rate is accepted (of 0.5—1.0 MPa s^{-1} according to the ISRM suggestions, or 1.63 kPa s^{-1} according to Dreyer (1973), etc.). As will be shown below the mechanical properties of rocks are strongly rate-dependent, and the loading rate is therefore a major parameter to be considered. As an example, Fig. 2.2.1 shows the stress—strain curves for green *schist* (from Palazu Mare) obtained with two loading rates. The experiments started from a very small preloading stress of about 0.98 MPa. Several loading—unloading loops are also shown. Figure 2.2.1 yields several important conclusions which are typical for most rocks (see also Walsh and Brace (1966),

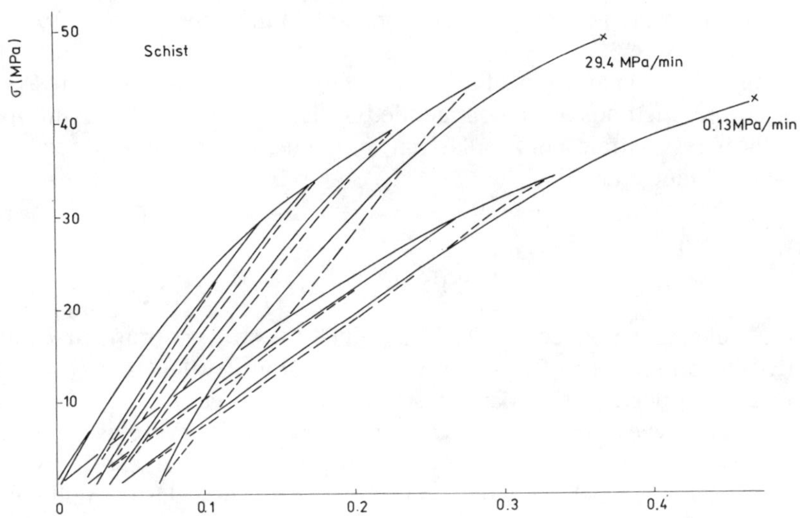

Figure 2.2.1. Stress—strain curves for schist during loading and unloading obtained with two loading rates.

Morgenstern and Tamuly Phukan (1969), Hansen *et al.* (1984), Tijani *et al.* (1983), Stavrogin and Protosenya (1985), etc.):

— the stress—strain curves are generally non-linear;

— the stress—strain curves are strongly loading-rate-dependent, starting from the smallest stresses;

— unloading reveals an irreversible (plastic) part of the strain;

— during the loading—unloading cycles significant hysteresis loops are evident;

— the failure stress is also loading-rate-dependent: an increase of the loading rate increases the stress at failure, but the strain at failure decreases (failure is shown by a cross on the curves in Fig. 2.2.1): see Chapter 9.

For various rocks a variety of concavities of the stress—strain curves is possible. Sometimes these stress—strain curves (for some limestones for instance) are nearly straight lines; but again, for distinct loading rates these straight lines are distinct, starting from the smallest stresses (see Chapter 9).

A conventional yield point cannot be distinguished on the stress—strain curves since irreversible strains occur even after small applied stresses, and because from the very beginning the stress—strain curve is rate-dependent. For these reasons, and taking into account also the experimental arguments given in subsequent chapters, it will be assumed that the *yield stress of most rocks is practically zero* (if very small stresses of no practical significance in mining applications are disregarded).

The *work-hardening* exhibited by most rocks when the stress is increased is mainly due to the closing of microcracks, combined with a slip with friction along some other microcracks. The significant hysteresis loop during the loading—unloading process can also be explained by the mechanism of the opening (with some retardation) of some microcracks during unloading, and reclosing during loading. The permanent closing of some microcracks as well as the opening of some new microcracks can explain the appearance of *irreversible strain*.

For the determination of the *elastic parameters* from the stress—strain curves, several methods are recommended in the literature. For a point of stress σ_1^* on the stress—strain curve a 'secant' modulus, a 'tangent' one, or a 'mean' one can be obtained from

$$E_s = \left.\frac{\Delta\sigma_1}{\Delta\varepsilon_1}\right|_{\sigma_1^*}, \qquad E_t = \left.\frac{d\sigma_1}{d\varepsilon_1}\right|_{\sigma_1^*}, \qquad E_{med} = \frac{\Delta\sigma_1}{\Delta\varepsilon_1} \tag{2.2.2}$$

with the meaning explained in Fig. 2.2.2. Since the stress—strain curves are not unique, the formulae (2.2.2) yield a great diversity of values for E. Generally, only $(2.2.2)_3$ applied to that part of the stress—strain curve where concavity is changing (and where most of the existing microcracks are closed while new microcracks are not yet open) may possibly furnish a value of E somehow asymptotically approaching what can be termed the 'elastic modulus' (see Chapter 6). The definition $(2.2.2)_1$ is sometimes used for practical reasons, when the non-linear and non-unique stress—strain curve is replaced by a unique

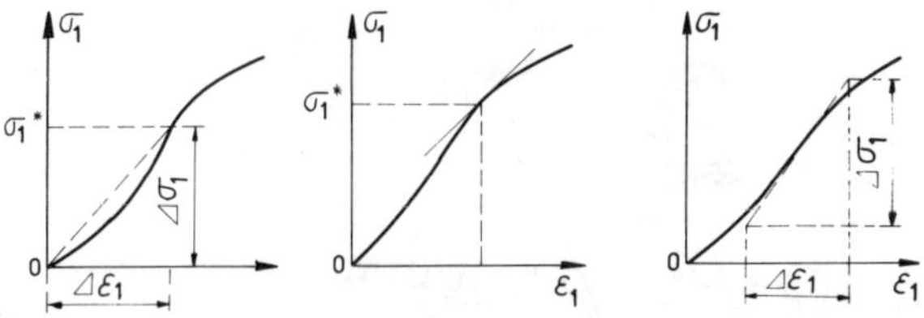

Figure 2.2.2. Notation used in formulae (2.2.2) for the determination of Young's modulus.

linear relationship in order to allow the use of simple formulae from linear elasticity theory.

If the lateral (diameter) strain ε_2 is also recorded, then from

$$\nu = -\frac{\varepsilon_2}{\varepsilon_1} \tag{2.2.3}$$

a certain kind of Poisson ratio can be determined. One can determine it directly from

$$\nu = -\frac{\text{slope of the axial stress—axial strain curve}}{\text{slope of the axial stress—diameter strain curve}}$$

by estimating these slopes according to a definition from (2.2.2). Thus the Poisson ratio is not uniquely determined and is generally of little use in determining a reasonably accurate constitutive equation for rocks.

Generally, the 'elastic' constants can also be determined during the unloading process following a loading process done at a slow loading rate. If the first portions of the unloading curves are used, the values thus determined are approaching what can be termed the 'elastic constants'. Formulae from elasticity,

$$K = \frac{E}{3(1-2\nu)}, \qquad G = \frac{E}{2(1+\nu)} \tag{2.2.4}$$

can further be used to determine the bulk modulus and the shear modulus. The dynamic procedures for determining the elastic constants (see Chapter 6) are also useful for the formulation of the mathematical model of the rock. Generally, *the last two are the only proper methods of determining the elastic constants.* The elastic parameters have also been obtained by back analysis of loading and unloading tests on large coal specimens (600 × 600 × 200 mm) before and after excavation of a tunnel (Kaiser and Maloney (1982)).

The elastic parameters for common rocks, as obtained by various authors with the methods just described, are reported, for instance, by Jumikis (1979), Lama and Vutukuri (1978), and Obert and Duvall (1967). Generally, higher values for the moduli are obtained with the dynamic methods. Typically, data points are widely scattered for most rocks. On the other hand, a particular value for a certain constant is not a characteristic of the type of rock considered. Thus the mechanical properties which can be expressed by numerical values of the elastic parameters cannot be a characteristic of a certain kind of rock.

The *temperature* has a complex influence on the mechanical properties of

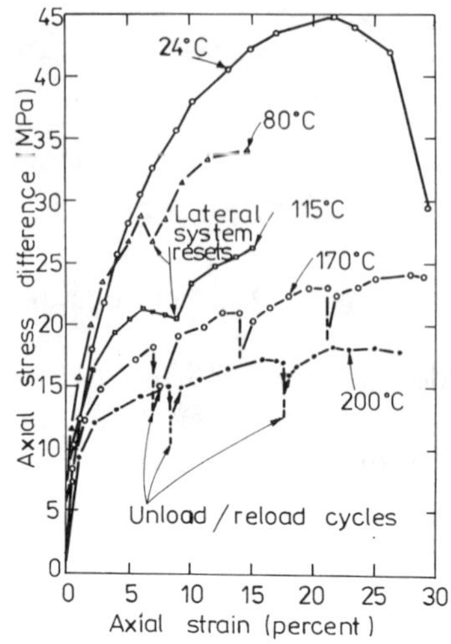

Figure 2.2.3. Typical triaxial compressive test results at various temperatures for salt (after Hansen and Carter (1980)).

rocks (see, for instance, Walsh (1973), Wong and Brace (1979), Hansen and Carter (1980), Hansen and Mellegard (1980), Heard and Page (1982), Heuze (1983), Tijani *et al.* (1983), and the references in these papers). As an example, Fig. 2.2.3 (after Hansen and Carter (1980)) shows the plot of differential (or relative — see §8.2) axial stress as a function of axial strain for triaxial (see §2.4) quasi-static compression experiments on Avery Island *salt*, at a constant confining pressure of 3.45 MPa, for temperatures of 24, 80, 115, 170, and 200°C. The axial stress is increased at a rate of 1.38 MPa min^{-1}. The authors point out that all quasi-static triaxial compression experiments deform qualitatively in a similar manner; *near-zero yield stress* followed by *strong work-hardening* at room temperature, decreasing towards a steady-state flow as the temperature approaches 200°C.

Generally, with an increase of temperature the work-hardening modulus decreases significantly; also, the elastic modulus decreases, but if a hydrostatic pressure is also superimposed, this decrease is smaller. The coefficient of linear expansion increases with the increase of temperature but decreases with the increase of pressure. The porosity similarly increases with temperature but decreases with pressure.

2.3. Other mechanical tests

'*Direct*' *tensile test* experiments are significant, since the tensile strength of the rock is an important parameter. As is well known, the tensile strength of most rocks is much smaller than the compressive strength. Since the tensile strength is involved in many applications (computations of the strength of underground structures, possible failure due to the reflection of compressive waves from free surfaces, etc.), it should be determined as exactly as possible. However, since 'direct' tensile tests are difficult to perform (see Obert and Duvall (1967), Vutukuri *et al.* (1974, Chapter 3)), in most laboratories *indirect* (or Brazilian) tests are replacing them. In these tests a circular disk is compressed by two concentrated forces at the two ends of a diameter. From the stress state which results, if linear elasticity is assumed, the tensile strength can be obtained from

$$\sigma_x = -\frac{F}{\pi t R}, \tag{2.3.1}$$

where F is the magnitude of the applied force at failure, t the thickness of the disk, and R its radius. It must be mentioned that besides a tensile stress (2.3.1), a compressive one

$$\sigma_y = \frac{F}{\pi t R}\,\frac{3R^2 + y^2}{R^2 - y^2} \tag{2.3.2}$$

is also present, as well as some local effects (where the platens of the testing machine impart pressure on the rock). This method, initially proposed by Carneiro (1947) was further developed by Wijk (1978), Sundaram and Corrales (1980), Loureiro-Pinto (1979), and others. See also Vutukuri *et al.* (1974, §3.3.3).

Another widely used test is the *shear strength test* in which the shear takes place along a predetermined shear plane, and a normal stress σ is applied normal to this plane (Fig. 2.3.1). If τ is the shear stress along the plane AB (in a global rough estimation), then at failure a linear relationship is found in the form

$$\tau = C + \sigma \tan \varphi \tag{2.3.3}$$

which is called *Coulomb's shear strength line* (Fig. 2.3.2). Here φ is the *angle of friction* and C the *cohesion*. After failure when $C = 0$, $\tan \varphi = f$ is the friction

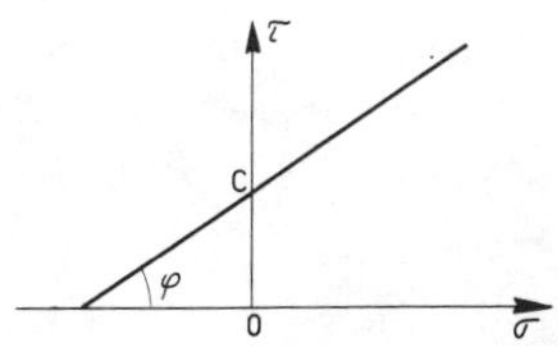

Figure 2.3.1. Shear test of a prismatic specimen.

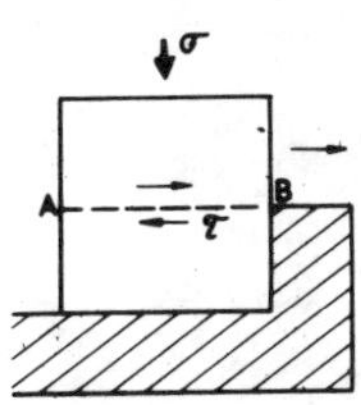

Figure 2.3.2. Coulomb's shear strength line.

coefficient between the two parts of the broken rock specimen. The values of φ, $\tan\varphi$ and the shear strength $\tau_f = C$ for various rocks according to several authors are reported by Jumikis (1979), Vutukuri *et al.* (1974), and others.

The friction between two rock blocks has been studied by many authors (see Byerlee (1975, 1978)). It was shown by Crawford and Curran (1981) that for harder rocks the frictional resistance decreases with increasing shear displacement rates greater than a variable critical velocity; conversely, for softer rocks, the resistance was observed to increase with increasing shear velocity, up to a critical shear displacement rate, and thereafter to remain essentially constant. The frictional properties are influenced by temperature and moisture (Shimamoto (1986)).

2.4. Triaxial tests

While all the tests previously described have only a limited value in the formulation of a constitutive equation, in triaxial tests the rock specimen can be subjected to any desired stress state, so that the loading conditions encountered around underground openings can be reproduced quite accurately in the laboratory. There are two kinds of triaxial test. In most laboratories, for 'triaxial' tests cylindrical rock specimens are loaded with an axial stress σ_1 and a lateral pressure $\sigma_2 = \sigma_3$; both can be varied independently, but always $\sigma_2 = \sigma_3$. Seldom can all three principal components of the stress be varied independently (Mogi (1971, 1979), Hunsche (1984*a*), Demiris (1985), Mier (1986) and Mier and Borst (1986)).

The first kind of 'triaxial' test is now described briefly (see Obert and Duvall (1967)). A more detailed analysis is due to Brady (1971), Wawersik (1975), and others: some technical recommendations concerning the best way to do the experiment can be found in the ISRM suggestions (1978).

Consider a cylindrical specimen, subjected to an axial force F and a lateral pressure p (Fig. 2.4.1). If A is the initial cross-section we have, with some approximation,

$$\sigma_1 = \frac{F}{A}, \qquad \sigma_2 = \sigma_3 = p. \tag{2.4.1}$$

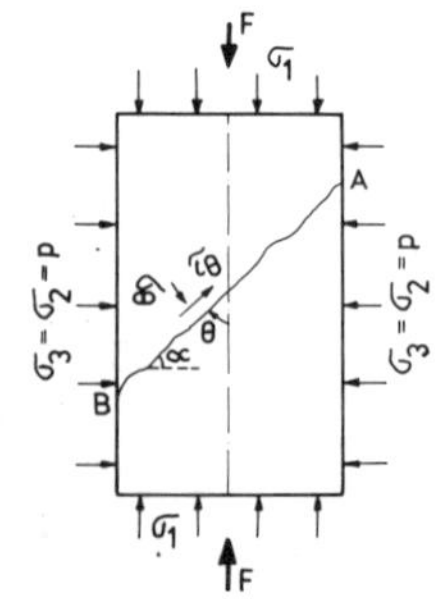

Figure 2.4.1. Cylindrical specimen subjected to a 'triaxial' test; notation used.

If the experiment is performed up to failure, let us denote by AB the plane of failure; then at failure the normal and tangential components on this plane are

$$\sigma_\theta = \frac{\sigma_1 + \sigma_3}{2} + \frac{\sigma_1 - \sigma_3}{2} \cos 2\alpha$$
$$\tau_\theta = \pm \frac{\sigma_1 - \sigma_3}{2} \sin 2\alpha. \tag{2.4.2}$$

These are the coordinates of point P on Mohr's circle:

$$\left(\sigma_\theta - \frac{\sigma_1 + \sigma_3}{2}\right)^2 + \tau_\theta^2 = \left(\frac{\sigma_1 - \sigma_3}{2}\right)^2 \tag{2.4.3}$$

(see Fig. 2.4.2). If the angle θ is measured after the experiment, then the coordinates of point P are known and so is the *angle of internal friction* $\varphi = \frac{1}{2}\pi - 2\theta$. For various lateral pressures σ_3 corresponds an axial stress σ_1 and an angle θ at failure. Thus we obtain another Mohr's circle and another point P on it which corresponds to failure, leading to the envelope $\sigma_\theta = f(\tau_\theta)$ of Mohr's circles which characterizes failure (broken line on Fig. 2.4.2). This relationship is sometimes written in principal components $\sigma_1 = g(\sigma_3)$. Sometimes this relationship is linear, as in (2.3.3).

In order to give an example of the results obtained in triaxial tests, Fig. 2.4.3 presents the stress—strain curves for Jefferson Island *salt*, obtained for the various confining pressures shown, for a load rate of 0.002 MPa s^{-1} at room temperature. The figure reveals that both ultimate strength and ductility increase with confining pressure and that the *yield strength is essentially zero*. The *unloading/reloading cycles are remarkably linear and time-independent*: the specimens neither recover nor creep when held at a constant stress of less than about 90% of the previous maximum. The elastic parameters at room temperature are determined from the slope of the unloading/reloading curves; the average values, nearly site-independent (from ten different sites in the United States) are $E = 30$ GPa and $\nu = 0.35$. Most of these properties are obtained for other rocks as well. For instance, for Westerly *granite* Schock and Duba (1972) found that the loading and unloading curves are strongly non-linear with a pronounced hysteresis and permanent volume changes. The hysteresis is

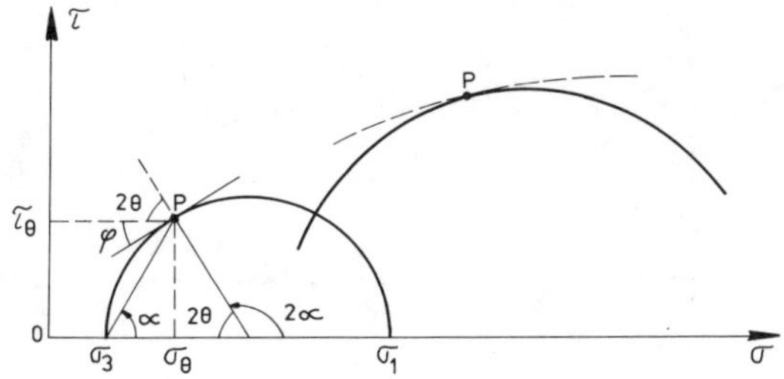

Figure 2.4.2. Mohr's circles.

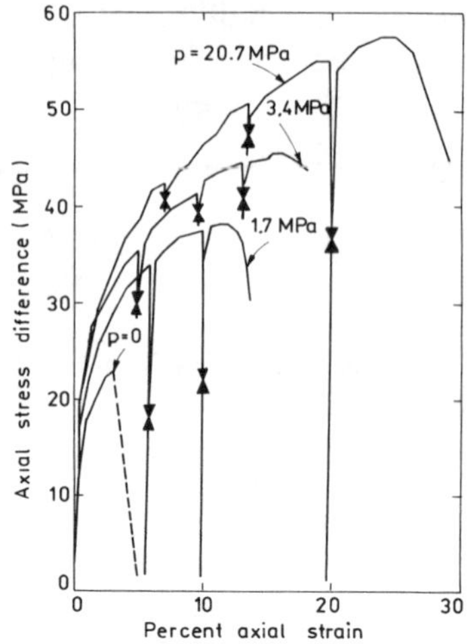

Figure 2.4.3. Axial stress difference as a function of percent axial strain for salt (after Hansen *et al.* (1984)).

absent at high confining pressures. At 0.2 GPa confining pressure the axial stress—strain curve becomes more linear and steeper, and shows little or no hysteresis. For Solenhofen *limestone*, however, linear stress—strain relationships are found with no hysteresis at atmospheric confining pressure.

The humidity influences the strength of the rock. Generally, an increase in humidity slightly decreases the compressive strength.

In order to illustrate the results which can be obtained in true triaxial tests, Fig. 2.4.4 shows the data of failure experiments on cubic samples of *rock salt* from a West German salt dome. First the mean stress is increased and then,

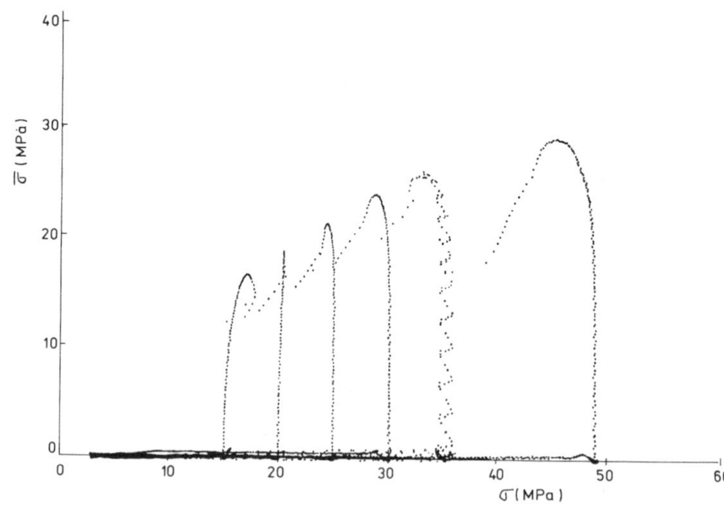

Figure 2.4.4. Stress trajectories in a $\bar{\sigma}$ 0 σ plane, with hydrostatic loadings followed by deviatoric loadings; the maxima point failure. (By courtesy of U. Hunsche.)

with σ = const., the deviatoric stress is increased (τ_0 is the octahedral shear stress $\sqrt{3}\ \tau_0 = \bar{\sigma}$, with $\bar{\sigma}$ the equivalent stress defined by (4.1.1)). The time interval between dots is 1 s. The six curves shown are obtained for 15 MPa $\leqslant \sigma \leqslant$ 48.4 MPa. Failure occurs at peak points. The stresses are corrected to true stresses. The results shown in Fig. 2.4.4 are obtained for the value $m = -1$ of the Lode parameter

$$m = \frac{2\sigma_2 - \sigma_1 - \sigma_3}{\sigma_1 - \sigma_3} .$$

In Fig. 2.4.5 (after Hunsche (1984*a*)) shows the results of the failure-strength experiments on Gorleben *rock salt*, plotted in the octahedral plane as function of m and σ. The curves shown are obtained for various constant values of σ. It is obvious that for relatively small values of σ these curves depend more significantly on m, i.e. on the third invariant of the stress tensor (see Exercise 8 of Chapter 4) besides the dependency on $\bar{\sigma}$ (the distance from the centre).

True triaxial tests reveal all the necessary information for the formulation of a triaxial constitutive equation, since besides the three stress components, all three strain components (see Fig. 2.4.6) are recorded, as well as the time elapsed between two successive readings. Even standard triaxial tests have a great importance for rock mechanics, since during tests both axial ε_1 and diametral strain ε_2 can be measured. Thus both axial stress—axial strain and axial stress—diametral strain relationships can be obtained for various laterally applied pressures. Thus we obtain significant information which will be used in subsequent chapters in order to formulate a constitutive equation. The first such stress—strain curves obtained at various lateral pressures, for Carrara *marble*, seem to be due to Kármán (1911). Today such tests are done in all rock mechanics laboratories.

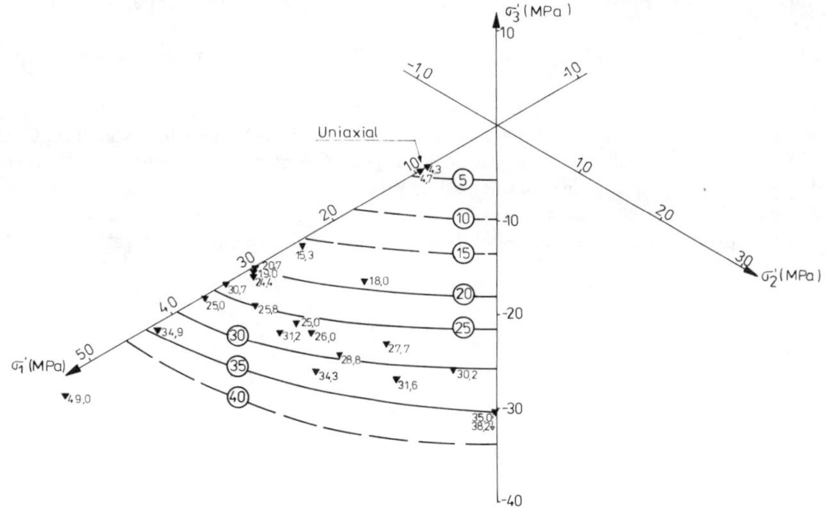

Figure 2.4.5. Failure—strength curves for Gorleben rock-salt, plotted in the octahedral plane, as function of m and σ (shown in circles).

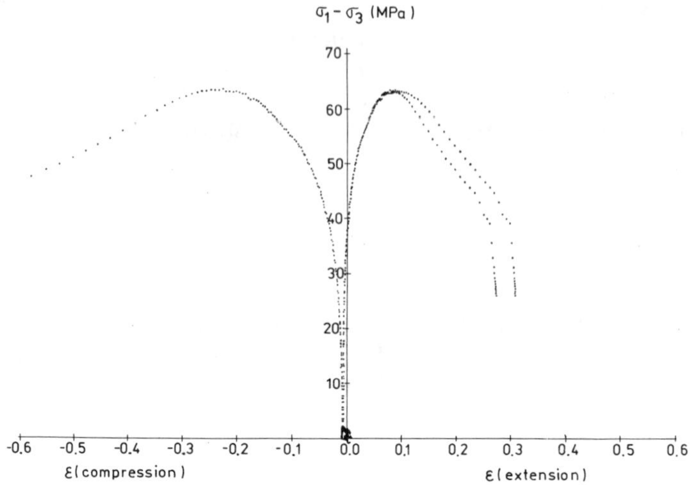

Figure 2.4.6. Stress—strain curves obtained for rock-salt (true stress—true strain curves) for $\sigma =$ 46 MPa and $m = -1$. (By courtesy of U. Hunsche.)

Exercises

1. Carry out uniaxial compression tests on a rock of your choice. First use a very low loading rate. Then increase the loading rate several times by orders of magnitude. Find out if the whole stress—strain curve is influenced by the loading rate. At small stresses, is there an elastic part of this curve independent of the loading rate?
2. Make a uniaxial compression test in which several unloadings are produced starting from various stress levels. Check how significant the irreversible parts of the strains are. Does the slope of the unloading curve vary with the level of maximum stress? Is this slope constant during a single unloading process?
3. Do several compression tests with various loading rates. Determine how the loading rate influences the stress and strain at failure.
4. Determine the Young's modulus during unloading tests using various definitions (2.2.2), and various stress intervals $\Delta\sigma_1$. Compare the results obtained for various intervals $\Delta\sigma_1$. Compare the results with those obtained by dynamic tests (see Chapter 6). Determine if there is any dependency of the unloading slopes on the loading rate.
5. Repeat all the above suggested tests, this time as triaxial tests, by successively increasing the lateral pressure. Compare the stress—strain curves obtained with various lateral pressures.

3

Creep of rocks

3.1. History of creep tests

Creep tests on rocks performed in the laboratory are very significant in mining and in improved design of underground structures, in order to ensure safety, to increase the amount of extracted ore, etc. The simplest creep tests are the unconfined ones, when the lateral surface of the rock specimen is stress-free. Confined creep tests performed in a 'triaxial' apparatus where the lateral surface of the specimen is subjected to a constant pressure are seldom carried out. Sometimes the creep tests are done at elevated temperatures or, on the contrary, at lowered temperatures. The devices used to study the creep of rocks are described by several authors (see, for instance, Lama and Vutukuri (1978, Vol. III), Wawersik and Preece (1984), Kartashov (1973, §8), Hardy *et al.* (1984), etc.). Generally, the devices used in creep tests are now highly sophisticated (Wawersik (1975)).

Many survey papers present the experimental results obtained for various kinds of rocks as well as several attempts to match the experimental data with various simple rheological models (Langer (1979, 1984), Carter and Kirby (1978), Carter *et al.* (1981), Krieg (1982), Carter and Hansen (1983), Horseman and Passaris (1984), Wawersik (1985, 1986), and Wawersik and Zeuch (1986)). The creep tests last a long time, generally several weeks or months. Tests lasting several years have also been reported (Kumagai and Itô (1970), Sasajima and Itô (1980), Itô (1983), Mirza (1984), and Russell and Lomenick (1984)).

In this chapter we present mainly uniaxial creep tests obtained in our laboratory. However, in the last section the triaxial tests will also be discussed, since without such tests it is not possible to formulate a general triaxial constitutive equation. Some of the data given below may also be found in Cristescu (1979) and Cristescu and Suliciu (1982, Chapter IX).

3.2. Uniaxial creep

Various types of apparatus have been used for the study of unconfined uniaxial creep of rocks. In some of them, the axial loading is obtained by mechanical devices; in others, by hydraulic or pneumatic devices. In all cases the initial dimensions of the specimens have been diameter $d = 5$ cm and height $h = 10$ cm. (The specimen size and history effect on creep of *salt* are discussed by Senseny (1984).) Descriptions of the apparatus used by various authors are given by Lama and Vutukuri (1978, §9.5), Kartashov (1973, §8), and others.

Generally in such tests an axial stress is applied to the specimen and is then kept constant for a long time interval. During this interval the variations of the axial strain ε_1 and of the diameter strain ε_2 are recorded. When after a certain time (several weeks or even months) the strains no longer increase (transient creep), the loading stress is increased again. This stepwise loading is continued until at a certain loading level the creep continues by secondary and tertiary creep, which is followed by failure. Typical axial strain-time creep curves for schist are shown in Fig. 3.2.1. The final points on these curves, marked by asterisks, correspond to failure. There is a limit stress with the following property: only stresses below this limit stress will produce a primary creep that ends at a final stable stress—strain state. By definition, all the points of equilibrium in the stress—strain plane, which are obtained at the end of creep tests done with various stress levels, belong to the '*creep stabilization boundary*' or '*equilibrium boundary*'. Stresses above this limit stress ultimately produce secondary and tertiary creep, finally ending in failure. This limit stress can be linked to a certain degree to the compressive strength of the rock. If we denote by σ_c the compressive short-term strength obtained with a conventionally chosen standard loading rate, then this limit stress ranges approximately between 0.6 σ_c and 0.7 σ_c, and is only occasionally slightly lower than this. Thus the loading level during creep tests is expressed by the ratio $\Delta = \sigma/\sigma_c$, called the *relative loading level*. We recall here that σ_c is not uniquely determined, since it depends on the loading rate and loading history (see Chapter 9); however, in conjunction with the definition of the relative loading level, σ_c is considered as a material constant obtained through a conventionally chosen compressive test.

The curves in Fig. 3.2.1 show at each reloading an initial 'instantaneous'

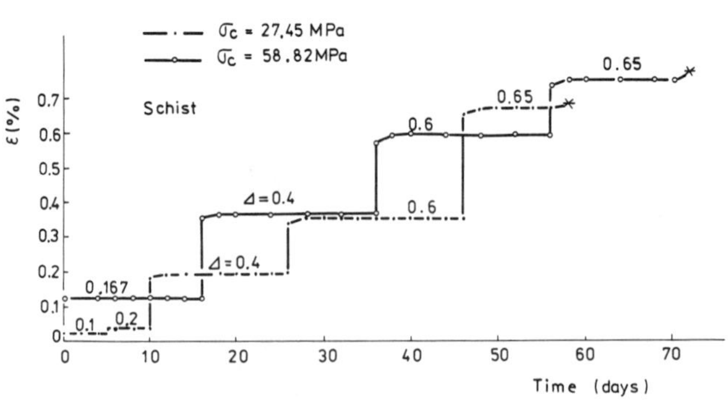

Figure 3.2.1. Creep curves for two kinds of schist, differing by the compressive strength.

increase of the strain, followed by an irreversible one, obtained after a long-time interval. However, if a sudden unloading is produced, another 'instantaneous' decrease of the strain is evident, followed by a retarded reversible one; a final irreversible deformation of the specimen is also observed for any previously applied loading level. It is interesting to note that the 'instantaneous' strain obtained during loading is always larger than the 'instantaneous' strain observed during unloading. The reason is that in conventional creep devices, the loading is applied only for a period of several tens of seconds or even minutes, a time interval during which the strain certainly increases by 'short-term' creep. That is the reason why we have studied, in specially devised tests, the short-term creep of all rock tested, i.e. creep taking place within a few minutes after reloading of the specimen.

3.3. Mathematical model

A unidimensional mathematical model able to describe the main features of deformation by creep exhibited by most rocks must be defined in a domain D in the stress—strain plane, bounded by three curves, all obtained by experiment (see Fig. 3.3.1 for *schist*, where three stress—strain curves obtained with various loading rates are also shown). Curve C_c is the *creep stabilization boundary*; the points on this curve are points of maximal strain which can be obtained in creep tests of relatively small or moderate loadings. Let $\sigma_1 = a$ be the maximal stress on this curve, or a horizontal asymptote of it. Further, the curve C_d is the curve

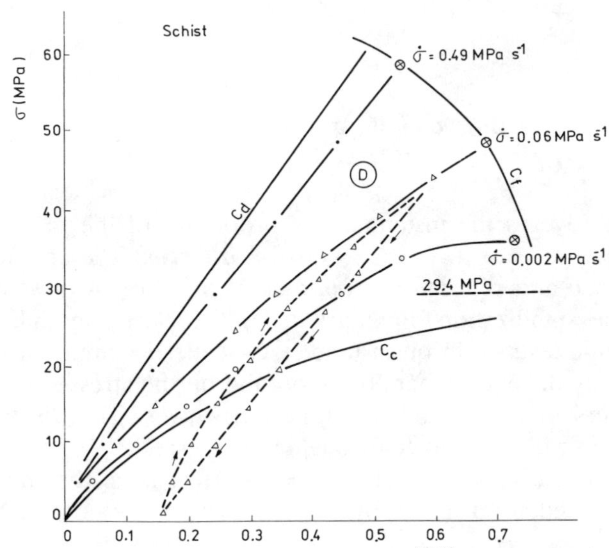

Figure 3.3.1. Constitutive domain for schist. C_c is the creep stabilization boundary, C_f the boundary where failure takes place, and C_d the instantaneous response curve. Also shown are three stress—strain curves obtained with three loading rates.

of the instantaneous response of the rock, i.e. the rock response obtained in the fastest possible test starting from $\sigma_1 = 0$, $\varepsilon_1 = 0$. Finally, C_f is the boundary where failure takes place; this curve is not uniquely determined (see Chapter 9), but in the present chapter C_f will be the locus of stress and strain states where failure takes place in creep tests done with a single loading step.

The stabilization boundary introduced above is not uniquely determined either, since it depends on the loading history and the loading rate. This aspect will be discussed in later chapters (see Chapter 8).

The three curves previously mentioned bound the domain D of all stress–strain states which can be obtained by creep tests. We observe that for stress-states in the neighbourhood of $\sigma = a$ (in the present section we use the simplified notation $\sigma_1 = \sigma$) the peculiarities of curves C_c and C_f are difficult to determine by experiment since very long time intervals would be necessary. Thus it is not known if these two curves have, say, a common horizontal asymptote. Similar constitutive domains have been found for *cement* by Rusch (1960) using strain rates ranging between $2 \times 10^{-6}\,\text{s}^{-1}$ and $8 \times 10^{-10}\,\text{s}^{-1}$ and for *halitic salt* by Dreyer (1973). See also the experimental data for *rock salt* by Wawersik and Hannum (1980) and for *granite* by Genevois and Prestininzi (1979), and Schmidtke and Lajtai (1985).

Next we present the uniaxial model (Cristescu (1979)); the general triaxial model will be presented in subsequent chapters.

Consider a general form of the constitutive equation (initially proposed for metals by Cristescu (1963))

$$\dot{\varepsilon} = \left(\frac{1}{E} + F(\varepsilon, \sigma, \operatorname{sgn}\dot{\sigma})\right)\dot{\sigma} + \Psi(\varepsilon, \sigma) \tag{3.3.1}$$

with

$$F(\varepsilon, \sigma, \operatorname{sgn}\dot{\sigma}) = \begin{cases} \Phi(\varepsilon, \sigma) & \text{if } \dot{\sigma} \geqslant 0 \\ 0 & \text{if } \dot{\sigma} < 0, \end{cases} \tag{3.3.2}$$

where $\Phi(\varepsilon, \sigma)$ governs the instantaneous properties of the rock and is a non-negative function of class C^1 on $\varepsilon \geqslant 0$, $0 \leqslant \sigma \leqslant E\,\varepsilon$. The coefficient $\Psi(\varepsilon, \sigma)$ governs the creep and relaxation properties of the rock, with $\Psi(\varepsilon, \sigma) = 0$ the equation of the stabilization boundary. E is the elastic modulus determined either in dynamic tests or in quasistatic tests at the beginning of the unloading processes. E may depend (generally, smoothly) on the stress-state as well. Dots above the letters in (3.3.1) indicate time derivatives. The displacements and material rotations will be assumed throughout to be small.

The 'instantaneous' response of the rock starting from the state $\varepsilon = \sigma = 0$ can be approximated, for instance, by

$$\varepsilon = \alpha\left(\frac{\sigma}{E_0}\right)^{\beta}, \tag{3.3.3}$$

with $\alpha > 0$ and $\beta > 0$ the material dimensionless constants of the rock, while

E_0 is a well-defined value of E, for instance the dynamic value. Thus the function Φ can be defined, for instance, by

$$\Phi(\sigma) = \frac{\alpha\beta}{E_0}\left(\frac{\sigma}{E_0}\right)^{\beta-1} - \frac{1}{E} \tag{3.3.4}$$

for any stress state in D.

Concerning the coefficient $\Psi(\varepsilon, \sigma)$, if the equation of the stabilization boundary is $\sigma = \sigma_\infty(\varepsilon)$, we assume for Ψ the property

$$\begin{aligned} &\Psi(\varepsilon, \sigma) > 0 \quad \text{if } \sigma > \sigma_\infty(\varepsilon) \\ &\Psi(\varepsilon, \sigma) = 0 \quad \text{if } \sigma \leqslant \sigma_\infty(\varepsilon). \end{aligned} \tag{3.3.5}$$

Since for stresses above the limit $\sigma = a$ the deformation by creep proceeds to failure, the equation of the curve C_c was chosen in the form

$$\sigma = a[1 - \exp(-b\varepsilon)], \tag{3.3.6}$$

with $b > 0$ a dimensionless constant. Sometimes it is useful to express the coefficient a as

$$a = \Delta_s\, \sigma_c, \tag{3.3.7}$$

where Δ_s is the maximal value of the relative loading level still producing stabilization of the deformation by creep.

Generally the stabilization boundary will be written as

$$\sigma = \sigma_c f(\varepsilon) \tag{3.3.8}$$

with a possible linear form

$$\frac{\sigma}{\sigma_c} = h\varepsilon, \tag{3.3.9}$$

where h is a constant.

Thus the coefficient $\Psi(\varepsilon, \sigma)$ involved in the constitutive equation can be chosen as

$$\Psi(\varepsilon, \sigma) = \begin{cases} \dfrac{k(\sigma)}{E^m}\{\sigma - \sigma_c\,\Delta_s[1-\exp(-b\varepsilon)]\}^m & \text{if } \sigma > \sigma_c\,\Delta_s[1-\exp(-b\varepsilon)] \\ 0 \quad \text{if } \sigma \leqslant \sigma_c\,\Delta_s[1-\exp(-b\varepsilon)] \end{cases} \tag{3.3.10}$$

with $m > 0$ a constant, or in the form

$$\Psi(\varepsilon, \sigma) = \begin{cases} \dfrac{k(\sigma)}{E}[\sigma - \sigma_c\, h\varepsilon] & \text{if } \sigma > \sigma_c\, h\varepsilon \\ 0 & \text{if } \sigma \leqslant \sigma_c\, h\varepsilon. \end{cases} \tag{3.3.11}$$

Various other variants are mentioned by Cristescu and Suliciu (1982).

In the above formulae, $E/k(\sigma)$ is a *viscosity coefficient* of the rock in uniaxial creep tests; it can also be called the coefficient of apparent viscosity. Generally, it may be expected that $k(\sigma)$ is variable; for simplicity, a dependence on the stress alone will be assumed. The viscosity coefficient can be determined in creep tests when $\sigma =$ const. at various time intervals. For instance, let us denote by σ_0 the value of the constant applied stress, by ε_i the strain at the moment t_i which is considered the 'initial' moment for the creep period, and by ε_f the 'final' strain at time t_f. The time interval (t_i, t_f) can be chosen in an arbitrary manner and, in particular, it can be the whole period of creep which takes place for a certain constant stress level. Thus when using (3.3.10) with $m = 1$, the value of k is obtained from the experimental data by using the formula

$$(\varepsilon_f - \varepsilon_i) + \frac{1}{b} \ln \left| \frac{\sigma_0 - a + a \exp(-b\varepsilon_f)}{\sigma_0 - a + a \exp(-b\varepsilon_i)} \right| = \frac{k}{E} (\sigma_0 - a)(t_f - t_i). \quad (3.3.12)$$

Similarly, if formula (3.3.11) is used, we get

$$\frac{1}{\sigma_c h} \ln \left| \frac{\sigma_0 - \sigma_c h \varepsilon_i}{\sigma_0 - \sigma_c h \varepsilon_f} \right| = \frac{t_f - t_i}{E} k. \quad (3.3.13)$$

These formulae have been used at various stress levels and for various time intervals during creep at each stress level. The units expressing E/k are Ns m^{-2} (= 10 P), while k is given in units of (time)$^{-1}$.

3.4. Examples

The numerical values of various coefficients involved in the constitutive equation given above have been determined for a variety of rocks. Some of these values are listed in Table 3.1 for several rocks. The numerical values given in this table are only illustrative. Within a single kind of rock significant scatter of these values is seen. The coefficient k has been determined for various stress levels and time intervals. Its value for hard rocks ranges between 0.02 d^{-1} and 0.7 d^{-1}, so that the viscosity coefficient is of the order 10^{15} P. Softer rocks have smaller viscosity coefficients of the order 10^{12}—10^{13} P for *rock salt*, 10^{13} for *sandstone*, 10^{13}—10^{14} for *chalk*, etc. Generally, determinations *in situ* yield

Table 3.1. Values of constitutive constants for various rocks

No.	Rock	σ_c (MPa)	E (GPa)	a (MPa)	b
1	Schist	19.6— 98.0	16.2—36.3	9.8— 58.8	200—750
2	Dolomite	49.0—166.6	27.4—41.2	34.3—116.6	366—856
3	Limestone	20.0— 78.4	12.5—31.8	14.2— 54.9	140—240
4	Gneiss	24.5— 68.6	17.6—29.4	12.2— 35.7	650—800
5	Rock salt	19.6	15.6	5.9	75
6	Chalk	3.9— 7.8	3.4— 4.9	1.4— 4.3	157.7—495.1
7	Sandstone	17.8	2.45	9.8	

Table 3.1. Cont.

No.	h	α	β	Ek^{-1} (P) (3.3.13)	Ek^{-1} (P) (3.3.12)
1	12—181	10— 2.10⁴	1.2—4.5	3.10^{14}—2.10^{15}	5.10^{14}—6.10^{15}
2	92—169	1— 2.10^{5}	1 — 2.7	7.10^{14}—5.10^{15}	1.10^{15}—1.10^{16}
3	61— 97	1	1	2.10^{14}—1.10^{15}	4.10^{14}—3.10^{15}
4	105—157	1—65	1 —2.2	7.10^{14}—2.10^{15}	1.10^{15}—3.10^{15}
5	—	1611	1.76	2.10^{12}—3.10^{13}	—
6	—	41—99	1	4.10^{12}—7.10^{14}	—
7	79.4	—16	1.8	2.10^{13}—4.10^{13}	—

higher viscosity coefficients. For instance (some authors use a different definition for this coefficient from that given in the above model), Obert and Duvall (1967, p. 479), obtain for *limestone* $Ek^{-1} = 7 \times 10^{18}$—$10^{19}$ P. Several values can be found for *rock salt*; thus Griggs (1939) for a single monocrystal specimen finds 2.6×10^{17} P, Albrecht and Langer (1974) give values of the order 10^{16}—10^{17} P (see also Lama and Vutukuri (1978, Vol. III, p. 245)), while Baar (1977, p. 93), reports a quite wide interval, 10^{11}—10^{18} P, for various kinds of *salt* and various experimental procedures. For *granite, gabbro*, and *sandstone* tested at rates of deformation of 10^{-4}—10^{-8} s^{-1}, Volarovich (1978) reports viscosity coefficients ranging between 10^{12} and 10^{14} P, while for *marble*, Repko (see Kartashov (1973, p. 85)) gives 1.6×10^{14} P. Long-term creep tests lasting more than twenty years have been done by Itô (1983); for very small rates of deformations of 10^{-12}—10^{-15} s^{-1}, the viscosity coefficient for *granite* is 10^{20} P and for *gabbro*, 10^{18}—10^{21} P. It should be mentioned that geophysical studies suggest that the viscosity coefficient of the lower mantle ranges between 1×10^{22} P and 4×10^{22} P (Yuen *et al.* (1982)); sometimes a variable viscosity coefficient is accepted depending on the depth, ranging between 10^{25} P and 10^{20} P (Yang and Toksöz (1981)).

It should be stressed here again that various authors have used distinct rheological models and therefore distinct definitions for the viscosity coefficient. It seems that the viscosity coefficient determined by uniaxial creep tests done in the laboratory is much smaller than that determined *in situ*. In the solutions of engineering problems, only the order of magnitude of the viscosity coefficient is significant.

3.5. Creep in triaxial stress-state

This is the test which brings the rock very close to the stress state existing around underground openings. The creep of rock around an underground opening is a very important practical problem, mainly for deep openings and for soft rocks. However, because of their complexity, triaxial creep tests are performed only in a few laboratories (see Wawersik and Preece (1984), Langer (1979, 1982), Serata *et al.* (1972), Hunsche (1984*a*), Mellegard *et al.* (1981),

Hansen and Mellegard (1980), Natau *et al.* (1983), and Farmer and Gilbert (1984)). In a triaxial creep test a cylindrical rock specimen is first subjected to hydrostatic pressure; the pressure on the lateral surface of the specimen is then held constant, and finally the axial stress is increased up to a desired value. Furthermore, this axial stress is held constant for a very long time interval during which the variations of axial and diameter strains are recorded. The difference between the axial stress and the constant hydrostatic pressure is called *stress difference* or *relative stress* (see §8.2).

Sometimes, creep in rocks is studied directly *in situ* (Ladanyi and Gill (1983), Hunsche and Plischke (1984)). The creep of *rock salt* in triaxial tests has been studied by several authors who report primary, secondary, and tertiary creep (see Hendron (1968), Galperin and Shafarenko (1977, Chapter VII), Erzhanov and Bergman (1977), Proskuryakov *et al.* (1973, Chapter XI), Wawersik and Hannum (1980), Hansen (1977), Hansen and Mellegard (1980), Mellegard *et al.* (1981), Borchert *et al.* (1979), Wawersik *et al.* (1982), Gangi (1983), etc.).

The *influence of temperature on creep* has also been studied (Handin and Carter (1979), Hansen and Carter (1980), Savage and Swolfs (1980), Langer (1982), Wai and Lo (1982), Carter *et al.* (1981), Gangi *et al.* (1984), Hunsche (1984*b*, 1984*c*), Lindner and Brady (1984), Munson and Dawson (1984), Vouille *et al.* (1984)). Generally, a rise in temperature increases the strain rate during creep. As an example showing the influence of the temperature on the creep curves, Fig. 3.5.1 presents such curves for Avery Island *salt* according to Hansen and Carter (1980), obtained for various temperatures, on initial confining pressure of 3.5 MPa, and two stress differences (relative axial stresses). Increasing the temperature above a threshold value, introduces thermal cracks that arise from differential thermal expansion of adjacent crystals in most instances (Carter *et al.* (1981)). Temperature greatly influences the creep rate. It was also found that the influence of the stress difference on creep behaviour

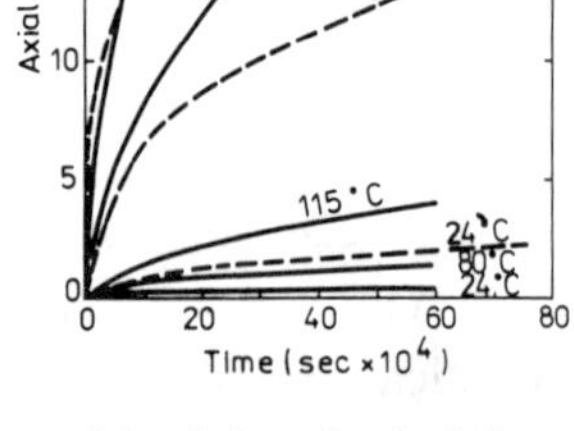

Figure 3.5.1. Creep curves for Avery Island domal salt deformed at various temperatures shown and two stress differences; initial confining pressure 3.5 MPa (after Hansen and Carter (1980)).

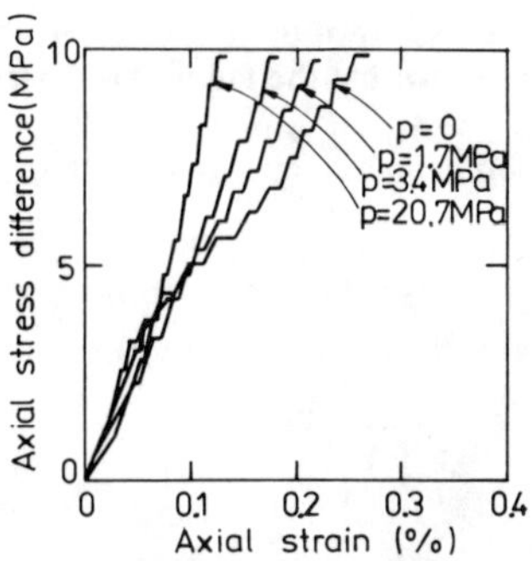

Figure 3.5.2. Details of low axial stress difference and low axial strain for Jefferson Island salt (after Hansen *et al.* (1984)).

is similar to the influence of temperature: a high stress difference produces higher creep rates. However, according to the author's knowledge, no Zener–Hollomon type of parameter relating the variation of the rate of strain to that of temperature, or a generalization thereof, has yet been found. The influence of water content, temperature, and loading conditions on the creep of *bischofite* was studied by Eekelen *et al.* (1984).

Another characteristic of *rock salt* pointed out by Hansen *et al.* (1984) is that the *yield strength is essentially zero*. This is shown in Fig. 3.5.2, where deformation is not linearly elastic at the onset; even at very low strains and stresses, time-dependent deformation occurs. Akagi (1985) found also that 'according to nearly all existing experimental data for rock, the plastic creep behaviour is observed even under the low stress levels, and it is almost *impossible to find the definite yield point* which may be observed on metal'. This fundamental property, found by the present author for many rocks, is used in the formulation of the models developed in Chapters 7 and 8.

Generally, all the above-mentioned experimental conclusions are used in the formulation of the mathematical models for rocks given in the following chapters.

Exercises

1. Perform uniaxial compressive creep tests on a rock of your choice. Determine the creep curves for various constant stress levels. Determine the limit stress for which the creep stabilizes. (after a reasonably long time-interval). Plot the stabilization boundary.

2. After reaching a certain stress level in creep tests using incremental loading, do unloading. Determine the magnitude of the instantaneous reversible strain component, the retarded irreversible in time strain component, and the irreversible component of the strain. Study the influence of the loading rate and loading history on these components.

3. Study the creep of rocks in uniaxial compressive tests using a universal testing machine (incremental loading in steps), in the first minutes after loading. Compare the results obtained for various rocks. Compare the magnitude of the strain obtained in a few minutes, with that obtained in one hour, one day, one week, etc.

4. Determine the three boundaries shown in Fig. 3.3.1 for a rock of your choice. By changing the loading rate or loading history show that the failure boundary is not unique (see also Chapter 9).

5. Starting from a set of experimental creep curves, determine a mathematical model of the form (3.3.1). See if any of the terms is negligible. Determine the viscosity coefficient using a formula of the type (3.3.12) or (3.3.13). Compare the magnitude of this coefficient for various rocks.

4

Volume deformation

4.1. Dilatancy and/or compressibility

The essential distinction between the mechanical properties of rocks and those of other materials (such as metals or plastics) is the characteristic of the volume deformation. Before discussing the formulation of a constitutive equation for a rock or a rock-like material, let us first discuss the characteristics of the volume deformation when a certain load is applied. Since the rock possesses voids, microcracks, etc., non-linear volume deformation is to be expected. A part of this deformation is irreversible. In what follows it will be assumed that for most rocks (of relatively small porosity) and stress-states of interest in mining applications, the irreversible volumetric strain is mainly due to microcracks. In other words we assume that the *irreversible volumetric strain can be related mainly to a measure of microcrack evolution* (microcrack density change, microcrack formation and/or closure, etc.). Sometimes, however, pores also play a significant role in the irreversible volume change.

In order to investigate the properties of the volume deformation, two kinds of loading path will be considered. The first kind of loading is called *hydrostatic* if at any time $\sigma_1 = \sigma_2 = \sigma_3$, or *quasi-hydrostatic*, if with some approximation $\sigma_1 \simeq \sigma_2 \simeq \sigma_3$. In the second kind of loading path, one starts from various initial hydrostatic stress levels and then increases only the equivalent stress to failure. This second kind of loading could be called *deviatoric* for convenience, although in the experiments done in most laboratories it is not exactly so. By combining these two kinds of loading, we try to disclose the constitutive relationship for the volume deformation of the rock for all possible loading histories of interest for engineering applications.

In the laboratory the variation of the volume of rocks is studied using either a 'triaxial' apparatus or a real triaxial one, or perhaps various other devices in which lateral displacement of the specimen is prevented. Generally, all tests in which the lateral surface of the specimen is subjected to a certain stress are called *confined* tests. Let us discuss first what kind of information such tests can reveal. In Fig. 4.1.1 two domains can be identified in the compressive stresses range. Domain D_c, containing a certain segment OP of the straight line $\sigma_1 =$

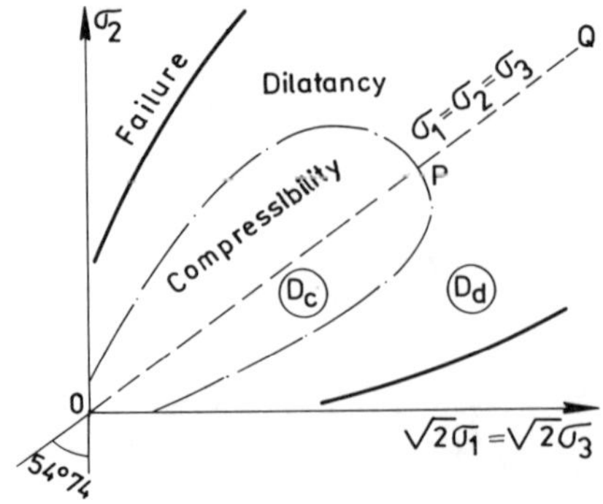

Figure 4.1.1. Domains of compressibility D_c and of dilatancy D_d in a meridian plane. The border line (—.—.) is the compressibility/dilatancy boundary.

$\sigma_2 = \sigma_3$ and corresponding to relatively small equivalent stress (the stress intensity), i.e.

$$\bar{\sigma}^2 = \sigma_1^2 + \sigma_2^2 + \sigma_3^2 - \sigma_1\sigma_2 - \sigma_2\sigma_3 - \sigma_3\sigma_1, \tag{4.1.1}$$

is the domain where, in the dissipation process during irreversible deformation of the volume, the hydrostatic pressure is the dominant factor compared with the stress deviator (expressed for instance by the difference $\sigma_1 - \sigma_3$, etc.). The domain D_d is located outside this first domain and is characterized either by large values of the equivalent stress compared to the mean stress

$$\sigma = \tfrac{1}{3}(\sigma_1 + \sigma_2 + \sigma_3), \tag{4.1.2}$$

or by very great mean stresses. In domain D_c, when σ increases, the microcracks close and that is why this domain is called the *domain of compressibility*; a part of the volume compressibility is irreversible. In domain D_d, called the *domain of dilatancy*, the microcracks open when $\bar{\sigma}$ increases and their number increases. The boundary between the two domains will be called the *compressibility/dilatancy boundary* (C/D boundary), or the dilatancy threshold boundary, or finally the onset of dilatancy boundary.

For an arbitrary mean stress satisfying $\sigma < \sigma_P$, these two domains are shown in Fig. 4.1.2, in a deviatoric plane. If in the plane $\sigma =$ const. the deviatoric stress components increase according to the path OAB, then from O to A the volume decreases while from A to B the volume increases; finally at B, failure occurs. For $\sigma \geqslant \sigma_P$ the volume compressibility is reversible for stress-states in the immediate neighbourhood of the 0σ-axis (the line PQ in Fig. 4.1.1), since at these pressures practically all microcracks are closed; however, for stress-states which are far from this line an irreversible dilatancy will occur.

The relationships describing the volume deformation are obtained both from quasi-hydrostatic tests which reveal the behaviour for stress-states in the neighbourhood of the $\sigma_1 = \sigma_2 = \sigma_3$ line, and from 'triaxial' tests which reveal the shape of the C/D boundary and the dilatancy behaviour of the rock for stress-states far from the hydrostatic line. The shape of the C/D boundary may

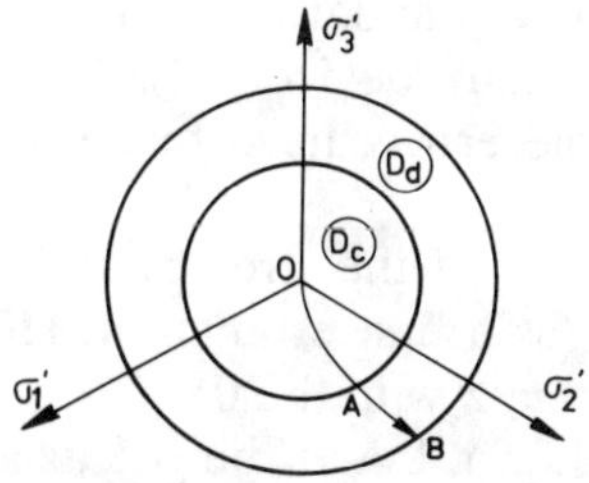

Figure 4.1.2. Domains of compressibility D_c and of dilatancy D_d in a deviatoric plane. OAB is an arbitrary loading path in the deviatoric plane.

be quite different for different rocks. The equation of the C/D boundary must follow from the constitutive equation.

4.2. Volume compressibility

The compressibility of the volume, especially at high pressures, is studied in quasi-hydrostatic tests in which small cylindrical specimens are compressed by a piston inside a rigid metallic cylinder. Various devices manufactured for this purpose are described by Bridgman (1952), Swenson (1960), Baidyuk (1963), Stephens *et al.* (1970), Volarovich *et al.* (1974, 1977), Volarovich (1978), Heard and Duba (1978), Levykin and Vavakin (1978), and others.

Variants of such devices have also been manufactured at the University of Bucharest. For instance, in Fig. 4.2.1 the device made by M. Constantinescu (1981) is shown schematically. The cylindrical specimen (7) is compressed under two pistons (6) inside a multisheet cylinder (5—4—3). The pistons are made of cermet, while all other pieces are of hard-tool steel. The openings (9) are used for experiments on ultrasonic wave propagation through the highly compressed rock. The loading is applied through the platens (1) using a

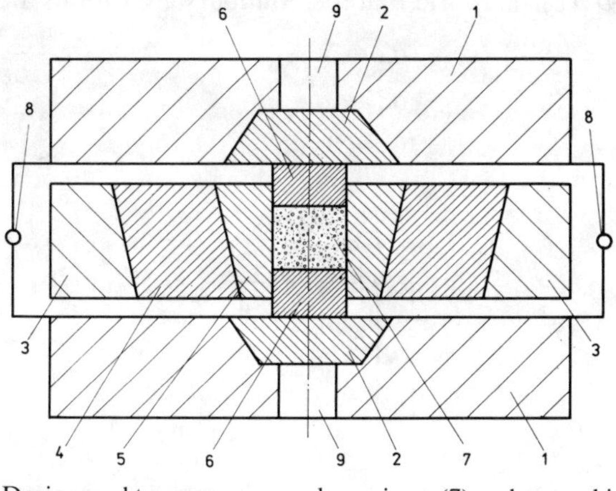

Figure 4.2.1. Device used to compress a rock specimen (7) under very high pressures.

universal testing machine or a standard creep test apparatus. Only the axial strain is measured using various devices (8). Due to the very high stresses involved, the various possible errors have to be analysed by experimental or theoretical methods.

An approximate distribution of the stresses and displacements in the specimen can be obtained using (as a first estimation) Hooke's law. Let us consider the cylindrical coordinates r, θ, z with the $0z$-axis coinciding with the symmetry axis of the specimen (Fig. 4.2.2). Let σ_1 be the axial stress applied at the end $z = \ell$ of the specimen, and τ be the shearing stress at the lateral surface $r = R$ of the specimen due to friction. It will be assumed that τ is independent of z and θ. ℓ is the length of the specimen. Axial symmetry will be assumed.

In the absence of body forces the balance of linear momentum yields

$$\begin{aligned} &\frac{\partial \sigma_{rr}}{\partial r} + \frac{\partial \sigma_{zr}}{\partial z} + \frac{1}{r}(\sigma_{rr} - \sigma_{\theta\theta}) = 0 \\ &\frac{\partial \sigma_{rz}}{\partial r} + \frac{\partial \sigma_{zz}}{\partial z} + \frac{1}{r}\sigma_{rz} = 0, \end{aligned} \tag{4.2.1}$$

since $\sigma_{r\theta} = \sigma_{z\theta} = 0$. The strain components are

$$\varepsilon_{rr} = \frac{\partial u_r}{\partial r}, \qquad \varepsilon_{\theta\theta} = \frac{u_r}{r}, \qquad \varepsilon_{zz} = \frac{\partial w}{\partial z}, \qquad 2\varepsilon_{rz} = \frac{\partial w}{\partial r} + \frac{\partial u_r}{\partial z}, \tag{4.2.2}$$

and Hooke's law

$$\begin{aligned} &\varepsilon_{rr} = \frac{1}{E}[\sigma_{rr} - \nu(\sigma_{\theta\theta} + \sigma_{zz})], \qquad \varepsilon_{\theta\theta} = \frac{1}{E}[\sigma_{\theta\theta} - \nu(\sigma_{zz} + \sigma_{rr})] \\ &\varepsilon_{zz} = \frac{1}{E}[\sigma_{zz} - \nu(\sigma_{rr} + \sigma_{\theta\theta})], \qquad \varepsilon_{rz} = \frac{1+\nu}{E}\sigma_{rz}, \end{aligned} \tag{4.2.3}$$

with E the Young's modulus and ν the Poisson ratio.

In order to get a simple solution it will be assumed that $u_r = 0$ everywhere, that each stress component depends on a single variable only (either on r as $\sigma_{rz}(r)$, or on z as $\sigma_{zz}(z)$, etc.), and that the boundary conditions are

$$\begin{aligned} &r = R, \quad z \in [0, \ell]: \quad \sigma_{rz} = \tau \\ &z = \ell, \quad r \in [0, R]: \quad \sigma_{zz} = \sigma_1 \\ &z = 0, \quad r = R \quad\;\; : \quad w = 0. \end{aligned} \tag{4.2.4}$$

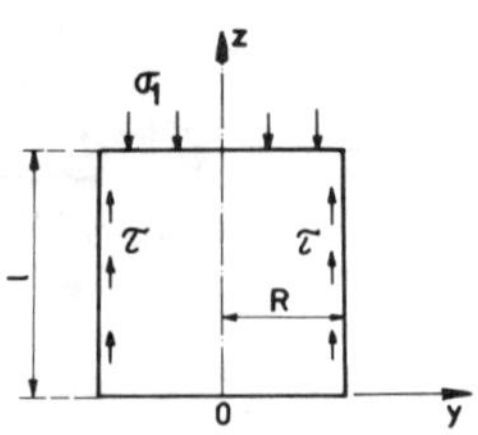

Figure 4.2.2. Rock specimen; notations used.

With these assumptions we get the solution

$$\sigma_{rr} = \sigma_{\theta\theta} = \frac{\nu}{1-\nu}\left[\sigma_1 + \frac{2\tau}{R}(\ell - z)\right]$$

$$\sigma_{zz} = \sigma_1 + \frac{2\tau}{R}(\ell - z) \tag{4.2.5}$$

$$\sigma_{rz} = \frac{\tau}{R} r$$

and the mean stress

$$\sigma = \frac{1}{3}\frac{1+\nu}{1-\nu}\left[\sigma_1 + \frac{2\tau}{R}(\ell - z)\right]. \tag{4.2.6}$$

Hence it results that if the friction is small, the radius of the specimen is larger than its length ($R > \ell$), and/or if the value of ν is close to $\frac{1}{2}$, then the mean stress is not too different from σ_1.

These approximate formulae have been used for the calculation of the dimensions of the specimen only. They can also be used for an estimation of the frictional force by considering specimens of various lengths or by measuring the force necessary to pull out the specimen from the cylinder (during or after the experiment).

For the displacement we easily get

$$w = \frac{1+\nu}{E}\frac{\tau}{R}(r^2 - R^2) + \frac{1-\nu-2\nu^2}{E(1-\nu)} \times$$
$$\times\left[\left(\sigma_1 + \frac{2\tau}{R}\ell\right)z - \frac{\tau}{R}z^2\right]. \tag{4.2.7}$$

Thus according to this solution we have to accept that the displacement on the symmetry axis $r = 0$ is larger than that along the lateral surface $r = R$, and therefore, that the surfaces $z = 0$ and $z = \ell$ of the pistons will deform according to this formula.

A more general and more accurate solution has to take into account that the lateral surface of the cylinder will also deform during the experiment. Such a solution has been given by Massier (D. Massier, personal communication).

Other authors have done similar experiments in which the specimen has been covered with a thin sheet of a ductile metal in order to subject the specimen to hydrostatic pressure (Stephens *et al.* (1970)).

From the experiments described above the relationship between the axial stress σ_1 and the axial strain ε_1 can be determined, both during loading (when $\dot{\sigma} > 0$) and during unloading (when $\dot{\sigma} < 0$). These relationships will suggest the volume compressibility laws for stress-states which are in the neighbourhood of the hydrostatic line $\sigma_1 = \sigma_2 = \sigma_3$.

The length and density of the specimens before and after the experiments have also been measured and compared with the data obtained during the experiments. Various loading rates have been used. Volumetric creep tests have also been performed.

Several rocks have been tested with the method described above: *rock salt, schist*, natural *chalk, cement concrete*, etc. As an example, the $\sigma_1 - \varepsilon_1$ curves obtained with three loading rates (see Table 4.1) are given for *cement concrete* (Constantinescu (1981)) in Fig. 4.2.3. The first non-linear portion of these curves may be due to the setting of the specimen in the testing device, to the crushing of some asperities or pores, to the exceeding of the static friction, etc. For higher stresses the curves have the tendency to become vertical. Further-

Table 4.1. Dimensions of the specimens used

No.	Global rate of deformation (s^{-1})	Heights (mm) initial	Heights (mm) final	Diameters (mm) initial	Diameters (mm) final	Volume (cm^3) initial	Volume (cm^3) final	Density ($g\ cm^{-3}$) initial	Density ($g\ cm^{-3}$) final
1	6.4×10^{-4}	17.99	15.75	19.92	20.08	5.603	4.988	2.167	2.434
2	1.67×10^{-3}	17.98	15.64	19.95	20.08	5.620	4.934	2.155	2.455
3	2.55×10^{-3}	17.98	15.59	19.94	20.08	5.615	4.937	2.167	2.465

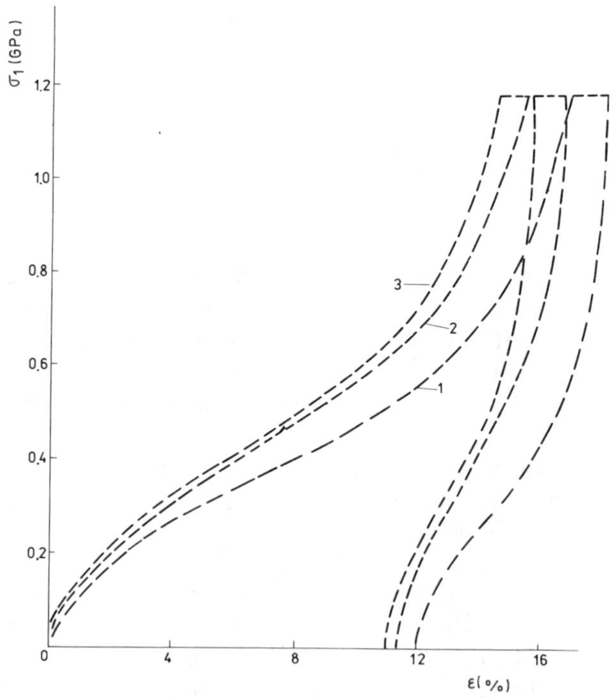

Figure 4.2.3. Volumetric loading/unloading curves for cement concrete showing also volumetric creep and irreversible deformation.

more, at $\sigma_1 = 1.166$ GPa, the stress was kept constant for a certain interval of time (102 minutes, 87 minutes, and 70 minutes, respectively), and the slow variation of strain by creep was recorded. Finally, the unloading was started with a nearly vertical slope. The strong non-linearity at the end of the unloading process may be due to the opening of microcracks, etc. Similar results for the hydrostatic compressibility of *rock salt* have been obtained by Wawersik and Hannum (1980).

4.3. Mathematical models for the hydrostatic compressibility of volume

Two variants will be presented. In the first model it is assumed that at high hydrostatic stresses the volumetric strains tend toward a limit value. In the second model it is assumed that at very high hydrostatic stresses the behaviour of the volumetric response is essentially elastic, i.e. it is defined by the elasticity of the rock mineral constituents. In the present subsection a constitutive equation for the volumetric compressibility will be given. However, after the formulation of a general constitutive equation, this relationship must be obtained from the general one for the particular case of hydrostatic loading.

4.3.1. The first variant of the model (Cristescu (1979), see also Cristescu and Suliciu (1982)) is based on the following three assumptions:

— there is a limit strain ε_* which cannot be exceeded no matter how much the stress is increased (a 'locking' model);

— there is a stabilization boundary for the volumetric creep and $\varepsilon = \varepsilon_*$ is the vertical asymptote of this curve;

— in the fastest possible loading tests starting from $\sigma = 0$ and $\varepsilon = 0$, the 'instantaneous' response of the rock can be described by an 'instantaneous response curve'.

Let us denote by σ the mean stress and by ε the mean strain. The previous assumptions have been suggested by the experimental evidence. If we apply a loading which is increased in steps, the result shown schematically in Fig. 4.3.1 is obtained. At each increase in loading we get an instantaneous response of the rock which is followed by a volumetric creep taking place in the long subse-

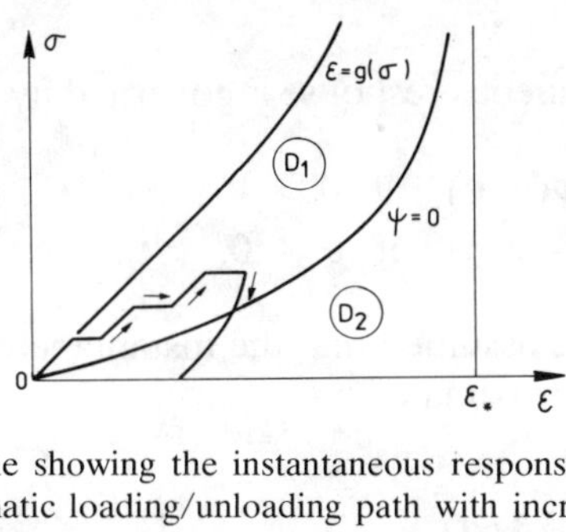

Figure 4.3.1. Stress-strain plane showing the instantaneous response curve $\varepsilon = g(\sigma)$, relaxation boundary $\psi = 0$ and a schematic loading/unloading path with incremental loading followed by volumetric creep and afterwards by unloading.

quent time interval. If an experiment is carried out with a very small global loading rate (the smallest possible one for the kind of applications we have in mind) and very small loading steps, then the stress—strain relationship will closely follow the stabilization boundary for volumetric creep. The equation of such a boundary can be written, for instance, in the form

$$\varepsilon = \varepsilon_* \left[1 - \exp\left(- \frac{\sigma}{\sigma_*} \right) \right], \tag{4.3.1}$$

where σ_* is a material constant. The boundary (4.3.1) has the following property: if from a point $(\varepsilon, \sigma) \in D_1$ the stress is kept constant (or is decreased), then the boundary (4.3.1) is reached in an infinite (or finite, respectively) time interval. The states represented by points on the boundary are stable states, providing that the mean stress is kept constant.

According to our third assumption, in the fastest possible test the 'instantaneous' response of the rock can be expressed by a relationship of the form

$$\varepsilon = g(\sigma), \tag{4.3.2}$$

if the tests start from the states of zero stress and zero strain. In particular, (4.3.2) can be a straight line. Here, $g(\sigma)$ is a non-negative continuous function in the domain $\sigma > 0, 0 < \varepsilon < \varepsilon_*$. Furthermore the domain D_1 is defined by

$$\sigma > 0, \qquad g(\sigma) < \varepsilon < \varepsilon_* \left[1 - \exp\left(- \frac{\sigma}{\sigma_*} \right) \right],$$

while the domain D_2 is defined by

$$\sigma > 0, \qquad \varepsilon_* \left[1 - \exp\left(- \frac{\sigma}{\sigma_*} \right) \right] \leqslant \varepsilon < \varepsilon_* .$$

In order to describe the compressibility of the rocks in hydrostatic loading, a rate-type constitutive equation of the form

$$\dot{\varepsilon} = \left[\frac{1}{3K(\sigma)} + f(\varepsilon, \sigma, \operatorname{sgn}\dot{\sigma}) \right] \dot{\sigma} + \psi(\varepsilon, \sigma) \tag{4.3.3}$$

will be used. The instantaneous response is governed by

$$f(\varepsilon, \sigma, \operatorname{sgn}\dot{\sigma}) = \begin{cases} \varphi(\varepsilon, \sigma) & \text{if } \dot{\sigma} \geqslant 0 \\ 0 & \text{if } \dot{\sigma} < 0. \end{cases} \tag{4.3.4}$$

For many rocks it can be assumed that the instantaneous response starting from any state in D_1 can be approximated by

$$\varphi(\sigma) = \frac{\mathrm{d}g(\sigma)}{\mathrm{d}\sigma} - \frac{1}{3K(\sigma)}. \tag{4.3.5}$$

Here $K(\sigma)$ (possibly a gently varying function of σ) is the elastic modulus describing unloading, and therefore has to be determined in unloading tests. In many engineering applications one can assume K = const. and the instantaneous response is therefore linearly elastic.

The function $\psi(\varepsilon, \sigma)$ which is related to the slow deformation in time, i.e. to the volumetric creep, is defined by

$$\psi(\varepsilon, \sigma) = \begin{cases} k_v \left\{ \varepsilon_* \left[1 - \exp\left(-\dfrac{\sigma}{\sigma_*} \right) \right] - \varepsilon \right\} & \text{if } \varepsilon < \varepsilon_* \left[1 - \exp\left(-\dfrac{\sigma}{\sigma_*} \right) \right] \\ 0 & \text{if } \varepsilon \geqslant \varepsilon_* \left[1 - \exp\left(-\dfrac{\sigma}{\sigma_*} \right) \right], \end{cases} \tag{4.3.6}$$

where k_v is a viscosity coefficient for the volume change, measured in units of $(\text{time})^{-1}$. Generally it is to be expected that k_v is variable, but in many cases, if stresses and strains do not vary too much, the previous assumption concerning the constancy of k_v seems reasonable.

In order to apply the model mentioned above to the computation of the stress and strain distribution in the rock surrounding an underground opening, one can use a simplified linearized version of this model. First, the stabilization boundary will be written in the linearized form

$$\sigma = 3K_0\, \varepsilon, \tag{4.3.7}$$

with $K_0 = \sigma_*/(3\varepsilon_*)$. Then the instantaneous response will be assumed also to be linearly elastic, i.e. $f = 0$ and K = const. Thus the simplified variant of the model can be written as

$$\dot{\varepsilon} = \frac{\dot{\sigma}}{3K} + k_v \left\langle \frac{\sigma}{3K_0} - \varepsilon \right\rangle, \tag{4.3.8}$$

with $\langle Z \rangle = \frac{1}{2}(Z + |Z|)$.

If the stabilization boundary is already determined from experimental data, the value of the viscosity coefficient k_v can be determined using the data obtained during uniaxial strain volumetric creep, when σ_1 = const. If the creep test is done in the time interval $t_f - t_i$, then by integration of the constitutive equation we get

$$k_v(\sigma_1) = \frac{1}{t_f - t_i} \ln \left[\frac{\varepsilon_* \left[1 - \exp\left(-\dfrac{\sigma_1}{\sigma_*} \right) \right] - \varepsilon_{1f}}{\varepsilon_* \left[1 - \exp\left(-\dfrac{\sigma_1}{\sigma_*} \right) \right] - \varepsilon_{1i}} \right], \tag{4.3.9}$$

where ε_{1i} is the initial axial strain and ε_{1f} is the final axial strain corresponding to times t_i and t_f, respectively. A global value for $k_v(\sigma_1)$ can be obtained if the integration interval is extended over the whole period when σ_1 = const. Generally, this integration can be done over any time interval during a creep test.

Numerical examples have been obtained for various rocks. For instance, for *schist* (from Palazu Mare) we get $\varepsilon_* = 0.0108$, $\sigma_* = 0.392$ GPa, $k_v = 43\ \mathrm{d}^{-1}$, while for *chalk* (from Dobrogea) which is a soft rock with $\sigma_c = 4.41$ MPa and density $\rho = 1.67\ \mathrm{g\ cm}^{-3}$, we obtain $\varepsilon_* = 0.07$, $\sigma_* = 53.94$ MPa, $k_v = 3.80 \pm 1.07\ \mathrm{d}^{-1}$.

More details will be given for *cement concrete*. The σ_1—ε_1 relationship is shown in Fig. 4.3.2 for two tests done with a rate of strain of the order $10^{-4}\ \mathrm{s}^{-1}$. The stabilization boundary has been written in the form

$$\varepsilon = \varepsilon_* \left[1 - \exp\left(- \frac{\sigma - \sigma_0}{\sigma_* - \sigma_0} \right) \right], \tag{4.3.10}$$

where σ_0 has been introduced, since the first non-linear portion of the curve at small stresses has been disregarded (in another variant of the model it was not disregarded; see §8.7). We get $\varepsilon_* = 0.22$, $\sigma_* = 0.445$ GPa, $\sigma_0 = 0.13$ GPa. Furthermore, the mean value for the viscosity coefficient obtained for various

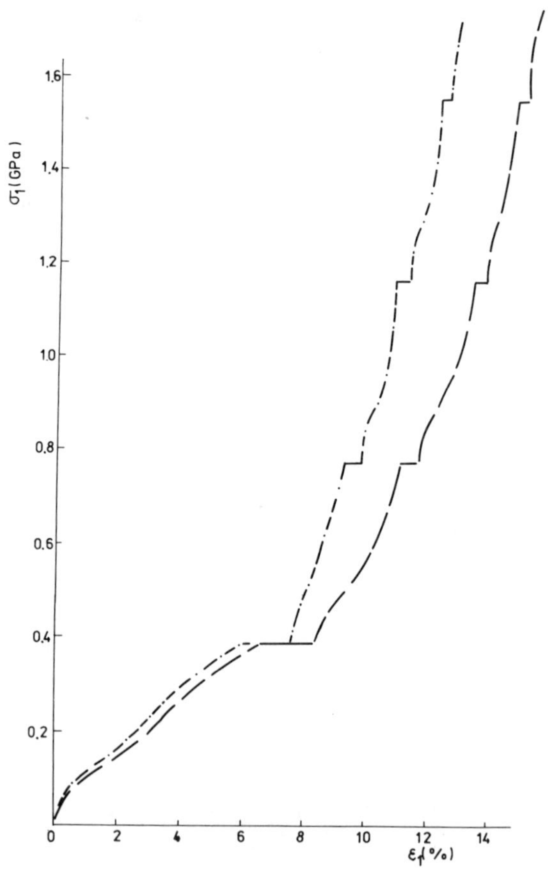

Figure 4.3.2. Incremental confined compression creep curves for cement concrete.

stress levels is between 3.9×10^{-2} min^{-1} and 0.39×10^{-2} min^{-1}. These values have been obtained from tests done with eleven specimens. Although in engineering applications the viscosity coefficient may be assumed with some approximation to be constant, the tests show that k_v depends on the loading history: if the loading preceding the creep tests is done with a higher loading rate, then during the creep tests we get a higher value for k_v, and this value of k_v is smaller for higher values of σ. In other words, k_v generally depends on both the loading rate and on the loading history (order of successive loading applications). The toes which appear in Fig. 4.3.2 after each reloading may be due to the crushing of additional pores or to the surpassing of the lateral friction, or to other causes. It is to be mentioned here that the cement concrete has an initial porosity of 25% and that during the tests described above the volume decreased about 10—14%.

The instantaneous response of the cement concrete, i.e. the determination of the relation (4.3.2), can be obtained by making compressive quasi-hydrostatic tests with relatively high rates of strains. In the present tests the mean rate of strain $\dot{\varepsilon} = 5.6 \times 10^{-3}$ s^{-1} has been used, while the instantaneous response has been determined as

$$\sigma - \sigma_i = a\varepsilon^b,$$

where the constant stress σ_i is an initial static preloading applied to the specimen. Thus we get $\sigma_i = 0.19$ GPa, $a = 31.548$ GPa, and $b = 1.8$. In the first approximation the instantaneous response can be considered to be linearly elastic, with the elastic moduli determined by dynamic procedures or in dynamic processes.

4.3.2. The second variant of the model has been formulated (Cristescu (1985*d*)) starting from the more realistic assumption suggested by the experimental evidence that only at relatively small pressures is the rock compressible (partially irreversible), but at very high pressures, after the closing of all microcracks, the volume variation of the rock obeys an elastic (reversible, nearly linear) law. Returning to Fig. 4.1.1, along the segment OP the compressibility of the rock is partially irreversible, while for $\sigma > \sigma_P$, no irreversible compressibility is to be found in small cyclic loading/unloading tests. In order to introduce the mathematical model, we first present various kinds of hydrostatic compressibility behaviour, as discussed by Brace (1978). For rocks of very small porosity, due mainly to microcracks, the volumetric compressibility is shown schematically in Fig. 4.3.3*a*. During loading the curve is initially strongly non-linear, but for pressures $\sigma > \sigma_P$, this curve is nearly linear following the elastic mineral response, and thus a small loading/unloading cycle shows reversibility. For $\sigma < \sigma_P$, unloading is generally non-linear, always showing an irreversible part of the volumetric strain. For rocks of relatively high porosities (over 5%, say), the irreversible compaction of the volume is more significant (Fig. 4.3.3*b*). For rocks of even higher porosities, due mainly to pores and not to microcracks, the compressibility curve has the aspect shown in Fig. 4.3.3*c*. In this latter case, when the pressure is greatly increased, the pores first collapse

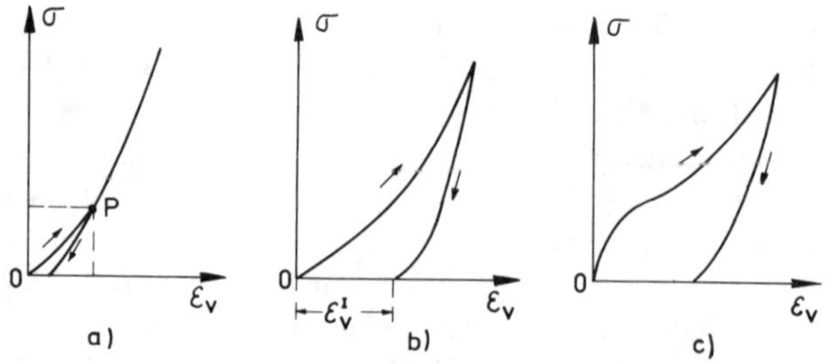

Figure 4.3.3. Loading/unloading compressibility curves for rocks of various initial porosities (schematic).

and thus microcracks are formed. This is revealed by the non-linear unloading pattern. In what follows we consider mainly the first two cases, those for which a certain pressure σ_P (at point P), which corresponds to the closure of most microcracks, can be perceived clearly.

The compaction of rocks and the mechanism of closing and opening of microcracks have been studied by several authors (see Brace (1965, 1978), Walsh (1965, 1980), and the references cited in these papers).

Walsh (1980) suggests an experimental method for the determination of the volume V_c of the microcracks or of the 'microcrack porosity' $\Delta V_c/V$, from the formula

$$\frac{\Delta V_c}{V} = \int_0^\sigma \beta \mathrm{d}\sigma - \int_0^\sigma \beta_s \mathrm{d}\sigma, \tag{4.3.11}$$

with $\beta = \mathrm{d}V/V\mathrm{d}\sigma$ the compressibility of the rock and β_s the compressibility of the minerals from which the skeleton is made. Integration is done up to a pressure $\sigma > \sigma_P$. The meaning of various terms is shown in Fig. 4.3.4.

For the formulation of the model given below the experimental data obtained for *granite* by Volarovich *et al.* (1974, p. 91) and Brace (1965) will be used. Because of their nature these experiments are performed for relatively long time intervals, so that to a first approximation it will be assumed that the obtained data can be used to determine the creep stabilization boundary. The experi-

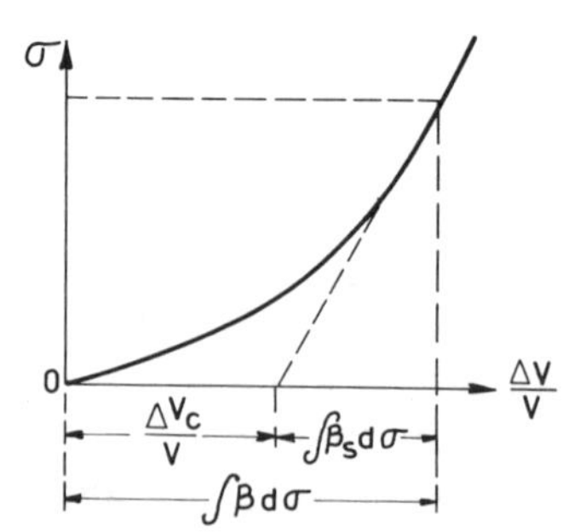

Figure 4.3.4. Experimental determination of microcrack porosity.

mental data are given in Table 4.2 for the pressure interval $0 \leqslant \sigma \leqslant 3.2$ GPa. The variation of ε_v with σ in the interval $0 \leqslant \sigma \leqslant 1$ GPa is shown in Fig. 4.3.5 by crosses. An equation which approximate these data is

$$\varepsilon_v = -p + q\sqrt{\sigma + r} \quad \text{if } 0 \leqslant \sigma \leqslant \sigma_0$$
$$\varepsilon_v = \frac{\sigma + K_2}{K_0} \quad \text{if } \sigma_0 \leqslant \sigma. \tag{4.3.12}$$

Therefore the last portion of the curve is approximated by a straight line. The values of various constants are

$$p = 0.0017469, \qquad q = 0.020112\ (\text{GPa})^{-1/2}, \qquad r = \frac{p^2}{q^2}$$
$$K_0 = 59\ \text{GPa}, \qquad K_2 = 0.2565\ \text{GPa}, \qquad \sigma_0 = 0.3445\ \text{GPa}. \tag{4.3.13}$$

In the interval $0 \leqslant \sigma \leqslant \sigma_0$ an unloading of the form

$$\dot{\sigma} = K(\sigma)\,\dot{\varepsilon}_v \tag{4.3.14}$$

will be assumed, in which the modulus $K(\sigma)$ is monotonically increasing with σ,

Table 4.2. Compressibility data for granite (Volarovich *et al.* (1974, p. 91))

σ (GPa)	ε_v	ρ (g cm^{-3})
0.000098	0	2.621
0.03	0.00215	2.620
0.06	0.00348	2.630
0.1	0.00467	2.633
0.4	0.01110	2.650
1.0	0.0223	2.681
1.4	0.0296	2.701
2.0	0.0403	2.731
2.4	0.0472	2.751
3.2	0.0609	2.791

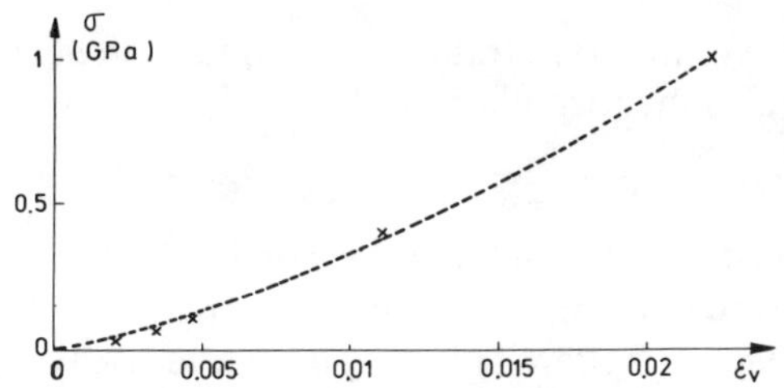

Figure 4.3.5. Hydrostatic compressibility curve for granite.

the maximal value of K being K_0. For this function the following empirical formula has been used:

$$K(\sigma) = \begin{cases} K_0 - K_1 \left(1 - \dfrac{\sigma}{\sigma_0}\right)^4 & \text{for } 0 \leqslant \sigma \leqslant \sigma_0 \\ K_0 & \text{for } \sigma_0 \leqslant \sigma, \end{cases} \tag{4.3.15}$$

with $K_1 = 48$ GPa. If unloading is approximated by an elastic straight line, then in (4.3.15), K must be interpreted as being a function of the maximally reached pressure, the one from which the unloading process begins. While the constant K_0 can be determined quite accurately, the determination of K_1 is less precise from the existing data. Since for $\sigma \leqslant \sigma_0$ the unloading slope is larger than the one during loading, we have for K_1 the restriction

$$K_1 < K_0 - \frac{2p}{q^2}.$$

Generally, smaller values for K_1 mean more significant irreversible volumetric strain.

In the present model the instantaneous response is assumed to be elastic with the slope K given above. We recall that for $\sigma \geqslant \sigma_0$, loading and unloading take place along the straight line $(4.3.12)_2$, without any time effects and without irreversible deformation. The viscosity coefficient E/k is determined as in uniaxial creep tests.

4.4. Volume dilatancy

Dilatancy is a specific property of rocks and of other porous materials. It is an increase in the volume of the rock, quite evident mainly for stress-states approaching failure, but generally present in the whole domain D_d shown in Fig. 4.1.1 and Fig. 4.1.2. In order to study this phenomenon, let us first consider the uniaxial unconfined tests in which the evolution of both longitudinal ε_1 and diameter strain ε_2 are recorded. Then, using the formula

$$\varepsilon_v = \frac{\Delta V}{V_0} = \varepsilon_1 + 2\varepsilon_2, \tag{4.4.1}$$

the variation of the volumetric strain can be determined, in the case of small strains. For finite strains the formula

$$\varepsilon_v = 1 - (1 - \varepsilon_1)(1 - \varepsilon_2)^2 \tag{4.4.2}$$

will be used. A typical result which can be obtained with such tests is shown in Fig. 4.4.1 for *sandstone* (from Siriu). For this rock the standard compressive strength is $\sigma_c = 98.04$ MPa. The σ_1—ε_1 curve is typical, showing three portions: an initial non-linear one, attributed to the closing of the microcracks, followed by a nearly linear portion, and the final non-linear portion with

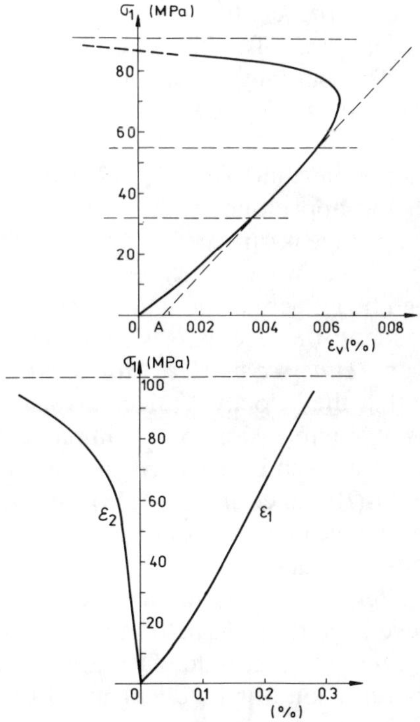

Figure 4.4.1. Stress—strain curves for sandstone obtained in uniaxial tests showing dilatancy.

opposed concavity, where the microcracks are opening and their number is increasing quite fast. The volume first decreases at lower stresses and then increases at higher stresses. The volume increase which is observed on the last portion of the σ_1—ε_v curve is called *dilatancy.* At intermediate stress values, between the levels C and I, there is a portion of the σ_1—ε_v curve which can be approximated by a straight-line segment with the slope approaching to some extent the magnitude of the elastic bulk modulus K. On this portion small loading—unloading cycles show reversibility, i.e. the irreversible part of ε_v is negligible. The magnitude of the segment OA on the deformation axis is a measure of the initial microcrack porosity of the rock (cf. Walsh (1965), and Brace *et al.* (1966)).

As a measure of dilatancy one can use the *difference between the actual volume of the rock and the volume which would be obtained for the same stress level from the linear elastic law*, represented by the broken line passing through the point A shown in Fig. 4.4.1 (i.e. the elastic behaviour after the closing of most microcracks). The dilatancy precedes failure and thus can be used as a warning of imminent failure. For this reason alone the dilatancy phenomenon is of great significance in mining or oil engineering, in earthquake prediction, etc.

Since the unloading process, at least a partial one, is assumed to take place according to a linear elastic law with the slope approximately determined from the slope of the mid —nearly linear— portion of the σ_1—ε_v curve during loading, dilatancy is by definition the *irreversible increase of the volume produced during a small loading–unloading cycle.* This 'local' definition of dilatancy is used in the following chapters. The partially reversible dilatancy, which is sometimes observed after full unloading in repeated loading—unloading cycles, is disregarded here.

It was found by Tapponnier and Brace (1976) that the onset of dilatancy roughly coincides with the appearance of *new* cracks and that crack density approximately doubles up to peak stress. The dilatancy threshold is evident also by the decrease of the seismic wave velocities V_s and V_p (Soga *et al.* (1978)); this decrease can be significant across the fault when failure is imminent, and it is also strain-rate-dependent: the decrease is more significant for lower strain rates. In confined tests on *Tennessee marble, Ruhr sandstone*, and *Greek marble*, Rummel (1974) found that the velocity V_p decreases starting from the onset of dilatation. During biaxial compression of a *granite* cube the travel time of ultrasonic waves first decreases and then starts increasing just from the moment of the dilatancy threshold (Rummel *et al.* (1978)). It was found, however, that while in granite the wave velocity V_p during progressive fracture increases monotonically, a velocity anomaly is found for *serpentinite*: a distinct velocity increase of about 2% after a previous monotonic velocity decrease (Rummel and Frohn (1982)). Several authors have found that the compressional wave velocity in the radial direction begins to decrease at about half the fracture stress, while in the axial direction it is hardly changed by stress (see Brace *et al.* (1966)).

Sudden intensification of the acoustic emission during loading marks the beginning of the dilatancy process (Rummel *et al.* (1978), Cristescu (1984, Chapter 9); see also §8.8).

A detailed study of *granite* dilatancy is due to Brace *et al.* (1966). For this rock the uniaxial compressive strength is of the order 0.23 ± 0.01 GPa (1 kbar $= 10^8$ Pa $= 10^8$ N m^{-2} = 987 atmospheres). The axial stress was increased to 95% of the compressive strength followed by unloading. The loading rate $\dot{\sigma}_1$ was 0.01 GPa s^{-1} to about half of fracture stress. Afterwards the stress was increased in steps, and after each increase the stress was held constant for several minutes. During this time interval a *significant creep phenomenon* was monitored on both stress—strain curves. The steps due to creep became broader as the stress was increased. Unloading was produced also in increments, but *this time the rock did not creep.* After full unloading, a significant irreversible diameter strain ε_2, but only a small axial irreversible strain ε_1, were evident. The variation of the volumetric strain was obtained using (4.4.1). After unloading, a significant irreversible increase of the volume was observed. *Dilatancy* is defined globally as the increase of the volume relative to elastic changes caused by deformation.

Similar results have been obtained in tests done under confining pressure. With steadily higher lateral pressures the initial non-linear portion on the various stress—strain curves steadily disappears, i.e. the non-linear compressibility of the volume is observed only when the specimen is not subjected to a lateral pressure, or maybe if this pressure is small. Two loading rates have been used; it was found that with higher loading rates the stress corresponding to the beginning of dilatancy (dilatancy threshold) is significantly higher. Dilatancy seems to be due to the opening of many cracks in planes oriented nearly parallel to the specimen axis. This conclusion has been pointed out by many authors using various experimental evidence: by measuring the variation of the velocity of propagation of the seismic waves in various directions, by observing the cracks with the electron microscope, with X-rays (see Brace (1978)), etc. Acoustic emission experiments carried out in our laboratory have shown that at the dilatancy threshold the frequency of the microseismic events increases quite suddenly. As the stress states further surpass the dilatancy threshold, the irreversible increase of the volume becomes increasingly significant. Other peculiar properties have also been reported by various authors. For some rocks dilatancy apparently disappears at very high pressures (Brace (1978)). Gowd and Rummel (1980) have also found that for porous *sandstone* (porosity 15%) the dilatancy is less pronounced if the lateral pressure is higher, while for $\sigma_2 \geqslant 100$ MPa, the volume seems to be non-linearly compressible only. Michelis (1981) has found that for *marble* and *granular rock*, dilatancy increases with confining pressure. A statistical study of the stress—volumetric strain behaviour of Westerly *granite*, during hydrostatic and triaxial compression experiments, is due to Constantino (1978). Chen *et al.* (1979) have shown that for *gabbro* the development of microcracks is dependent on the dimensions of the crystal grains. Concerning the influence of pore pressure, Gowd and Rummel (1977) have shown for *sandstone* tested under triaxial compression, that: '(1) if the pore pressure in elastically deformed specimens is increased, the axial stress at the specimen boundary will increase; (2) on the other hand, the boundary stress will decrease, if the pore pressure applied to a specimen is undergoing inelastic deformation.'

One should also mention that crack growth in compression produces not a mechanical instability as it does in tension, but a *dilatancy-hardening by cracking*, since the cracks grow and then become stable (Brace *et al.* (1966)) (the concept of work-hardening for rocks will be defined in later chapters). Work-hardening is due to the effective stresses, i.e. the applied stress minus the pore pressure. With increasing dilatancy and if the total volume of the pore fluid is fixed, the pore pressure decreases, thus increasing the effective stresses which produce work-hardening. For stress-states very close to failure, the final crack producing macroscopic failure along a strip of instability is formed from some of the microcracks which appeared during dilatancy.

Attempts to describe the dilatancy phenomenon by simple empirical formulae have been made by various authors (Bordia (1971, 1972), Brady (1970), etc.). In the next few chapters this phenomenon will be described by the general constitutive equations for rocks which have to deal with both dilatancy and compressibility phenomena.

Dilatancy has been studied also by *proper triaxial* test devices, i.e. those in which all three principal components of the stress tensor can be varied independently and the evolution of all three corresponding strains are recorded (Mogi (1977) and Hunsche (1984*a*)). Concerning the main dilatancy features, the results obtained with such tests are similar to the previous ones. Depending on the loading orientation, a certain anisotropy of dilatancy is also found.

The study of dilatancy both *before* and *after failure*, when the specimen is subjected also to a lateral pressure, has also been made (Rummel *et al.* (1978)). It has been shown that dilatancy continues well after failure, during decrease of the relative stress.

In the case of *cyclic loadings* the dilatancy is evident at each reloading (Scholz and Kranz (1974), Zoback and Byerlee (1975)), while the irreversible deformation of the volume seems to increase steadily after each cycle (mainly at low confining pressures); the dilatancy threshold is apparently not affected by the repeated cycling under triaxial conditions, and at high pressures, dilatancy is stable, increasing very little from cycle to cycle. Stevens and Holcomb (1980) and Holcomb and Stevens (1980) have shown that cyclic dilatancy, with huge hysteresis loops, is difficult to explain by existing models, and have developed a new model to describe this phenomenon. They found that the dilatant rock exhibits 'memory' of the maximum stress difference since the last complete unloading.

A *temperature variation* influences significantly both the dilatancy and the compressibility of rocks. This influence is complex, since on the one hand, cracking is a thermally activated process, but on the other hand, temperature may affect the pore pressure, may produce grain boundary cracks due to the difference in thermal expansion of minerals, may enhance stress corrosion processes at the crack tip, etc. The influence of temperature has been studied by

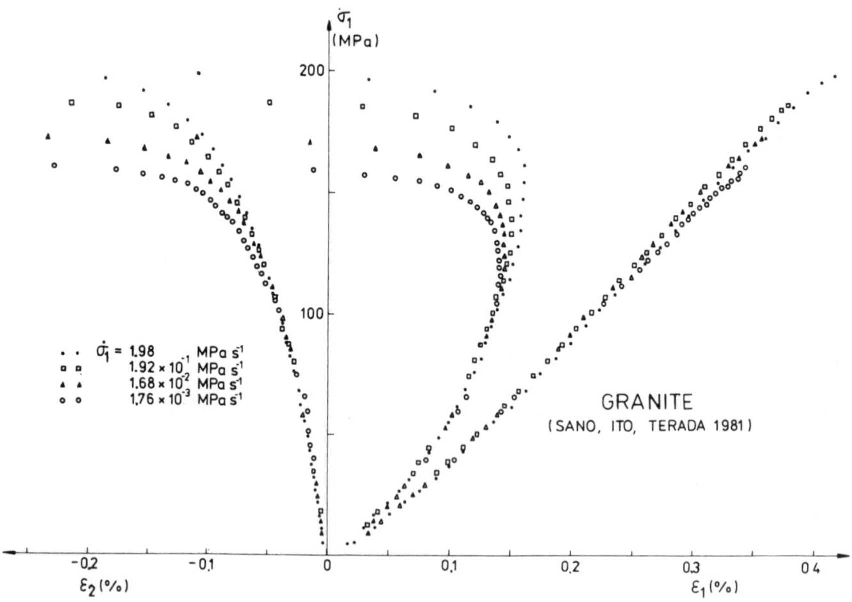

Figure 4.4.2. Influence of the loading rate on the stress—strain curves, and on dilatancy, for granite.

several authors: Houpert and Homand-Etienne (1979), Hansen and Carter (1980), Wawersik and Hannum (1980), and others.

The influence of the *rate of strain* on dilatancy was also studied and the results are revealing. In order to find the influence of the rate of strain on dilatancy we start by looking for this influence on the $\sigma_1 - \varepsilon_1$ and $\sigma_1 - \varepsilon_2$ curves. Most authors have observed a rise in both these curves when the rate of strain is increased (see for instance Perkins *et al.* (1970)).

The influence of the loading rate on dilatancy has been studied for *granite* by Sano *et al.* (1981). Four compressive loading rates were considered, and the results are shown in Fig. 4.4.2. If the loading rate does not have a very marked influence on the $\sigma_1 - \varepsilon_1$ curves, the effect on the $\sigma_1 - \varepsilon_2$ and $\sigma_1 - \varepsilon_v$ curves is remarkable. An increase of the loading rate increases significantly the apparent dilatancy threshold.

Other results concerning the influence of loading rate on dilatancy are reported by several authors (see Chapter 8 and Lama and Vutukuri (1978, Vol. II §6.7.3)).

There are rock-like materials which do not exhibit a significant dilatancy in uniaxial or triaxial tests. For instance, *cement concrete* (Constantinescu and Cristescu (1983)) does not exhibit, up to failure, any dilatancy during uniaxial standard tests (performed with constant rate of strain) or during uniaxial creep tests. Let us mention that this rock-like material has a significant initial porosity of 25%, due mainly to pores.

4.5. Rock dilatancy during creep

Rock dilatancy during creep tests has also been studied (Kranz and Scholz (1977), Kranz (1979), and Cristescu (1981, 1982)), mainly in uniaxial tests. In such tests successive stress increments $\Delta\sigma_1$ of the axial stress are applied, the lateral surface of the specimen being stress-free, where $\sigma_2 = \sigma_3 = 0$. After each such increase $\Delta\sigma_1$ of the axial stress, the stress is subsequently kept constant for a long time-interval during which the variation in time of the axial strain $\varepsilon_1(t)$ and of the diameter strain $\varepsilon_2(t)$ are recorded. Thus, creep curves for both strain components are obtained and also for the volumetric strain ε_v. The results are represented in the $\sigma_1 - \varepsilon_1$ and $\sigma_1 - \varepsilon_2$ plane (see Fig. 4.5.1 which corresponds to *sandstone* from Siriu). At each reloading $\Delta\sigma_1$, 'instantaneous' variations of the strains are observed which are followed by slow strain variations of long time-intervals (weeks or even months). Thus, the broken lines in Fig. 4.5.1 are obtained, where the horizontal portions correspond to creep at various stress levels. The next reloading is applied only if a stabilization ($\dot{\varepsilon}_1 \simeq 0$, $\dot{\varepsilon}_2 \simeq 0$) of the creep is observed for several successive days.

A volumetric creep is thus also revealed, while after stabilization a rock dilatancy by creep is recorded. In Fig. 4.5.1 the stabilization boundary for the axial strain is shown by circles, that of the diameter strain by crosses, and that of the volumetric strain by triangles. In connection with the creep of the volumetric strain the experiments sometimes show (Cristescu (1982)) a certain

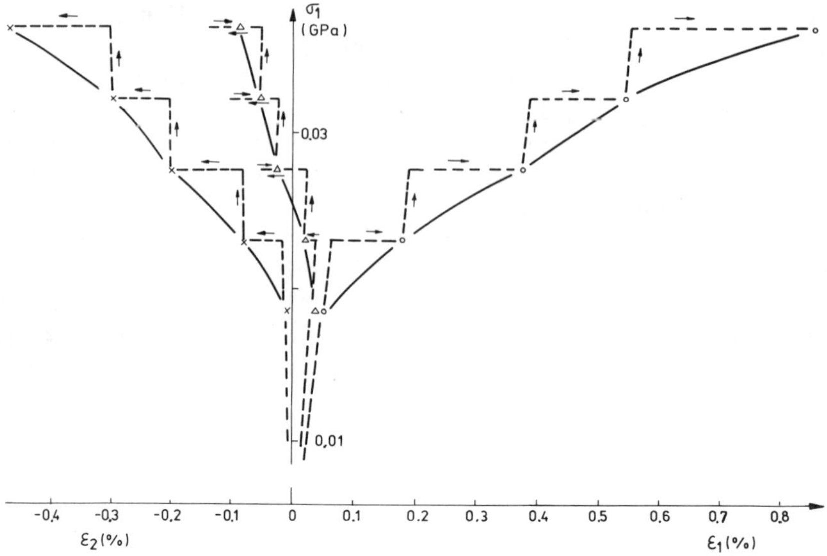

Figure 4.5.1. Incremental creep curves for sandstone showing oscillatory volumetric creep.

oscillation during creep: generally, in the first time-interval the volumetric strain decreases by creep and afterwards increases before creep stabilization. This oscillation is shown by small arrows in Fig. 4.5.1. Thus both phenomena, compressibility and dilatancy, are simultaneously present during creep, but are dominant in successive time-intervals. In other words, it seems that for sandstone, microcrack formation is generally a faster phenomenon than microcrack closure.

Crack growth and development in Barre *granite* as a function of time, under constant load, was studied by Kranz (1979) using the scanning electron microscope. He found that 'new stress-induced cracks were continuously generated with time' and that 'average crack lengths increased with time as much or more than they did upon loading'. Furthermore 'almost all cracks are tensile in nature' (see also Tapponnier and Brace (1976)); 'small amounts of shear were seen to occur only where cracks coalesce'. Several authors have shown (see Kranz (1979)) that 'as the time proceeds more and more of the cracks are roughly aligned with the maximum stress direction' so that there will be more accumulated inelastic strain in the perpendicular direction in a creep test.

Further details concerning dilatancy by creep, and mathematical models describing this phenomenon, are given in subsequent chapters.

Dilatancy during creep has also been studied in *triaxial tests*. *Rock salt* has been widely studied: Wawersik and Hannum (1980), Hansen and Mellegard (1980), Mellegard *et al.* (1981). The qualitative results are similar to those given above. It is shown that creep occurs even for the smallest applied stresses. The threshold of dilatancy increases with confining pressure and the irreversible component of the volumetric strain is significant. Unloading and reloading are nearly perfectly linearly elastic.

Observations. It was shown above that volumetric creep in hydrostatic compression tests always ends in a stable state, while in 'triaxial' tests it ends in a stable state only if the applied relative stresses are not too high. An important question arises: is the stabilization boundary thus unique? Certainly not, since it may depend on the loading history. Generally, for a fixed stress level, if the number of steps in the incremental loading history increases, then the ultimate strain at stabilization is more significant. Thus, if engineering problems are to be solved, the most disadvantageous case is a continuous loading with very small incremental loadings following by a creep period. In various cases the dependence of the stabilization boundary on the loading history is more or less significant. The stabilization boundary will be introduced in subsequent chapters using the concept of stress-work.

Exercises

1. Give a detailed derivation of the formulae (4.2.5)—(4.2.7).

2. Choose a soft rock and make hydrostatic compression tests (either in a 'triaxial' test or in a device of the kind shown in Fig. 4.2.1). Make incremental creep tests and try to determine a stabilization boundary. Find an empirical simple equation of this boundary.

3. After solving the previous exercise, try to determine the viscosity coefficient k_v by writing a formula of the form (4.3.9) corresponding to the stabilization boundary you have determined. Compare the magnitudes of k_v obtained for various stress levels.

4. Repeat the previous two exercises for a harder rock and then for one possessing a great percentage of voids besides microcracks. Compare the results with the previous ones. Formulate mathematical models of your own for each kind of rock.

5. Starting from the experimental data given in Table 4.2 find another equation for the volumetric compressibility than (4.3.12). Keep it as simple as possible.

6. From experimental data of your own obtained during unloading in hydrostatic compression tests, try to determine the dependence of the bulk modulus on the mean stress (see (4.3.15)). Compare the results with those obtained by dynamic tests (see Chapter 6). Try to determine the pressure σ_0 which would completely close the microcracks.

7. Make uniaxial unconfined creep compression tests on a rock of your choice with incremental increase of the stress followed by incremental decrease of the stress (see Fig. 4.5.1). Find out if the volume is dilatant and if volumetric creep can be observed during loading and unloading.

8. Let σ_i ($i = 1, 2, 3$) be the principal components of the stress tensor and $\mathbf{e}_i$ the unit vector in the principal directions. Consider the 'octahedral' plane of normal

$$\mathbf{n}_0 = \sum_{i=1}^{3} \frac{1}{\sqrt{3}} \mathbf{e}_i.$$

(a) Show that the projection of the stress vector corresponding to $\mathbf{n}_0$, $\mathbf{t}(\mathbf{x}, \mathbf{n}_0) = 1/\sqrt{3}\, \sigma_i \mathbf{e}_i$ on $\mathbf{n}_0$ is the mean stress

$$\mathbf{t}(\mathbf{x}, \mathbf{n}_0) \cdot \mathbf{n}_0 = \sigma = \tfrac{1}{3} \operatorname{tr}(\boldsymbol{\sigma}).$$

(b) Show that the projection of the same stress vector on the octahedral plane is a vector

$$\mathbf{t}_\tau(\mathbf{x}, \mathbf{n}_0) = \tfrac{1}{3}\,[(\sigma_1 - \sigma)\mathbf{e}_1 + (\sigma_2 - \sigma)\mathbf{e}_2 + (\sigma_3 - \sigma)\mathbf{e}_3]$$

of magnitude

$$|\mathbf{t}_\tau(\mathbf{x}, \mathbf{n}_0)| = \frac{\sqrt{2}}{3}\,\bar{\sigma} = \sqrt{\frac{2}{3}\,\mathrm{II}_{\sigma'}}$$

with $\bar{\sigma}$ the equivalent stress (see (4.1.1)) and $\mathrm{II}_{\sigma'}$ the second invariant of the stress deviator.

(c) Let $\mathbf{e}_i'$ be the projections of $\mathbf{e}_i$ on the octahedral plane. Show that

$$\cos(\mathbf{e}_i', \mathbf{e}_i) = \sqrt{\tfrac{2}{3}} \qquad \text{(fixed } i \text{ with } i = 1, 2, 3).$$

Now project $\mathbf{t}(\mathbf{x}, \mathbf{n}_0)$ on various directions $\mathbf{e}_i'$ and equate with the projection of $\bar{\sigma}$ on the same direction. Show that

$$\cos 3\alpha = \left(\sqrt{\frac{3}{\mathrm{II}_{\sigma'}}}\right)^3 \frac{\mathrm{III}_{\sigma'}}{2},$$

where $\alpha \in [0, \tfrac{1}{3}\pi]$ defines the orientation of $\mathbf{t}_\tau(\mathbf{x}, \mathbf{n}_0)$ in the octahedral plane and is the angle between $\mathbf{t}_\tau(\mathbf{x}, \mathbf{n}_0)$ and that of the unit vector $\mathbf{e}_i'$ which corresponds to the maximum σ_i'.

5

Classical constitutive equations

5.1. The linear elastic model

The linear elastic model for isotropic and homogeneous bodies is still widely used in rock mechanics. The design and strength of underground structures are now computed mainly with the help of linear elasticity. Even in those cases in which time-dependent phenomena such as creep, relaxation, etc. are involved and obviously significant, linear elasticity may still be useful in providing a first approximation to the stress, strain, and displacement distribution around the underground openings. The linear elastic model is also used as a theoretical background for some experimental techniques (photoelasticity, strain gauges, dynamic tests, etc.). Besides their mathematical simplicity, the solutions and methods developed with the assumption of linear elasticity can sometimes be easily generalized to some other cases of interest for rock mechanics: anisotropic elasticity (stratified rocks), non-homogeneous elasticity, linear visco-elasticity, etc. For the purpose of the present work, the 'linear elastic solution' is used as a first approximation to a more complete solution obtained with a more complex rheological model.

Let us assume that the rock is, at least locally, homogeneous and isotropic. A linear elastic constitutive equation is then written in the form of Hooke's law:

$$\varepsilon_{xx} = \frac{1}{E}\left[\sigma_{xx} - \nu(\sigma_{yy} + \sigma_{zz})\right], \ldots, \qquad \varepsilon_{xy} = \frac{\sigma_{xy}}{2G}, \ldots, \tag{5.1.1}$$

with E, G, K, and ν elastic moduli satisfying

$$G = \frac{E}{2(1+\nu)} = \frac{3KE}{9K-E} = \frac{3K(1-2\nu)}{2(1+\nu)}. \tag{5.1.2}$$

The determination of the elastic moduli from experimental data is not an easy task (see Chapter 2 and Chapter 6). Within the framework of the present chapter, the elastic moduli must be determined as mean values obtained from the slopes of the stress—strain curves, in circumstances when these stress—strain curves can somehow be approximated by straight lines.

Sometimes Hooke's law is written using deviators:

$$\boldsymbol{\sigma}' = 2G\boldsymbol{\varepsilon}', \qquad \sigma = 3K\varepsilon \tag{5.1.3}$$

with

$$\begin{aligned} \boldsymbol{\sigma}' &= \boldsymbol{\sigma} - \sigma\mathbf{1}, \qquad & \boldsymbol{\varepsilon}' &= \boldsymbol{\varepsilon} - \varepsilon\mathbf{1} \\ \sigma &= \tfrac{1}{3}\,\mathrm{tr}\,\boldsymbol{\sigma}, \qquad & \varepsilon &= \tfrac{1}{3}\,\mathrm{tr}\,\boldsymbol{\varepsilon}. \end{aligned} \tag{5.1.4}$$

Some useful formulae, besides (5.1.2), which relate the values of the elastic constants are:

$$\begin{aligned} K &= \frac{GE}{3(3G - E)} = \frac{2G(1+\nu)}{3(1-2\nu)} = \frac{E}{3(1-2\nu)} \\ E &= 2G(1+\nu) = \frac{9KG}{3K+G} = 3K(1-2\nu) \\ \nu &= \frac{E-2G}{2G} = \frac{3K-2G}{3(3K+G)} = \frac{3K-E}{6K}\,. \end{aligned} \tag{5.1.5}$$

According to the linear elastic model the volume response is always linearly compressible and reversible. It follows that this model cannot be used in all cases in which non-linear compressibility or dilatancy, possibly irreversible, are non-negligible phenomena.

The linear elastic constitutive equation is used when the body under consideration has a natural stress-free and strain-free reference configuration. Since rocks do not have such a configuration, Hooke's law will be written in *relative* quantities for solving various problems: relative stresses (i.e. the difference between the actual stresses and those existing *in situ* before excavation) and relative strains (deformation with respect to the configuration before excavation). In other words, in rock mechanics the *reference configuration is the state existing* in situ *before excavation* (see Chapter 10).

For most rocks the displacements and material rotations are small during their deformation. Therefore, the actual configuration is not too far from the reference configuration. Thus in a general form, for anisotropic bodies the relationship between the Cauchy stress $\boldsymbol{\sigma}$ (coinciding with the Piola—Kirchhoff one) and the infinitesimal strain tensor $\boldsymbol{\varepsilon}$ is

$$\boldsymbol{\sigma} = \mathbf{C}[\boldsymbol{\varepsilon}] \quad \text{or} \quad \sigma_{ij} = C_{ijk\ell}\varepsilon_{k\ell}, \quad i, j, k, \ell = 1, 2, 3, \tag{5.1.6}$$

where the 81 components $C_{ijk\ell}$ of the fourth-rank tensor $\mathbf{C}$ are material constants; due to the symmetry of the stress tensor $C_{ijk\ell} = C_{jik\ell}$ and due to the symmetry of the strain tensor $C_{ijk\ell} = C_{ij\ell k}$. Thus there are only 36 independent constants $C_{ijk\ell}$. If, further, the fourth-rank tensor $\mathbf{C}$ is symmetric, $\mathbf{C} = \mathbf{C}^{\mathrm{T}}$, i.e. $C_{ijk\ell} = C_{k\ell ij}$, the number of these constants reduces to 21.

When Hooke's law is used for anisotropic bodies, sometimes 'contracted' notation is preferred to the tensorial one according to the following rule for the indices:

tensorial	11	22	33	23 or 32	31 or 13	12 or 21
contracted	1	2	3	4	5	6

Thus (5.1.6) is written in contracted notation

$$\sigma_i = C_{ij}\varepsilon_j \quad (i, j = 1, \ldots, 6), \tag{5.1.7}$$

where C_{ij} is *stiffness matrix* with 36 independent constant components. The stress and strain components will be denoted by

$$\sigma_{11} = \sigma_1, \ldots, \sigma_{12} = \sigma_6 \quad \text{and} \quad \varepsilon_{11} = \varepsilon_1, \ldots, 2\varepsilon_{12} = \varepsilon_6.$$

According to Green the elastic model (often called hyperelastic) can be introduced with a scalar potential $W(\boldsymbol{\varepsilon})$. Thus a material is elastic in the Green sense, or hyperelastic if there is a scalar function $W(\boldsymbol{\varepsilon})$ playing the role of an elastic potential, so that the constitutive equation can be written in the form

$$\sigma_{ij} = \frac{\partial W(\boldsymbol{\varepsilon})}{\partial \varepsilon_{ij}} \quad \text{or} \quad \sigma_i = \frac{\partial W(\boldsymbol{\varepsilon})}{\partial \varepsilon_i}, \tag{5.1.8}$$

where W must be symmetric with respect to ε_{ij} and ε_{ji}. $W(\boldsymbol{\varepsilon})$ is the energy of deformation per unit volume in the undeformed reference configuration.

W is generally considered to be a quadratic form in ε_i, i.e. in contracted notation:

$$W = \tfrac{1}{2} C_{ij}\varepsilon_i\varepsilon_j \quad (i, j = 1, \ldots, 6). \tag{5.1.9}$$

We could also add linear terms to this expression, for instance

$$C_0 + C_1\varepsilon_1 + C_2\varepsilon_2 + \cdots + C_6\varepsilon_6. \tag{5.1.10}$$

However, if we prefer to have $W = 0$ for the stress-free strain-free state, then $C_0 = 0$. Furthermore, if we wish to use (5.1.8) for rocks *in situ*, we can use the linear terms from (5.1.10) as well, just to take into account in the constitutive equation the primary stresses already existing *in situ* when $\varepsilon_i = 0$. In this case the stresses obtained from (5.1.8) are secondary stresses (see Chapter 10).

More often the state *in situ* before excavation is used as a reference state. Thus the stresses involved in the constitutive equation are 'relative' (or 'difference') stresses, i.e. the actual stress minus the stress existing *in situ* before excavation. In this case all linear terms (5.1.10) are missing in the expression of W. From (5.1.9) we have

$$\frac{\partial^2 W}{\partial \varepsilon_i \partial \varepsilon_j} = C_{ij} \tag{5.1.11}$$

and furthermore, the symmetry of $\mathbf{C}$: $C_{ij} = C_{ji}$.

Hooke's law, written in the general form (5.1.6), corresponds to general elastic anisotropy. If the body has in each point a plane of symmetry and if this plane is chosen to be the $z = 0$ plane, then the mapping

$$x' = x, \qquad y' = y, \qquad z' = -z \tag{5.1.12}$$

must leave invariant the elastic potential. It follows that the general form of Hooke's law is written as

$$\begin{bmatrix} \sigma_1 \\ \sigma_2 \\ \sigma_3 \\ \sigma_4 \\ \sigma_5 \\ \sigma_6 \end{bmatrix} = \begin{bmatrix} C_{11} & C_{12} & C_{13} & 0 & 0 & C_{16} \\ C_{12} & C_{22} & C_{23} & 0 & 0 & C_{26} \\ C_{13} & C_{23} & C_{33} & 0 & 0 & C_{36} \\ 0 & 0 & 0 & C_{44} & C_{45} & 0 \\ 0 & 0 & 0 & C_{45} & C_{55} & 0 \\ C_{16} & C_{26} & C_{36} & 0 & 0 & C_{66} \end{bmatrix} \begin{bmatrix} \varepsilon_1 \\ \varepsilon_2 \\ \varepsilon_3 \\ \varepsilon_4 \\ \varepsilon_5 \\ \varepsilon_6 \end{bmatrix}. \tag{5.1.13}$$

The stiffness matrix now has only 13 independent components.

If in each point of the body there are two orthogonal symmetry planes, then there is also a third orthogonal symmetry plane passing through the same point. The symmetry with respect to three mutually orthogonal planes is called *orthotropy*, and the stiffness matrix reduces to

$$\begin{bmatrix} C_{11} & C_{12} & C_{13} & 0 & 0 & 0 \\ C_{12} & C_{22} & C_{23} & 0 & 0 & 0 \\ C_{13} & C_{23} & C_{33} & 0 & 0 & 0 \\ 0 & 0 & 0 & C_{44} & 0 & 0 \\ 0 & 0 & 0 & 0 & C_{55} & 0 \\ 0 & 0 & 0 & 0 & 0 & C_{66} \end{bmatrix} \tag{5.1.14}$$

with only nine independent components.

Let us consider now the mapping

$$\begin{aligned} x' &= x \cos \theta + y \sin \theta \\ y' &= -x \sin \theta + y \cos \theta \\ z' &= z. \end{aligned} \tag{5.1.15}$$

If the elastic potential is invariant with respect to such mapping for any angle θ, the body is *transversely isotropic*. The planes z = const. are isotropy planes. The stiffness matrix

$$\begin{bmatrix} C_{11} & C_{12} & C_{13} & 0 & 0 & 0 \\ C_{12} & C_{11} & C_{13} & 0 & 0 & 0 \\ C_{13} & C_{13} & C_{33} & 0 & 0 & 0 \\ 0 & 0 & 0 & C_{44} & 0 & 0 \\ 0 & 0 & 0 & 0 & C_{44} & 0 \\ 0 & 0 & 0 & 0 & 0 & \frac{1}{2}(C_{11} - C_{12}) \end{bmatrix} \tag{5.1.16}$$

possesses only five independent components. Transverse isotropy is quite important for rock mechanics, since very often the rock texture is layered. In this case the $0z$-axis is chosen normal to the planes of layers.

In order to determine the numerical values of C_{ij} from experimental data, it is useful to express the stiffness matrix components in terms of the so-called engineering constants (Lekhnitskii (1977); see also Amadei (1983)). For transverse isotropy we have

$$C_{11} = C_{22} = \frac{1 - \nu_{32}\nu_{23}}{E_3 E_1 C}, \qquad C_{33} = -\frac{1 - \nu_{21}^2}{E_1^2 C},$$

$$C_{12} = C_{21} = \frac{\nu_{21} + \nu_{32}\nu_{23}}{E_1^2 C}, \qquad C_{32} = C_{31} = \frac{\nu_{23}(1 + \nu_{21})}{E_1^2 C},$$

$$C_{66} = \frac{C_{11} - C_{21}}{2} = \frac{1 - \nu_{21} - 2\nu_{32}\nu_{23}}{2E_1 E_3 C}, \qquad C_{44} = C_{55} = G_{13} = G_{23}, \tag{5.1.17}$$

$$C = \frac{(1 + \nu_{21})(1 - \nu_{21} - 2\nu_{32}\nu_{23})}{E_1^2 E_3},$$

where E_1 is the elastic modulus in any one of the directions contained in the planes z = const., E_3 is the modulus in the $0z$-direction, $\nu_{ij} = -\varepsilon_j/\varepsilon_i$ are Poisson ratios, where ε_i is the longitudinal strain in the i-direction, and ε_j is the strain in the corresponding transverse direction j. From the symmetry of the stress tensor, the so-called reciprocal relations follow:

$$\frac{\nu_{ij}}{E_i} = \frac{\nu_{ji}}{E_j}, \quad i, j = 1, 2, 3 \text{ (no summation).} \tag{5.1.18}$$

From all these constants the shear modulus G_{13} is the most difficult to determine from experimental data. One can use some procedures employed in the mechanics of composite materials to determine G_{13} (see Jones (1975, §2.8.2), Cristescu (1983*a*, §3.3)). A cylindrical specimen can be cut with its 0ξ-axis making an angle of $\frac{1}{4}\pi$ with the $0z$-axis; from a uniaxial test in the direction 0ξ one can determine

$$E_\xi = \frac{P}{A\varepsilon_\xi},$$

with P the applied force and A the area of the cross-section. Introducing E_ξ together with E_1, E_3, and ν_{13} assumed to be already determined, in the standard formula

$$G_{13} = \left(\frac{4}{E_\xi} - \frac{1}{E_1} - \frac{1}{E_3} + \frac{2\nu_{13}}{E_1}\right)^{-1}, \tag{5.1.19}$$

one can obtain G_{13}. The so-called 'Iosipescu specimen' may also be useful for this purpose (see Walrath and Adams (1983)), but the cutting of a rock specimen in this shape is not an easy task.

Numerical values for the constants (5.1.17) have been determined for various anisotropic rocks by several authors (see Chong *et al.* (1980), Podio *et al.* (1968), Lama and Vutukuri (1978, vol. II, §6.7 and §6.8), etc.). Generally with rocks the elastic moduli (5.1.17) are not really constant since their magnitude is, as a rule, stress-state-dependent. The elastic rock anisotropy is discussed in detail by Amadei (1983).

In the case of *isotropic* bodies any plane is a symmetry plane and any axis is a symmetry axis. The stiffness matrix is in this case

$$\begin{bmatrix} C_{11} & C_{12} & C_{12} & 0 & 0 & 0 \\ C_{12} & C_{11} & C_{12} & 0 & 0 & 0 \\ C_{12} & C_{12} & C_{11} & 0 & 0 & 0 \\ 0 & 0 & 0 & \frac{1}{2}(C_{11}-C_{12}) & 0 & 0 \\ 0 & 0 & 0 & 0 & \frac{1}{2}(C_{11}-C_{12}) & 0 \\ 0 & 0 & 0 & 0 & 0 & \frac{1}{2}(C_{11}-C_{12}) \end{bmatrix}. \tag{5.1.20}$$

Therefore, there are only two independent elastic constants. Hooke's law is written in the form

$$\sigma_{ij} = \lambda\varepsilon_{kk}\delta_{ij} + 2\mu\varepsilon_{ij} \tag{5.1.21}$$

with λ and μ the Lamé constants, and therefore $C_{11} = \lambda + 2\mu$ and $C_{12} = \lambda$. (5.1.21) is certainly equivalent to (5.1.1) and (5.1.3), if we take into account that

$$\lambda = \frac{\nu E}{(1+\nu)(1-2\nu)}, \qquad \mu = G = \frac{E}{2(1+\nu)}. \tag{5.1.22}$$

Further, we recall another well known result from linear elasticity: the necessary and sufficient condition for the displacements to be unique in a simply connected body is that the so-called Saint-Venant compatibility conditions be satisfied:

$$\begin{aligned} &\frac{\partial^2\varepsilon_{xx}}{\partial y^2} + \frac{\partial^2\varepsilon_{yy}}{\partial x^2} = 2\frac{\partial^2\varepsilon_{xy}}{\partial x\partial y}, \ldots, \\ &\frac{\partial^2\varepsilon_{xx}}{\partial y\partial z} = \frac{\partial}{\partial x}\left(-\frac{\partial\varepsilon_{yz}}{\partial x} + \frac{\partial\varepsilon_{zx}}{\partial y} + \frac{\partial\varepsilon_{xy}}{\partial z}\right), \ldots. \end{aligned} \tag{5.1.23}$$

From (5.1.23) all six compatibility equations can be obtained by circular permutations. Only three of them are independent. Sometimes it is useful to express these conditions in the stresses using Hooke's law and the equilibrium equations

$$\sigma_{ji,j} + \rho b_i = 0 \tag{5.1.24}$$

with **b** the body forces. We get

$$\nabla^2 \sigma_{xx} + \frac{3}{1+\nu} \frac{\partial^2 \sigma}{\partial x^2} = -\frac{\nu}{1-\nu} \left(\frac{\partial(\rho b_x)}{\partial x} + \frac{\partial(\rho b_y)}{\partial y} + \right.$$

$$\left. + \frac{\partial(\rho b_z)}{\partial z} \right) - 2 \frac{\partial(\rho b_x)}{\partial x}, \ldots$$

$$\nabla^2 \sigma_{xy} + \frac{3}{1+\nu} \frac{\partial^2 \sigma}{\partial y \partial z} = -\left(\frac{\partial(\rho b_y)}{\partial z} + \frac{\partial(\rho b_z)}{\partial y} \right), \ldots \tag{5.1.25}$$

with

$$\nabla^2 = \frac{\partial^2}{\partial x^2} + \frac{\partial^2}{\partial y^2} + \frac{\partial^2}{\partial z^2}, \tag{5.1.26}$$

the Laplace operator. This system of equations is used when the boundary conditions are expressed in terms of stresses.

Sometimes the equilibrium conditions are written in displacement components, using Hooke's law:

$$\nabla^2 u_x + \frac{1}{1-2\nu} \frac{\partial}{\partial x} \left(\frac{\partial u_x}{\partial x} + \frac{\partial u_y}{\partial y} + \frac{\partial u_z}{\partial z} \right) + \frac{\rho b_x}{G} = 0, \ \ldots \tag{5.1.27}$$

in those cases when the boundary conditions are also expressed in displacements.

Many engineering problems of rock mechanics are *plane strain problems* (Teodorescu (1966)), when $u_z = 0$ (say) and u_x and u_y are independent of z. In this case Hooke's law becomes

$$\begin{aligned} \sigma_{xx} &= \lambda(\varepsilon_{xx} + \varepsilon_{yy}) + 2G\varepsilon_{xx} \\ \sigma_{yy} &= \lambda(\varepsilon_{xx} + \varepsilon_{yy}) + 2G\varepsilon_{yy} \\ \sigma_{xy} &= 2G\varepsilon_{xy} \end{aligned} \tag{5.1.28}$$

and the unique compatibility condition is

$$\frac{\partial^2 \varepsilon_{xx}}{\partial y^2} + \frac{\partial^2 \varepsilon_{yy}}{\partial x^2} = 2 \frac{\partial^2 \varepsilon_{xy}}{\partial x \partial y}. \tag{5.1.29}$$

To these relations two balance-of-momentum equations must be added. Since $\varepsilon_{zz} = 0$, (5.1.1) also yields

$$\sigma_{zz} = \nu(\sigma_{xx} + \sigma_{yy}). \tag{5.1.30}$$

If the strain components are eliminated between (5.1.28) and (5.1.29) and if the equilibrium conditions are also used, we obtain the compatibility condition in stresses as

$$\left(\frac{\partial^2}{\partial x^2} + \frac{\partial^2}{\partial y^2}\right)(\sigma_{xx} + \sigma_{yy}) = -\frac{1}{1-\nu}\left[\frac{\partial(\rho b_x)}{\partial x} + \frac{\partial(\rho b_y)}{\partial y}\right]. \qquad (5.1.31)$$

In the absence of the body forces this equation becomes

$$\nabla^2(\sigma_{xx} + \sigma_{yy}) = 0, \qquad (5.1.32)$$

where the Laplace operator is now two-dimensional.

It is also well known that in the case of plane problems, and in the absence of body forces, the equilibrium conditions are satisfied if the stress components are obtained from

$$\sigma_{xx} = \frac{\partial^2\Phi}{\partial y^2}, \qquad \sigma_{yy} = \frac{\partial^2\Phi}{\partial x^2}, \qquad \sigma_{xy} = -\frac{\partial^2\Phi}{\partial x\partial y}, \qquad (5.1.33)$$

where the scalar function $\Phi(x, y)$ is the Airy function. By combining (5.1.33) with (5.1.31) it follows that the Airy function is a biharmonic one:

$$\nabla^2(\nabla^2\Phi) = 0. \qquad (5.1.34)$$

In Cartesian coordinates this equation becomes

$$\frac{\partial^4\Phi}{\partial x^4} + 2\frac{\partial^4\Phi}{\partial x^2\partial y^2} + \frac{\partial^4\Phi}{\partial y^4} = 0. \qquad (5.1.35)$$

A similar procedure is used for the *plane stress* problems, when the only non-zero stress components are σ_{xx}, σ_{yy}, and σ_{xy}, and they do not depend on the coordinate z. Now Hooke's law is

$$\varepsilon_{xx} = \frac{1}{E}(\sigma_{xx} - \nu\sigma_{yy}), \qquad \varepsilon_{yy} = \frac{1}{E}(\sigma_{yy} - \nu\sigma_{xx})$$
$$\varepsilon_{zz} = -\frac{\nu}{E}(\sigma_{xx} + \sigma_{yy}), \qquad \varepsilon_{xy} = \frac{\sigma_{xy}}{2G}. \qquad (5.1.36)$$

The equilibrium conditions and the compatibility condition for strains are the same as above, while the compatibility conditions for stresses become

$$\nabla^2(\sigma_{xx} + \sigma_{yy}) = -(1+\nu)\left[\frac{\partial(\rho b_x)}{\partial x} + \frac{\partial(\rho b_y)}{\partial y}\right]. \qquad (5.1.37)$$

Thus the plane stress problems yield the same kind of mathematical formulations as the plane strain problems. From the first problem one can pass to the second one by replacing E by $E/(1-\nu^2)$ and ν by $\nu/(1-\nu)$.

When Hooke's law is accepted for rock-like materials, one has to define the domain in a stress-space where this constitutive equation holds. Generally, the boundary of this domain is defined by a failure condition formulated in the classic sense, i.e. a condition formulated in terms of stress invariants (see Chapter 9). Thus within such a modelling we have to accept that the rock stays elastic up to failure. In the following chapters a linear elastic constitutive equation written in differential form will be used to describe the 'instantaneous' response of the rocks. Thus, besides 'elasticity', it will be assumed that rocks exhibit many other mechanical properties which have to be described by a much more complex constitutive equation.

The *influence of the pore pressure*, and how this influence is described by the equations which must be satisfied by the assumption that the rock is elastic, have been considered by several authors (see Stagg and Zienkienwicz (1968, pp. 84—89), Serafim (1968)). It is well known that the pressure of the water in the pores can be quite high, particularly at greater depths. The pore pressure can be modelled by additional body forces. According to one of the approaches, first the concept of *total stress* $\bar{\boldsymbol{\sigma}}$ is introduced; this is the ratio of all forces transmitted through an area to the area itself as its magnitude tends towards zero (preserving the proportion of voids). The equilibrium equations are

$$\begin{aligned}
&\frac{\partial \bar{\sigma}_{xx}}{\partial x} + \frac{\partial \bar{\sigma}_{xy}}{\partial y} + \frac{\partial \bar{\sigma}_{xz}}{\partial z} = 0 \\
&\frac{\partial \bar{\sigma}_{yx}}{\partial x} + \frac{\partial \bar{\sigma}_{yy}}{\partial y} + \frac{\partial \bar{\sigma}_{yz}}{\partial z} = 0 \\
&\frac{\partial \bar{\sigma}_{zx}}{\partial x} + \frac{\partial \bar{\sigma}_{zy}}{\partial y} + \frac{\partial \bar{\sigma}_{zz}}{\partial z} = -\gamma,
\end{aligned} \tag{5.1.38}$$

with γ the weight of the saturated material. The total stress can be decomposed into two parts: $\bar{\boldsymbol{\sigma}} - p\mathbf{1}$ corresponds to a stress with zero pore pressure, while $p\mathbf{1}$ corresponds to a pressure acting in the pores (the principle of superposition of forces is accepted). The first component will deform the rock as it would be without fluid in the pores, with the elastic constants for the rock being E and ν. The second component subjects the solid skeleton to a hydrostatic compression of compressive strain $p/3K_c$, with K_c the average bulk modulus for the solid phase. Therefore, the strains are obtained from

$$\begin{aligned}
\varepsilon_{xx} &= \frac{1}{E}\left[\bar{\sigma}_{xx} - p - \nu(\bar{\sigma}_{yy} + \bar{\sigma}_{zz} - 2p)\right] + \frac{p}{3K_c} \\
&= \frac{1}{E}\left[\bar{\sigma}_{xx} - \nu(\bar{\sigma}_{yy} + \bar{\sigma}_{zz})\right] - p\left[\frac{1-2\nu}{E} - \frac{1}{3K_c}\right], \ldots \\
\varepsilon_{xy} &= \frac{1}{2G}\,\bar{\sigma}_{xy} = \frac{2(1+\nu)}{E}\,\bar{\sigma}_{xy}, \ldots .
\end{aligned} \tag{5.1.39}$$

The equations for the problem just formulated are similar to the equations of thermoelasticity if we use the notation

$$\alpha T = -p\left[\frac{1-2\nu}{E} - \frac{1}{3K_c}\right] = -p\left[\frac{1}{3K} - \frac{1}{3K_c}\right], \tag{5.1.40}$$

where K is the bulk modulus of the porous body and α has the meaning of a certain expansion coefficient. If the porosity tends towards zero, i.e. $K \to K_c$, the bracket in (5.1.40) vanishes and therefore, the effect of pore pressure in Hooke's law disappears. If a boundary-value problem is formulated, (5.1.38) and (5.1.39) can be used to solve the problem.

Another approach uses the concept of *effective stress* $\boldsymbol{\sigma}^e$ defined by

$$\boldsymbol{\sigma}^e = \bar{\boldsymbol{\sigma}} - \eta p \mathbf{1}, \tag{5.1.41}$$

where η is the *effective area* with $0 \leqslant \eta \leqslant 1$ (see also Barnabe (1986, 1987)). The equilibrium equations are now

$$\frac{\partial \sigma^e_{xz}}{\partial x} + \frac{\partial \sigma^e_{yz}}{\partial y} + \frac{\partial \sigma^e_{zz}}{\partial z} + \frac{\partial(\eta p)}{\partial z} = -\gamma, \tag{5.1.42}$$

and Hooke's law is

$$\varepsilon_{xx} = \frac{1}{E}\left[\sigma^e_{xx} - \nu(\sigma^e_{yy} + \sigma^e_{zz})\right] - p\left[\frac{(1-2\nu)(1-\eta)}{E} - \frac{1}{3K_c}\right], \ldots \tag{5.1.43}$$

The problem can now be interpreted as an elastic problem with body forces $\partial(\eta p)/\partial x$, $\partial(\eta p)/\partial y$, $\partial(\eta p)/\partial z$, and a 'thermal expansion'

$$\alpha T = -p\left[\frac{(1-2\nu)(1-\eta)}{E} - \frac{1}{3K_c}\right]. \tag{5.1.44}$$

In the particular case

$$\eta = 1 - \frac{K}{K_c},$$

suggested by Geertsma (see Stagg and Zienkiewicz (1968, p. 89)), the term (5.1.44) corresponding to the thermal expansion is missing.

5.2. Plane strain elasticity in cylindrical coordinates

Quite often rock mechanics problems have to be solved in *cylindrical coordinates* r, θ, z. Since most of these problems are plane-strain problems, the displacement vector **u** has a radial component u (z = const., θ = const.) and a circumferential one v (z = const., r = const.) only, i.e.

$$\mathbf{u} = u\mathbf{e}_r + v\mathbf{e}_\theta, \tag{5.2.1}$$

and the only non-zero components of the infinitesimal strains are

$$\begin{aligned} \varepsilon_{rr} &= \frac{\partial u}{\partial r}, \qquad \varepsilon_{\theta\theta} = \frac{1}{r}\frac{\partial v}{\partial \theta} + \frac{u}{r}, \\ \varepsilon_{r\theta} &= \frac{1}{2}\left(\frac{1}{r}\frac{\partial u}{\partial \theta} + \frac{\partial v}{\partial r} - \frac{v}{r}\right). \end{aligned} \tag{5.2.2}$$

The compatibility equation for strains is obtained quite easily as

$$2\frac{\partial^2 (r\varepsilon_{r\theta})}{\partial\theta\partial r} = \frac{\partial^2 \varepsilon_{rr}}{\partial\theta^2} + r\frac{\partial^2 (r\varepsilon_{\theta\theta})}{\partial r^2} - r\frac{\partial \varepsilon_{rr}}{\partial r}. \tag{5.2.3}$$

The equilibrium equations are

$$\begin{aligned} &\frac{\partial \sigma_{rr}}{\partial r} + \frac{1}{r}\frac{\partial \sigma_{r\theta}}{\partial \theta} + \frac{1}{r}(\sigma_{rr} - \sigma_{\theta\theta}) + \rho b_r = 0 \\ &\frac{\partial \sigma_{r\theta}}{\partial r} + \frac{1}{r}\frac{\partial \sigma_{\theta\theta}}{\partial \theta} + \frac{2\sigma_{r\theta}}{r} + \rho b_\theta = 0, \end{aligned} \tag{5.2.4}$$

where the last terms are body forces.

Hooke's law for the plane strain case is added to these equations, in cylindrical coordinates:

$$\begin{aligned} &\sigma_{rr} = \lambda(\varepsilon_{rr} + \varepsilon_{\theta\theta}) + 2G\varepsilon_{rr}, \qquad \sigma_{\theta\theta} = \lambda(\varepsilon_{rr} + \varepsilon_{\theta\theta}) + 2G\varepsilon_{\theta\theta}, \\ &\sigma_{r\theta} = 2G\varepsilon_{r\theta}, \qquad \sigma_{zz} = \nu(\sigma_{rr} + \sigma_{\theta\theta}) \end{aligned} \tag{5.2.5}$$

which can be obtained quite easily, since in the orthogonal curvilinear coordinates the physical components of a tensor at a point are just the Cartesian components in a local Cartesian system with its axes tangent to the coordinate curves.

The inverse of (5.2.5) is

$$\begin{aligned} &\varepsilon_{rr} = \frac{1+\nu}{E}[(1-\nu)\sigma_{rr} - \nu\sigma_{\theta\theta}], \qquad \varepsilon_{\theta\theta} = \frac{1+\nu}{E}[-\nu\sigma_{rr} + (1-\nu)\sigma_{\theta\theta}] \\ &\varepsilon_{r\theta} = \frac{1}{2G}\sigma_{r\theta}. \end{aligned} \tag{5.2.6}$$

The compatibility condition (5.2.3) can be expressed also in stresses. This is obtained if (5.2.6) is introduced in (5.2.3)

$$2\left(r\frac{\partial^2\sigma_{r\theta}}{\partial\theta\partial r}+\frac{\partial\sigma_{r\theta}}{\partial\theta}\right)$$

$$=(1-\nu)\left(\frac{\partial^2\sigma_{rr}}{\partial\theta^2}+2r\frac{\partial\sigma_{\theta\theta}}{\partial r}+r^2\frac{\partial^2\sigma_{\theta\theta}}{\partial r^2}-r\frac{\partial\sigma_{rr}}{\partial r}\right)-$$

$$-\nu\left(\frac{\partial^2\sigma_{\theta\theta}}{\partial\theta^2}+2r\frac{\partial\sigma_{rr}}{\partial r}+r^2\frac{\partial^2\sigma_{rr}}{\partial r^2}-r\frac{\partial\sigma_{\theta\theta}}{\partial r}\right), \tag{5.2.7}$$

and, further, the equilibrium equations (5.2.4) are written as

$$\frac{\partial^2\sigma_{rr}}{\partial r^2}-\frac{1}{r^2}\frac{\partial\sigma_{r\theta}}{\partial\theta}+\frac{1}{r}\frac{\partial^2\sigma_{r\theta}}{\partial r\partial\theta}-\frac{\sigma_{rr}-\sigma_{\theta\theta}}{r^2}+$$

$$+\frac{1}{r}\frac{\partial(\sigma_{rr}-\sigma_{\theta\theta})}{\partial r}=0$$

$$\frac{1}{r}\frac{\partial^2\sigma_{r\theta}}{\partial\theta\partial r}+\frac{1}{r^2}\frac{\partial^2\sigma_{\theta\theta}}{\partial\theta^2}+\frac{2}{r^2}\frac{\partial\sigma_{r\theta}}{\partial\theta}=0. \tag{5.2.8}$$

If the term $(\sigma_{rr}-\sigma_{\theta\theta})/r^2$ from $(5.2.8)_1$ is replaced by the same expression obtained from $(5.2.4)_1$, we get

$$\frac{\partial^2\sigma_{rr}}{\partial r^2}+\frac{1}{r}\frac{\partial^2\sigma_{r\theta}}{\partial r\partial\theta}+\frac{1}{r}\frac{\partial\sigma_{rr}}{\partial r}+\frac{1}{r}\frac{\partial(\sigma_{rr}-\sigma_{\theta\theta})}{\partial r}=0. \tag{5.2.9}$$

Further, by adding (5.2.9) and $(5.2.8)_2$, one obtains

$$2\left(\frac{\partial\sigma_{r\theta}}{\partial\theta}+r\frac{\partial^2\sigma_{r\theta}}{\partial r\partial\theta}\right)=-r^2\frac{\partial^2\sigma_{rr}}{\partial r^2}-\frac{\partial^2\sigma_{\theta\theta}}{\partial\theta^2}-$$

$$-2r\frac{\partial\sigma_{rr}}{\partial r}+r\frac{\partial\sigma_{\theta\theta}}{\partial r}. \tag{5.2.10}$$

Finally, from (5.2.10) and (5.2.7) we obtain

$$\left(\frac{\partial^2}{\partial r^2}+\frac{1}{r^2}\frac{\partial^2}{\partial\theta^2}+\frac{1}{r}\frac{\partial}{\partial r}\right)(\sigma_{rr}+\sigma_{\theta\theta})=0, \tag{5.2.11}$$

with

$$\nabla^2=\frac{\partial^2}{\partial r^2}+\frac{1}{r^2}\frac{\partial^2}{\partial\theta^2}+\frac{1}{r}\frac{\partial}{\partial r} \tag{5.2.12}$$

the Laplace operator in polar coordinates.

In the absence of body forces, an Airy function $\Phi(r, \theta)$ in polar coordinates can be introduced as

$$\sigma_{rr} = \frac{1}{r}\frac{\partial\Phi}{\partial r} + \frac{1}{r^2}\frac{\partial^2\Phi}{\partial\theta^2}, \qquad \sigma_{\theta\theta} = \frac{\partial^2\Phi}{\partial r^2},$$
$$\sigma_{r\theta} = \frac{1}{r^2}\frac{\partial\Phi}{\partial\theta} - \frac{1}{r}\frac{\partial^2\Phi}{\partial r\partial\theta}. \tag{5.2.13}$$

The equilibrium equations are satisfied by the components (5.2.13), and further, from (5.2.11) and (5.2.13) we have

$$\nabla^2(\nabla^2\Phi) = 0, \tag{5.2.14}$$

i.e. the Airy function is a biharmonic one.

A general solution of equation (5.2.14) was given by Michell (1899) (see also Timoshenko and Goodier (1961), Obert and Duvall (1967)), using the method of separation of variables, i.e. the function Φ is written as

$$\Phi(r, \theta) = R(r)\Psi(\theta). \tag{5.2.15}$$

If (5.2.15) is introduced in (5.2.14), we get

$$\Psi\left(\frac{d^4R}{dr^4} + \frac{2}{r}\frac{d^3R}{dr^3} - \frac{1}{r^2}\frac{d^2R}{dr^2} + \frac{1}{r^3}\frac{dR}{dr}\right) +$$
$$+ \frac{d^2\Psi}{d\theta^2}\left(\frac{2}{r^2}\frac{d^2R}{dr^2} - \frac{2}{r^3}\frac{dR}{dr} + \frac{4R}{r^4}\right) +$$
$$+ \frac{R}{r^4}\frac{d^4\Psi}{d\theta^4} = 0. \tag{5.2.16}$$

Various particular cases can be considered. If the stress components are independent of θ then $\Psi = \text{const.}$ and (5.2.16) becomes

$$\frac{1}{r}\frac{d}{dr}\left\{r\frac{d}{dr}\left[\frac{1}{r}\frac{d}{dr}\left(r\frac{dR}{dr}\right)\right]\right\} = 0, \tag{5.2.17}$$

with the general solution of the form

$$R(r) = A_0 r^2 + B_0 r^2 \ln r + C_0 \ln r + D_0, \tag{5.2.18}$$

depending on four arbitrary integration constants. The stress components in this case are

$$\sigma_{rr} = \frac{1}{r}\frac{\partial\Phi}{\partial r} = 2A_0 + B_0(1 + 2\ln r) + \frac{C_0}{r^2},$$
$$\sigma_{\theta\theta} = \frac{\partial^2\Phi}{\partial r^2} = 2A_0 + B_0(3 + 2\ln r) - \frac{C_0}{r^2}, \tag{5.2.19}$$
$$\sigma_{r\theta} = 0.$$

The previous solution can be used in the case of axisymmetrical boundary conditions.

For other boundary conditions (half-space loaded by arbitrary forces),

$$\Psi = A_0 \theta, \tag{5.2.20}$$

and we find quite easily

$$\Phi(r, \theta) = E_0 \theta r^2 + F_0 \theta r^2 \ln r + G_0 \theta \ln r + H_0 \theta. \tag{5.2.21}$$

Other boundary value problems (bending of circular beams, etc.) suggest solutions of the form

$$\Psi = a \sin \theta \quad \text{or} \quad \Psi = b \cos \theta. \tag{5.2.22}$$

In this case the Airy function is of the form

$$\Phi(r, \theta) = \left(A_1 r + \frac{B_1}{r} + C_1 r^3 + D_1 r \ln r \right) \sin \theta + \\ + \left(E_1 r + \frac{F_1}{r} + G_1 r^3 + H_1 r \ln r \right) \cos \theta. \tag{5.2.23}$$

The problem of concentrated forces acting on a half-plane suggests the expressions of the form

$$\Psi = \theta \sin \theta \quad \text{or} \quad \Psi = \theta \cos \theta, \tag{5.2.24}$$

and the Airy function is

$$\Phi(r, \theta) = (J_1 r + K_1 r \ln r)(\theta \sin \theta) + (L_1 r + M_1 r \ln r)(\theta \cos \theta). \tag{5.2.25}$$

The important case of circular holes in an elastic plate suggests for Ψ the expressions

$$\Psi = \sin n\theta \quad \text{or} \quad \Psi = \cos n\theta, \tag{5.2.26}$$

with $n > 1$ an integer. The Airy function is in this case

$$\Phi(r, \theta) = (\sin n\theta)[A_n r^n + B_n r^{-n} + C_n r^{2-n} + D_n r^{2+n}] + \\ + (\cos n\theta)[E_n r^n + F_n r^{-n} + G_n r^{2-n} + H_n r^{2+n}]. \tag{5.2.27}$$

Therefore, adding up all the solutions given above, we obtain the solution (due to Michell), as

$$
\begin{aligned}
\Phi(r, \theta) = {} & A_0 r^2 + B_0 r \ln r + C_0 \ln r + D_0 + \\
& + (E_0 r^2 + F_0 r^2 \ln r + G_0 \ln r + H_0)\theta + \\
& + (A_1 r + B_1 r^{-1} + C_1 r^3 + D_1 r \ln r) \sin\theta + \\
& + (E_1 r + F_1 r^{-1} + G_1 r^3 + H_1 r \ln r) \cos\theta + \\
& + (J_1 r + K_1 r \ln r)\theta \sin\theta + (L_1 r + M_1 r \ln r)\theta \cos\theta + \\
& + \sum_{n=2}^{\infty} [A_n r^n + B_n r^{-n} + C_n r^{2-n} + D_n r^{2+n}] \sin n\theta + \\
& + \sum_{n=2}^{\infty} [E_n r^n + F_n r^{-n} + G_n r^{2-n} + H_n r^{2+n}] \cos n\theta.
\end{aligned}
\tag{5.2.28}
$$

For various specific problems a certain number of terms from (5.2.28) are chosen and the constants determined from the boundary conditions. The procedure for choosing the terms for a specific problem is discussed in Timoshenko and Goodier (1961). In the following chapters only solutions of relevance to rock mechanics will be presented.

5.3. Thick-walled tube subjected to internal and external pressures

Let us consider a thick-walled tube $a \leqslant r \leqslant b$ with the symmetry axis coinciding with the $0z$-axis. Let p_i be the internal pressure applied on the inner surface $r = a$ of the tube, and p_e be the external pressure applied on the outer surface $r = b$. The length of the tube is assumed to be infinite so that a plane strain state is assumed and polar coordinates are used. This problem was considered by Lamé (1852). From axial symmetry and plane strain follows $u_r = u$, $u_\theta = u_z = 0$, so that the (principal) components of the infinitesimal strains are

$$
\varepsilon_r = \frac{du}{dr}, \qquad \varepsilon_\theta = \frac{u}{r}, \qquad \varepsilon_{r\theta} = 0. \tag{5.3.1}
$$

Since two boundary conditions are available for the determination of the constants involved in (5.2.28), and since the problem is axisymmetrical and the stresses at infinity must be finite (i.e. $B_0 = 0$), we keep from expression (5.2.28) or (5.2.18) only two terms:

$$
\Phi(r) = A \ln r + Br^2, \tag{5.3.2}
$$

and therefore, the stresses are

$$\sigma_r = \frac{A}{r^2} + 2B, \qquad \sigma_\theta = -\frac{A}{r^2} + 2B, \qquad \sigma_{r\theta} = 0, \tag{5.3.3}$$

and σ_r, σ_θ, and σ_z are the principal stress components. Further, with the boundary conditions

$$\sigma_r|_{r=a} = p_i, \qquad \sigma_r|_{r=b} = p_e \tag{5.3.4}$$

we get

$$\begin{aligned} \sigma_r &= \frac{a^2b^2(p_i - p_e)}{b^2 - a^2}\frac{1}{r^2} + \frac{p_e b^2 - p_i a^2}{b^2 - a^2}, \\ \sigma_\theta &= -\frac{a^2b^2(p_i - p_e)}{b^2 - a^2}\frac{1}{r^2} + \frac{p_e b^2 - p_i a^2}{b^2 - a^2}, \\ \sigma_z &= \frac{2\nu(p_e b^2 - p_i a^2)}{b^2 - a^2}, \end{aligned} \tag{5.3.5}$$

where the component σ_z follows from $(5.2.5)_4$. The deformations and the displacements follow from Hooke's law (5.2.5) and from $(5.3.1)_2$, respectively:

$$\begin{aligned} \varepsilon_r &= \frac{1+\nu}{E}\left[\frac{a^2b^2(p_i - p_e)}{b^2 - a^2}\frac{1}{r^2} + (1 - 2\nu)\frac{p_e b^2 - p_i a^2}{b^2 - a^2}\right] \\ \varepsilon_\theta &= \frac{1+\nu}{E}\left[-\frac{a^2b^2(p_i - p_e)}{b^2 - a^2}\frac{1}{r^2} + (1 - 2\nu)\frac{p_e b^2 - p_i a^2}{b^2 - a^2}\right] \\ u &= \frac{1+\nu}{E}\left[-\frac{a^2b^2(p_i - p_e)}{b^2 - a^2}\frac{1}{r} + (1 - 2\nu)\frac{p_e b^2 - p_i a^2}{b^2 - a^2}\, r\right]. \end{aligned} \tag{5.3.6}$$

Let us observe that σ_z is constant along the tube and that the strain components satisfy the compatibility condition for strains (5.2.3).

In the particular case when the inner surface is stress-free, $p_i = 0$ and therefore

$$\sigma_r = \frac{b^2 p_e}{b^2 - a^2}\left(1 - \frac{a^2}{r^2}\right), \qquad \sigma_\theta = \frac{b^2 p_e}{b^2 - a^2}\left(1 + \frac{a^2}{r^2}\right),$$

$$\sigma_z = 2\nu \frac{b^2 p_e}{b^2 - a^2},$$

$$\varepsilon_r = \frac{1+\nu}{E}\frac{b^2 p_e}{b^2 - a^2}\left[1 - 2\nu - \frac{a^2}{r^2}\right], \tag{5.3.7}$$

$$\varepsilon_\theta = \frac{1+\nu}{E}\frac{b^2 p_e}{b^2 - a^2}\left[1 - 2\nu + \frac{a^2}{r^2}\right],$$

$$u = \frac{1+\nu}{E}\frac{b^2 p_e}{b^2 - a^2}\left[(1 - 2\nu)r + \frac{a^2}{r}\right].$$

Another noteworthy particular case is obtained if the sizes of the body are very large, i.e. when $b \to \infty$. This is the case with tunnels, which will be discussed in Chapter 12. The stresses, strains, and displacement field can be obtained from (5.3.5) and (5.3.6) when $b \to \infty$. We obtain

$$\sigma_r = p_e - (p_e - p_i)\frac{a^2}{r^2}, \qquad \sigma_\theta = p_e + (p_e - p_i)\frac{a^2}{r^2}, \qquad \sigma_z = 2\nu p_e,$$

$$\varepsilon_r = \frac{1+\nu}{E}\left[-(p_e - p_i)\frac{a^2}{r^2} + p_e(1 - 2\nu)\right],$$

$$\varepsilon_\theta = \frac{1+\nu}{E}\left[(p_e - p_i)\frac{a^2}{r^2} + p_e(1 - 2\nu)\right] \tag{5.3.8}$$

$$u = \frac{1+\nu}{E}\left[(p_e - p_i)\frac{a^2}{r} + p_e(1 - 2\nu)r\right].$$

From this solution it follows that at infinity when $r \to +\infty$ the far-field stresses and strains are

$$\sigma_r = \sigma_\theta = p_e, \qquad \sigma_z = 2\nu p_e$$
$$\varepsilon_r = \frac{(1+\nu)(1-2\nu)}{E} p_e, \qquad \varepsilon_\theta = \frac{(1+\nu)(1-2\nu)}{E} p_e, \tag{5.3.9}$$

and the displacement is infinite.

If we subtract (5.3.9) from (5.3.8), we obtain the so-called relative components (see Chapter 10):

$$\sigma_r^R = -(p_e - p_i)\frac{a^2}{r^2}, \qquad \sigma_\theta^R = (p_e - p_i)\frac{a^2}{r^2}, \qquad \sigma_z^R = 0,$$
$$\varepsilon_r^R = -\frac{1+\nu}{E}(p_e - p_i)\frac{a^2}{r^2}, \qquad \varepsilon_\theta^R = \frac{1+\nu}{E}(p_e - p_i)\frac{a^2}{r^2}, \tag{5.3.10}$$

and the corresponding displacement which follows from $(5.3.10)_4$ and $(5.3.1)_2$ is

$$u^R = \frac{1+\nu}{E}(p_e - p_i)\frac{a^2}{r}. \tag{5.3.11}$$

It is easy to see that all components (5.3.10) and (5.3.11) tend to zero as r becomes large.

The particular case just presented will be considered in detail in Chapter 11 and 12, while the meaning of relative components is presented in Chapter 10.

5.4. The general linear viscoelastic model

The general form of a rate-type linear viscoelastic constitutive equation for homogeneous and isotropic bodies is

$$\dot{\sigma} = 3K\dot{\varepsilon} - \eta_v(\sigma - 3K_0\varepsilon)$$
$$\dot{\boldsymbol{\sigma}}' = 2G\dot{\boldsymbol{\varepsilon}}' - \eta(\boldsymbol{\sigma}' - 2G_0\boldsymbol{\varepsilon}'), \tag{5.4.1}$$

or

$$\dot{\varepsilon} = -k_v\left(\varepsilon - \frac{\sigma}{3K_0}\right) + \frac{1}{3K}\dot{\sigma}$$
$$\dot{\boldsymbol{\varepsilon}}' = -k\left(\boldsymbol{\varepsilon}' - \frac{\boldsymbol{\sigma}'}{2G_0}\right) + \frac{1}{2G}\dot{\boldsymbol{\sigma}}', \tag{5.4.2}$$

with

$$k_v = \frac{K_0}{K}\eta_v, \qquad k = \frac{G_0}{G}\eta. \tag{5.4.3}$$

The constants K and G are the dynamic moduli, K_0 and G_0 are the static (in the relaxed state) moduli, while k_v and k (or η_v and η) are the constant viscosity coefficients.

If in (5.4.1) we put $K = K_0$ (an assumption which is generally accepted for metals) and if we choose as the initial state $\varepsilon = \varepsilon_0$ $\sigma = \sigma_0 = 3K_0\varepsilon_0$, then it follows that the mean stress and the mean strain are related by the relationship

$$\varepsilon = \frac{1}{3K_0}\sigma \tag{5.4.4}$$

for any deformation process of the volume, i.e. for any prescribed stress history $\sigma = \sigma(t)$. In this case we say that the volume of the material behaves elastically. The constitutive equations $(5.4.1)_2$ and (5.4.4) are known as the linear-standard viscoelastic model. This model, used mainly for metals and plastics, is recommended for rocks by Asszonyi and Richter (1979). We consider the more realistic model (5.4.1) for rocks which in the following chapters will be used to solve various engineering problems.

Let us assume that

$$K_0 \neq K \neq 0, \qquad K_0 \neq 0, \qquad G_0 \neq G \neq 0, \qquad G_0 \neq 0 \tag{5.4.5}$$

which is quite a realistic assumption for rock-like materials. In what follows it will be shown that the signs of various constants involved in (5.4.1) and a possible ordering by magnitude of the constants from (5.4.5), can be obtained from: the second law of thermodynamics, the stability properties of the material subjected to creep or relaxation, the conditions of existence of the waves which may propagate in rock and their order of propagation, and the general behaviour of the rock specimen subjected to simple standard tests.

The free energy ψ generated by the constitutive equation (5.4.1) (see Mihăilescu-Suliciu and Suliciu (1979), Gurtin *et al.* (1980)) is a function of class C^1, ψ: $R^6 \times R^6 \rightarrow R$ which satisfies the inequality (*the second law of thermodynamics*)

$$\boldsymbol{\sigma}' \cdot \dot{\boldsymbol{\varepsilon}}' + 3\sigma\dot{\varepsilon} - \dot{\psi}(\varepsilon, \boldsymbol{\varepsilon}', \sigma, \boldsymbol{\sigma}') \geqslant 0 \tag{5.4.6}$$

for any process $\varepsilon(t)$, $\boldsymbol{\varepsilon}'(t)$, $\sigma(t)$, $\boldsymbol{\sigma}'(t)$ which satisfies the constitutive equation (5.4.1). As a consequence of (5.4.6) and (5.4.1), ψ must be determined as a solution of the following problem:

$$\begin{gathered}
\frac{\partial\psi}{\partial\varepsilon} + 3K\frac{\partial\psi}{\partial\sigma} = 3\sigma \\
\frac{\partial\psi}{\partial\boldsymbol{\varepsilon}'} + 2G\frac{\partial\psi}{\partial\boldsymbol{\sigma}'} = \boldsymbol{\sigma}' \\
\eta_v\frac{\partial\psi}{\partial\sigma}(\sigma - 3K_0\varepsilon) + \eta\frac{\partial\psi}{\partial\boldsymbol{\sigma}'} \cdot (\boldsymbol{\sigma}' - 2G_0\boldsymbol{\varepsilon}') \geqslant 0.
\end{gathered} \tag{5.4.7}$$

The solution of the problem (5.4.7) was given by Podio-Guidugli and Suliciu (1984). The procedure is the following: first it is observed that

$$\psi(\varepsilon, \boldsymbol{\varepsilon}', \sigma, \boldsymbol{\sigma}') = \frac{\sigma^2}{2K} + \frac{\boldsymbol{\sigma}' \cdot \boldsymbol{\sigma}'}{4G} + \varphi_{\mathrm{v}}(\sigma - 3K\varepsilon) + \varphi(\boldsymbol{\sigma}' - 2G\boldsymbol{\varepsilon}') \quad (5.4.8)$$

identically satisfies $(5.4.7)_{1,2}$, where $\varphi_{\mathrm{v}}(\bar{\sigma})$ and $\varphi(\bar{\boldsymbol{\sigma}}')$ are arbitrary smooth functions of $\bar{\sigma} = \sigma - 3K\varepsilon$ and $\bar{\boldsymbol{\sigma}}' = \boldsymbol{\sigma}' - 2G\boldsymbol{\varepsilon}'$. The functions $\varphi_{\mathrm{v}}(\bar{\sigma})$ and $\varphi(\bar{\boldsymbol{\sigma}}')$ are determined from the condition that $(5.4.7)_3$ be satisfied. It is observed that (due to the particular form of the constitutive equation (5.4.1) and therefore of the inequality (5.4.6)) the finding of the solution of the problem (5.4.7) is equivalent to the solving of some one-dimensional problems (of the type discussed by Gurtin *et al.* (1980)). Let us consider, for instance, the case $\eta_{\mathrm{v}}(\sigma - 3K_0\varepsilon)\partial\psi/\partial\sigma = 0$. Since $\eta_{\mathrm{v}} \neq 0$, when $\eta_{\mathrm{v}}(\sigma - 3K_0\varepsilon) > 0$ it follows that $\partial\psi/\partial\sigma \geqslant 0$, while if $\eta_{\mathrm{v}}(\sigma - 3K_0\varepsilon) < 0$ we get $\partial\psi/\partial\sigma \leqslant 0$. Thus if $\sigma = 3K_0\varepsilon$ it follows that $\partial\psi/\partial\sigma = 0$. The equation $(5.4.7)_1$ with the condition $\partial\psi/\partial\sigma = 0$ has a unique solution and, as will be shown below, $\partial\psi/\partial\sigma = 0$ only if $\sigma = 3K_0\varepsilon$. Thus we obtain

$$\frac{\partial\psi}{\partial\sigma} = 0 \quad \text{if and only if } \sigma = 3K_0\varepsilon$$

$$\frac{\partial\psi}{\partial\sigma'_{ij}} = 0 \quad \text{if and only if } \sigma'_{ij} = 2G_0\varepsilon'_{ij}. \quad (5.4.9)$$

Formulae (5.4.9) are also the consequence of a more general result obtained by Gurtin *et al.* (1980) in the case when the solution $\psi(\varepsilon, \sigma)$ cannot be written explicitly.

From (5.4.8) and (5.4.9) follow

$$\frac{\partial\varphi_{\mathrm{v}}(-3(K - K_0)\varepsilon)}{\partial\bar{\sigma}} + \frac{3K_0}{K}\varepsilon = 0$$

$$\frac{\partial\varphi(-2(G - G_0)\boldsymbol{\varepsilon}')}{\partial\boldsymbol{\sigma}'} + \frac{G_0}{G}\boldsymbol{\varepsilon}' = 0. \quad (5.4.10)$$

If we change the arguments of the functions φ_{v} and φ,

$$\bar{\sigma} = -3(K - K_0)\varepsilon, \qquad \bar{\boldsymbol{\sigma}}' = -2(G - G_0)\boldsymbol{\varepsilon}' \quad (5.4.11)$$

we get

$$\varepsilon = -\frac{\bar{\sigma}}{3(K - K_0)}, \qquad \varepsilon'_{ij} = -\frac{\bar{\sigma}'_{ij}}{2(G - G_0)}. \quad (5.4.12)$$

Using (5.4.11) and (5.4.12) in (5.4.10), the functions φ_v and φ can uniquely be determined from

$$\frac{\partial \varphi_v(\bar{\sigma})}{\partial \bar{\sigma}} = \frac{K_0 \bar{\sigma}}{K(K - K_0)}, \qquad \varphi_v(0) = 0$$
$$\frac{\partial \varphi(\bar{\boldsymbol{\sigma}}')}{\partial \bar{\sigma}'_{ij}} = \frac{G_0 \bar{\sigma}'_{ij}}{2G(G - G_0)}, \qquad \varphi(\mathbf{0}) = \mathbf{0}, \tag{5.4.13}$$

choosing conventionally a zero energy for the stress-free and strain-free states. It follows that

$$\varphi_v(\bar{\sigma}) = \frac{K_0 \bar{\sigma}^2}{2K(K - K_0)}, \qquad \bar{\sigma} = \sigma - 3K\varepsilon,$$
$$\varphi(\bar{\sigma}') = \frac{G_0 \bar{\boldsymbol{\sigma}}' \cdot \bar{\boldsymbol{\sigma}}'}{4G(G - G_0)}, \qquad \bar{\boldsymbol{\sigma}}' = \boldsymbol{\sigma}' - 2G\boldsymbol{\varepsilon}', \tag{5.4.14}$$

and therefore

$$\psi(\varepsilon, \boldsymbol{\varepsilon}', \sigma, \boldsymbol{\sigma}') = \frac{\sigma^2}{2K} + \frac{\boldsymbol{\sigma}' \cdot \boldsymbol{\sigma}'}{4G} + \frac{K_0(\sigma - 3K\varepsilon)^2}{2K(K - K_0)} +$$
$$+ \frac{G_0}{4G(G - G_0)} (\boldsymbol{\sigma}' - 2G\boldsymbol{\varepsilon}') \cdot (\boldsymbol{\sigma}' - 2G\boldsymbol{\varepsilon}'). \tag{5.4.15}$$

It can be shown that ψ given by (5.4.15) is the unique solution of $(5.4.7)_{1,2}$ and (5.4.9), with $\psi(0) = 0$. It must still be shown that ψ determined by (5.4.15) satisfies $(5.4.7)_3$. We have

$$\frac{\eta_v}{K - K_0} (\sigma - 3K_0\varepsilon)^2 +$$
$$+ \frac{\eta}{2(G - G_0)} (\boldsymbol{\sigma}' - 2G_0\boldsymbol{\varepsilon}') \cdot (\boldsymbol{\sigma}' - 2G_0\boldsymbol{\varepsilon}') \geqslant 0. \tag{5.4.16}$$

Therefore the constitutive equations (5.4.1) accept an energy compatible with the second law of thermodynamics if and only if

$$\frac{\eta_v}{K - K_0} \geqslant 0, \qquad \frac{\eta}{G - G_0} \geqslant 0, \tag{5.4.17}$$

or, with the notation of (5.4.3),

$$\frac{K}{K_0} \frac{k_v}{K - K_0} \geqslant 0, \qquad \frac{G}{G_0} \frac{k}{G - G_0} \geqslant 0. \tag{5.4.18}$$

If in (5.4.17) or (5.4.18) the strict inequalities are satisfied, then the energy (5.4.15) is a positive-definite quadratic form in stresses and strains.

The above results are due to Suliciu. Another variant of the proof is due to Massier (D. Massier, personal communication).

Let us consider the *velocity of propagation* of the dilatational and shear waves as (Love (1944, Chapter XIII), Savarensky (1975)):

$$v_p^2 = \frac{1}{\rho}\left(K + \frac{4}{3}G\right), \qquad v_s^2 = \frac{G}{\rho}. \tag{5.4.19}$$

From the assumption that v_s is real it follows that

$$G > 0. \tag{5.4.20}$$

Since the natural order of propagation is $v_p > v_s$, from the condition that the bar velocity $(E/\rho)^{1/2}$ is real, it follows that

$$\frac{E}{\rho} = \frac{v_s^2(3v_p^2 - 4v_s^2)}{v_p^2 - v_s^2} > 0, \tag{5.4.21}$$

and therefore

$$v_p^2 > \tfrac{4}{3}v_s^2.$$

Thus,

$$K = \rho(v_p^2 - \tfrac{4}{3}v_s^2) > 0. \tag{5.4.22}$$

Let us consider now the *creep tests*, i.e. the tests done under constant stress. We assume that at moment t_0 we have $\boldsymbol{\varepsilon}(t_0) = \boldsymbol{\varepsilon}_0$, $\boldsymbol{\sigma}(t_0) = \boldsymbol{\sigma}_0$, and for $t \geqslant t_0$ the stress is kept constant: $\boldsymbol{\sigma}(t) = \boldsymbol{\sigma}(t_0) = \boldsymbol{\sigma}_0$. From (5.4.2) we have

$$\begin{aligned}\boldsymbol{\varepsilon}'(t) &= \left(\boldsymbol{\varepsilon}_0' - \frac{1}{2G_0}\boldsymbol{\sigma}_0'\right)\exp[-k(t-t_0)] + \frac{1}{2G_0}\boldsymbol{\sigma}_0' \\ \varepsilon(t) &= \left(\varepsilon_0 - \frac{1}{3K_0}\sigma_0\right)\exp[-k_v(t-t_0)] + \frac{1}{3K_0}\sigma_0.\end{aligned} \tag{5.4.23}$$

Therefore, from the condition of *stabilization of the deformation in creep tests* we must have, when $t \to \infty$

$$k > 0, \qquad k_v > 0. \tag{5.4.24}$$

In a similar way from the condition of *stabilization of the stress relaxation* under constant strain, when at $t = t_0$, $\boldsymbol{\varepsilon}(t_0) = \boldsymbol{\varepsilon}_0$, $\boldsymbol{\sigma}(t_0) = \boldsymbol{\sigma}_0$, and for $t \geqslant t_0$ the strain is kept constant, $\boldsymbol{\varepsilon}(t) = \boldsymbol{\varepsilon}(t_0) = \boldsymbol{\varepsilon}_0$. It follows that

$$K_0 > 0, \qquad G_0 > 0, \tag{5.4.25}$$

if (5.4.20), (5.4.22), and (5.4.24) are also taken into account. From (5.4.20), (5.4.22), (5.4.24), (5.4.25), and (5.4.18) we further get

$$K > K_0, \qquad G > G_0. \tag{5.4.26}$$

If inequalities (5.4.24) hold, for $t \to \infty$ the stabilization state satisfies

$$\boldsymbol{\varepsilon}'_\infty = \lim_{t \to \infty} \boldsymbol{\varepsilon}'(t) = \frac{1}{2G_0} \boldsymbol{\sigma}'_0$$
$$\varepsilon_\infty = \lim_{t \to \infty} \varepsilon(t) = \frac{\sigma_0}{3K_0} . \tag{5.4.27}$$

If the inequalities (5.4.24) and (5.4.26) are satisfied, the energy (5.4.15) is positive-definite.

Let us consider a *uniaxial compressive creep test* in which at $t = t_0 = 0$ a sudden loading is applied followed by a very long time interval in which the stress is kept constant,

$$\sigma_1 = \sigma_1^0 = \text{const.}, \qquad \sigma_2 = \sigma_3 = 0. \tag{5.4.28}$$

According to the model (5.4.2), the instantaneous response is

$$\varepsilon_1^0 - \varepsilon_2^0 = \frac{1}{2G} \sigma_1^0, \qquad \varepsilon_1^0 + 2\varepsilon_2^0 = \frac{1}{3K} \sigma_1^0. \tag{5.4.29}$$

These strain values are the initial data for the deformation by creep which follows, according to

$$\dot{\varepsilon}_1 - \dot{\varepsilon}_2 = -k \left[\varepsilon_1 - \varepsilon_2 - \frac{1}{2G_0} \sigma_1^0 \right]$$
$$\dot{\varepsilon}_1 + 2\dot{\varepsilon}_2 = -k_v \left[\varepsilon_1 + 2\varepsilon_2 - \frac{1}{3K_0} \sigma_1^0 \right] . \tag{5.4.30}$$

After integration we get

$$3\varepsilon_1 = \left(\frac{1}{G} - \frac{1}{G_0} \right) \sigma_1^0 \exp(-kt) +$$
$$+ \left(\frac{1}{3K} - \frac{1}{3K_0} \right) \sigma_1^0 \exp(-k_v t) + \left(\frac{1}{G_0} + \frac{1}{3K_0} \right) \sigma_1^0$$
$$3\varepsilon_2 = - \left(\frac{1}{2G} - \frac{1}{2G_0} \right) \sigma_1^0 \exp(-kt) +$$
$$+ \left(\frac{1}{3K} - \frac{1}{3K_0} \right) \sigma_1^0 \exp(-k_v t) + \left(- \frac{1}{2G_0} + \frac{1}{3K_0} \right) \sigma_1^0 . \tag{5.4.31}$$

Since during the uniaxial (unconfined) compressive creep test $\dot{\varepsilon}_1 > 0$ and $\dot{\varepsilon}_2 < 0$, it follows from the second condition (with (5.4.26)) that

$$k\left(\frac{1}{2G} - \frac{1}{2G_0}\right)\exp(-kt)$$

$$< k_v\left(\frac{1}{3K} - \frac{1}{3K_0}\right)\exp(-k_v t) < 0 \tag{5.4.32}$$

for any t. Therefore for $t \to 0$, the inequality (5.4.32) becomes

$$k\frac{G - G_0}{2GG_0} \geqslant k_v\frac{K - K_0}{3KK_0} > 0 \tag{5.4.33}$$

which is an additional inequality to be satisfied by the constitutive constants involved in the model (5.4.2). Alternatively, from (5.4.33) and (5.4.24), (5.4.18) follows. Rewriting (5.4.16) as

$$\frac{3k_v}{\dfrac{1}{3K_0} - \dfrac{1}{3K}}\left(\frac{\sigma}{3K_0} - \varepsilon\right)^2 + \frac{k}{\dfrac{1}{2G_0} - \dfrac{1}{2G}}\left(\frac{\boldsymbol{\sigma}'}{2G_0} - \boldsymbol{\varepsilon}'\right) \times$$

$$\times\left(\frac{\boldsymbol{\sigma}'}{2G_0} - \boldsymbol{\varepsilon}'\right) \geqslant 0, \tag{5.4.34}$$

inequality (5.4.33) establishes a relationship between the coefficients involved in (5.4.34) or in (5.4.15).

Further, from (5.4.32) it follows also for $t \to \infty$ that

$$k_v \geqslant k. \tag{5.4.35}$$

Furthermore, from (5.4.31) it follows that for $t \to \infty$ the stable states must satisfy the relationships

$$\varepsilon_1 = \frac{1}{3}\left(\frac{1}{G_0} + \frac{1}{3K_0}\right)\sigma_1^0, \qquad \varepsilon_2 = \frac{1}{3}\left(-\frac{1}{2G_0} + \frac{1}{3K_0}\right)\sigma_1^0. \tag{5.4.36}$$

Since the slope of $(5.4.36)_2$ is negative we get

$$2G_0 < 3K_0. \tag{5.4.37}$$

In a similar way from (5.4.29) it follows that the instantaneous curve

$$3\varepsilon_2^0 = \left(-\frac{1}{2G} + \frac{1}{3K}\right)\sigma_1^0$$

also has a negative slope, i.e.

$$2G < 3K. \tag{5.4.38}$$

Let us consider now the *uniaxial tests with constant loading rate* $\dot{\sigma}_1^0 > 0$:

$$\sigma_1(t) = \dot{\sigma}_1^0 t, \qquad \sigma_2 = \sigma_3 = 0. \tag{5.4.39}$$

Introducing (5.4.39) in (5.4.2), we get for this test

$$\dot{\varepsilon}_1 - \dot{\varepsilon}_2 = -k\left[\varepsilon_1 - \varepsilon_2 - \frac{1}{2G_0}\dot{\sigma}_1^0 t\right] + \frac{1}{2G}\dot{\sigma}_1^0$$

$$\dot{\varepsilon}_1 + 2\dot{\varepsilon}_2 = -k_v\left[\varepsilon_1 + 2\varepsilon_2 - \frac{1}{3K_0}\dot{\sigma}_1^0 t\right] + \frac{1}{3K}\dot{\sigma}_1^0.$$

By integrating with zero initial data, we obtain

$$\begin{aligned}
\varepsilon_1 - \varepsilon_2 &= \frac{\dot{\sigma}_1^0}{k}\left\{\frac{k}{2G_0}t + \left(\frac{1}{2G} - \frac{1}{2G_0}\right)[1 - \exp(-kt)]\right\} \\
\varepsilon_1 + 2\varepsilon_2 &= \frac{\dot{\sigma}_1^0}{k_v}\left\{\frac{k_v}{3K_0}t + \left(\frac{1}{3K} - \frac{1}{3K_0}\right)[1 - \exp(-k_v t)]\right\}.
\end{aligned} \tag{5.4.40}$$

Taking into account (5.4.39), we express the equation of the stress—strain curves as

$$\begin{aligned}
3\varepsilon_1 &= \left(\frac{1}{3K_0} + \frac{1}{G_0}\right)\sigma_1 + \\
&+ \left(\frac{1}{3K} - \frac{1}{3K_0}\right)\frac{\dot{\sigma}_1^0}{k_v}\left[1 - \exp\left(-k_v\frac{\sigma_1}{\dot{\sigma}_1^0}\right)\right] + \\
&+ \left(\frac{1}{G} - \frac{1}{G_0}\right)\frac{\dot{\sigma}_1^0}{k}\left[1 - \exp\left(-k\frac{\sigma_1}{\dot{\sigma}_1^0}\right)\right] \\
3\varepsilon_2 &= \left(\frac{1}{3K_0} - \frac{1}{2G_0}\right)\sigma_1 + \\
&+ \left(\frac{1}{3K} - \frac{1}{3K_0}\right)\frac{\dot{\sigma}_1^0}{k_v}\left[1 - \exp\left(-k_v\frac{\sigma_1}{\dot{\sigma}_1^0}\right)\right] - \\
&- \left(\frac{1}{2G} - \frac{1}{2G_0}\right)\frac{\dot{\sigma}_1^0}{k}\left[1 - \exp\left(-k\frac{\sigma_1}{\dot{\sigma}_1^0}\right)\right].
\end{aligned} \tag{5.4.41}$$

The slope of these stress—strain curves in the origin coincides with the slope of the instantaneous response straight lines, i.e.

$$\left.\frac{d\varepsilon_1}{d\sigma_1}\right|_{\sigma_1=0} = \frac{1}{3}\left(\frac{1}{3K} + \frac{1}{G}\right),$$

$$\left.\frac{d\varepsilon_2}{d\sigma_1}\right|_{\sigma_1=0} = \frac{1}{3}\left(\frac{1}{3K} - \frac{1}{2G}\right), \tag{5.4.42}$$

whence we can again derive (5.4.38). Further, if $k > 0$ and $k_v > 0$, from (5.4.41) it follows that the stress—strain curves have the oblique asymptotes

$$\varepsilon_1 = \frac{1}{3}\left(\frac{1}{3K_0} + \frac{1}{G_0}\right)\sigma_1 + \frac{1}{3}\left(\frac{1}{3K} - \frac{1}{3K_0}\right)\frac{\dot{\sigma}_1^0}{k_v} + $$
$$+ \frac{2}{3}\left(\frac{1}{2G} - \frac{1}{2G_0}\right)\frac{\dot{\sigma}_1^0}{k}$$

$$\varepsilon_2 = \frac{1}{3}\left(\frac{1}{3K_0} - \frac{1}{2G_0}\right)\sigma_1 + \frac{1}{3}\left(\frac{1}{3K} - \frac{1}{3K_0}\right)\frac{\dot{\sigma}_1^0}{k_v} - \tag{5.4.43}$$
$$- \frac{1}{3}\left(\frac{1}{2G} - \frac{1}{2G_0}\right)\frac{\dot{\sigma}_1^0}{k}.$$

When $\dot{\sigma}_1^0 \to \infty$, from (5.4.41) it follows that the stress—strain curves tend to the straight lines of the instantaneous response of slopes (5.4.42). For very slow tests when $\dot{\sigma}_1^0 \to 0$, the stress—strain curves tend to the creep stabilization boundaries (5.4.36). Generally, for an arbitrary finite value of $\dot{\sigma}_1^0$, the stress—strain curves are situated between the above-mentioned limit boundaries. A similar conclusion holds for the volumetric strain ε_v. However, with the restrictions used above, the volume is *always compressible*.

The viscosity coefficients k and k_v are determined, for instance, in uniaxial creep tests. If we denote by σ_1^0 the constant applied stress and denote by the subscript i the 'initial' value and by the subscript f the 'final' value of a certain parameter during the creep period, then from (5.4.2) we have

$$k = \frac{1}{t_f - t_i} \ln \left| \frac{\left(\varepsilon_1 - \varepsilon_2 - \dfrac{1}{2G_0} \sigma_1^0 \right)_i}{\left(\varepsilon_1 - \varepsilon_2 - \dfrac{1}{2G_0} \sigma_1^0 \right)_f} \right|$$

$$k_v = \frac{1}{t_f - t_i} \ln \left| \frac{\left(\varepsilon_1 + 2\varepsilon_2 - \dfrac{1}{3K_0} \sigma_1^0 \right)_i}{\left(\varepsilon_1 + 2\varepsilon_2 - \dfrac{1}{3K_0} \sigma_1^0 \right)_f} \right| . \tag{5.4.44}$$

These formulae are used to determine the coefficients k and k_v.

Examples

Let us give some examples to show how the constitutive constants involved in the constitutive equation (5.4.2) can be determined from experimental data. It should be pointed out that a linear viscoelastic constitutive equation can be fitted to a rock behaviour only for relatively small intervals of variation of stresses and strains, and that the above inequalities for the constitutive constants can be fulfilled only if the rock is compressible.

The procedure is the following. First, the dynamic moduli K and G are determined using the procedure described in §6.1 (see also the formulae (8.2.7) and Fig. 8.2.2). Then from the experimental data obtained in creep tests, such as those shown schematically in Fig. 8.2.2, one can determine the moduli G_0 and K_0 corresponding to the relaxed state. For this purpose the two relaxation boundaries are approximated by linear relationships

$$\varepsilon_1 - \varepsilon_2 = \frac{1}{2G_0} \sigma_1, \qquad \varepsilon_1 + 2\varepsilon_2 = \frac{1}{3K_0} \sigma_1 \tag{5.4.45}$$

(see (5.4.30)) where strains and stress may possibly be relative, if the creep tests are done using a triaxial apparatus. This linear approximation is possible only for relatively small stress intervals. Finally, from the creep data the two viscosity coefficients k and k_v are determined using the formulae (5.4.44). As a test we check if all constitutive inequalities are satisfied.

For instance, for a certain kind of *coal* (from Baraolt) we obtained: G = 210 MPa, K = 1954 MPa, E = 608 MPa, G_0 = 30.6 MPa, K_0 = 32.6 MPa, $k = 1.4 \times 10^{-6}$ s^{-1}, $k_v = 1.9 \times 10^{-6}$ s^{-1}, and the short-term tensile strength σ_t as determined in a Brazilian test ranges between 0.53 MPa and 1.24 MPa.

For another kind of *coal* (from Racoş) we have: G = 107.4 MPa, K = 1923.4 MPa, E = 316 MPa, G_0 = 39.2 MPa, K_0 = 53.9 MPa, $k = 1 \times 10^{-6}$ s^{-1}, $k_v = 1.23 \times 10^{-6}$ s^{-1}, and the short-term tensile strength σ_t (Brazilian) ranges between 0.56 MPa and 0.65 MPa.

Exercises

1. Starting from (5.1.1) and (5.1.4), write Hooke's law in the form (5.1.3).

2. Prove (5.1.13), starting from (5.1.7) and assuming the plane $z = 0$ to be a symmetry plane. Hint: the gradient of the mapping (5.1.12) is

$$Q_z = \begin{bmatrix} 1 & 0 & 0 \\ 0 & 1 & 0 \\ 0 & 0 & -1 \end{bmatrix},$$

and $\boldsymbol{\varepsilon}^* = \mathbf{Q}_z \boldsymbol{\varepsilon} \mathbf{Q}_z^{\mathrm{T}}$; find the restrictions for the components of $\mathbf{C}$ so that $\mathbf{Q}_z \boldsymbol{\sigma} \mathbf{Q}_z^{\mathrm{T}} = \mathbf{C}[\mathbf{Q}_z \boldsymbol{\varepsilon} \mathbf{Q}_z^{\mathrm{T}}]$ or $W = \frac{1}{2} C_{ij} \varepsilon_i \varepsilon_j = \frac{1}{2} C_{ij} \varepsilon_i^* \varepsilon_j^*$ is satisfied.

3. Show that if in each point of the body there are two orthogonal symmetry planes, then there is a third orthogonal plane of symmetry passing through this point. Hint: if $y = 0$ and $z = 0$ are symmetry planes, then $x = 0$ is also a symmetry plane since $Q_y Q_z = -Q_x$.

4. Show that in the case of orthotropy the stiffness matrix is (5.1.14).

5. Prove that for transversely isotropic elastic bodies the stiffness matrix is (5.1.16) (see Love (1944, §105), Cristescu (1983*a*, §2.5)).

6. Prove the compatibility conditions (5.1.23) for the components of the infinitesimal strains

$$\varepsilon_{ij} = \frac{1}{2}\left(\frac{\partial u_i}{\partial x_j} + \frac{\partial u_j}{\partial x_i}\right).$$

7. Find (5.1.25) starting from (5.1.23) and (5.1.24).

8. Prove formulae (5.1.27).

9. Carry out the details in establishing the formulae (5.1.28)—(5.1.35) of the plane-strain state.

10. Prove the compatibility condition (5.2.3) written in polar coordinates.

11. Show that for the particular case Ψ = const., (5.2.16) reduces to (5.2.17), and find the solution (5.2.18). Alternatively, change the variable by $r = \exp t$, etc.

12. Find the solution (5.2.23) of (5.2.16).

13. Show that for the case (5.2.24), the equation (5.2.16) reduces to

$$\theta \begin{pmatrix} \sin\theta \\ \cos\theta \end{pmatrix} \left[\frac{\mathrm{d}^4 R}{\mathrm{d}r^4} + \frac{2}{r}\frac{\mathrm{d}^3 R}{\mathrm{d}r^3} - \frac{3}{r^2}\frac{\mathrm{d}^2 R}{\mathrm{d}r^2} + \frac{3}{r^3}\frac{\mathrm{d}R}{\mathrm{d}r} - \frac{3R}{r^4} \right] +$$

$$+ \begin{pmatrix} 4\cos\theta \\ -4\sin\theta \end{pmatrix} \left[\frac{1}{r^2}\frac{\mathrm{d}^2 R}{\mathrm{d}r^2} - \frac{1}{r^3}\frac{\mathrm{d}R}{\mathrm{d}r} + \frac{R}{r^4} \right] = 0,$$

and that its general solution is (5.2.25).

14. Prove that the general solution of (5.2.16) for the case (5.2.26) is (5.2.27).

15. Introduce (5.3.1) into (5.2.5) and then the stress components in the equilibrium equation (5.2.4) written for the case of axial symmetry and absence of body forces. Show that the displacement is of the form

$$u = Ar + \frac{B}{r},$$

with A and B integration constants.

16. Carry out the details in establishing the formulae (5.3.5)–(5.3.11).

6

Rock 'elasticity' at high pressures

6.1. The elastic moduli

The determination of the elastic moduli arises in conjunction with the formulation of inelastic constitutive equations for rock-like materials if, as well as other properties, the rock also exhibits elastic properties. It has been mentioned in previous chapters that various authors have suggested that the elastic moduli have to be determined from the initial slopes of the stress—strain diagrams, assuming, by analogy with the behaviour of metals, that at relatively small stresses the behaviour of the rock is linearly elastic. However, as it has already been emphasized, the experimental evidence shows that even at relatively small stresses which produce small but non-negligible strains, the behaviour of the rock is not linearly elastic, but is in fact time-dependent. Thus, any method of determining the elastic moduli from the slopes of various portions of the stress—strain curves obtained during loading will not, strictly speaking, determine proper elastic moduli. With such a procedure, at best, a linearized version of the stress—strain relation is obtained for loading processes. Such a linearization is sometimes used for practical purposes, in order to make it possible to utilize simple strength-of-materials formulae.

The elastic moduli are sometimes determined during unloading processes. However, for such processes significant hysteresis loops may be revealed in unloading/reloading cycles. Thus, unloading does not always follow a proper straight line. The data show, however, that at the beginning of the unloading processes these curves are quite close to straight lines and therefore their slopes can be used for the determination of the elastic moduli. Similar conclusions follow from the σ—ε_v curve corresponding to the hydrostatic compressibility of the volume (see Chapter 4). Finally, for some rocks the unloading process can reasonably well be considered to be linearly elastic.

The terminology *static elastic moduli* is sometimes used for the values of these parameters as determined from the slopes of various portions of the stress—strain curves, mainly during loading processes; it is certainly assumed that on the corresponding portion of the curve the rock behaves 'elastically'. Let us recall (see Chapter 2) that the stress—strain curves are not even unique (since they are very sensitive to the loading rate) and neither are the static elastic moduli determined from them. The terminology *dynamic elastic moduli* is also used for the values of these moduli as obtained from the velocity of propagation of the longitudinal and shear waves, assuming that the waves propagating in the rock are 'seismic' waves (see below). Experience shows that the values of the dynamic elastic moduli are quite close to the moduli determined from the first part (approximately one-third) of the unloading curves, which can also be termed 'static elastic moduli'.

If the rock under consideration exhibits significant rheological properties, then before passing from loading to unloading the specimen must be kept under constant stress for a sufficiently long time interval for the specimen to approach, by creep, close enough to a stable state (see Chapter 8). The elastic moduli can further be determined during unloading. That is the only proper method of determining the static elastic moduli using, for instance, the simple formulae

$$E = \frac{\Delta\sigma_1}{\Delta\varepsilon_1^E}, \qquad K = \frac{1}{3}\,\frac{\Delta\sigma_1}{\Delta\varepsilon_1^E + 2\Delta\varepsilon_2^E}, \qquad G = \frac{1}{2}\,\frac{\Delta\sigma_1}{\Delta\varepsilon_1^E - \Delta\varepsilon_2^E}, \tag{6.1.1}$$

where $\Delta\sigma_1$ is the decrease of σ_1 during the first part of the unloading curve.

It is only with increasing hydrostatic pressure that the static elastic moduli determined during loading tests approach the dynamic moduli in magnitude.

In order to illustrate this, Fig. 6.1.1 shows the upper parts only of four stress—strain curves for *coal*. Between loading and unloading the maximum stress was kept constant for ten minutes, during which an obvious short-term creep takes place. The elastic constants are determined from the upper part of the unloading curves.

In the formulation of the constitutive equations which will be presented in the following two chapters, it will be assumed that 'unloading' is an elastic process. This assumption is not always entirely justified, due to the significant hysteresis loops which may be present, but it will still be made in the first stage of the formulation of reasonably simple models. Thus the 'elastic moduli' will be used in order to describe unloading, for the estimation of the magnitude of irreversible strains, for that of irreversible work, etc. Briefly, the elastic moduli

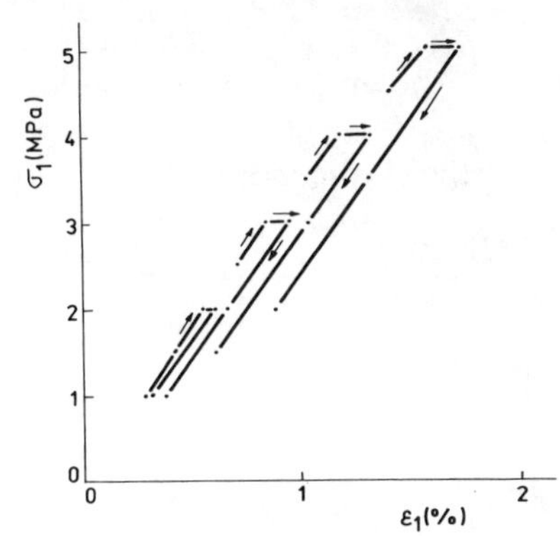

Figure 6.1.1. Beginning of unloading curves for coal, following a short-term creep of ten minutes, used to determine the static elastic constants.

play a major role in the formulation of the model and it is therefore desirable to determine them as exactly as possible. Ideally, elastic moduli should be determined by both methods: by dynamic methods, and during unloading following short-term creep in quasi-static tests. Since in dynamic tests the internal structure of the rock is not disturbed by the propagation of waves of small amplitudes, and since the dynamic tests are much more accurate, there is now an increasing tendency to use the dynamic procedure for the determination of elastic moduli (see for instance Volarovich *et al.* (1974), King (1970), Dortman (1976)), sometimes directly *in situ* (van Heerden (1985)).

6.2. Determination of elastic moduli by dynamic procedures

As already mentioned, an accurate method of determining the elastic moduli is the dynamic method based on the measurements of the velocities of propagation of various existing waves. For this purpose the specimen is previously subjected to an axial force and to a lateral hydrostatic pressure. The method uses the well-known technique of measurements of the transit times of the waves propagating in the stressed body. Besides various kinds of surface waves disregarded here, in extended solids with instantaneous elastic response, longitudinal and shear waves are propagated, while in thin bars longitudinal waves are of relevance in what follows. Their corresponding velocities of propagation are

$$
\begin{aligned}
v_P &= \left[\frac{E}{\rho}\,\frac{1-\nu}{(1+\nu)(1-2\nu)}\right]^{1/2} = \left(\frac{\lambda+2G}{\rho}\right)^{1/2} \\
v_S &= \left(\frac{G}{\rho}\right)^{1/2} = \left[\frac{E}{\rho}\,\frac{1}{2(1+\nu)}\right]^{1/2} \qquad (6.2.1) \\
v_B &= \left(\frac{E}{\rho}\right)^{1/2}.
\end{aligned}
$$

The first two velocities are known as *seismic velocities* and the last as *bar velocity*.

From (6.2.1) it follows for the elastic moduli that

$$
\begin{aligned}
E &= \rho\,\frac{v_S^2(3v_P^2 - 4v_S^2)}{v_P^2 - v_S^2} = \rho v_B^2 \\
K &= \rho(v_P^2 - \tfrac{4}{3}v_S^2) \\
G &= \rho v_S^2 \qquad (6.2.2) \\
\nu &= \frac{v_P^2 - 2v_S^2}{2(v_P^2 - v_S^2)} = \frac{v_B^2}{2v_S^2} - 1,
\end{aligned}
$$

where ρ is the current density. Let us observe that for the determination of ν only the ratio of two velocities is involved.

The experimental procedures used for the recording of the travel times of various waves are described by several authors: Kolsky (1963), Obert and Duvall (1967), King (1970), Volarovich *et al.* (1974), Schreiber *et al.* (1973), Lama and Vutukuri (1978, Chapter 7). Sometimes, as well as an axial force, the specimen is simultaneously subjected to a high hydrostatic pressure. Ultrasonic longitudinal and shear waves are transmitted through the specimen. For the history of the method see Bell (1973, §3.39).

It is well known (see Kolsky (1963)) that if the ratio between the radius of the specimen and the wave length is small, i.e. $R/\lambda < 0.3$, say, then the bar is considered 'thin' and the longitudinal waves are propagated with the bar velocity v_B. If however, $R/\lambda > 1.3$, then the bar is considered thick and the longitudinal waves are propagated with velocity v_P. If the frequency of the longitudinal waves can be changed, then both v_B and v_P can be measured on a single specimen. Since the velocities of propagation are quite large (several kilometres per second), and since the size of the specimen is generally small, the accuracy of the measurements must be high.

A practical method is the recording of the arrival time of the first and second group of waves arriving at the free end of the specimen—waves which were generated at the other end. The first to arrive are the longitudinal waves; the second group of waves arriving at the free end is due to the reflection of the longitudinal waves from the stress-free lateral surface of the specimen. Thus the second group of waves are shear waves. Let d be the diameter of the specimen and AB the front of the first longitudinal wave which is propagated from left to right, as shown by a heavy arrow in Fig. 6.2.1. Due to the reflection from the lateral surface, shear waves, shown as AC and BD, will be continuously generated. The direction of propagation is also shown by heavy arrows. The dotted arrows plotted along various front waves show the direction of motion of the material particles as a result of the passing of the corresponding front. The delay Δt with which the first shear wave arrives after the longitudinal one, depends only on the diameter d, on the two velocities of propagation, and on the Poisson ratio. This delay time does not depend on the length of the bar, since

$$\frac{BC}{d} = \text{cotan}\,\alpha = \left[\left(\frac{v_P}{v_S}\right)^2 - 1\right]^{1/2} = (1 - 2\nu)^{1/2}, \tag{6.2.3}$$

i.e.

$$\Delta t = \frac{d}{v_P}\left[\left(\frac{v_P}{v_S}\right)^2 - 1\right]^{1/2} = \frac{d}{v_P}(1 - 2\nu)^{1/2}. \tag{6.2.4}$$

If Δt is recorded, we also get

$$v_S = \frac{v_P d}{(d^2 + v_P^2(\Delta t)^2)^{1/2}}. \tag{6.2.5}$$

Therefore, with a single test in which both arrival times are recorded, the two velocities of propagation v_P and v_S can be determined, and therefore all elastic moduli.

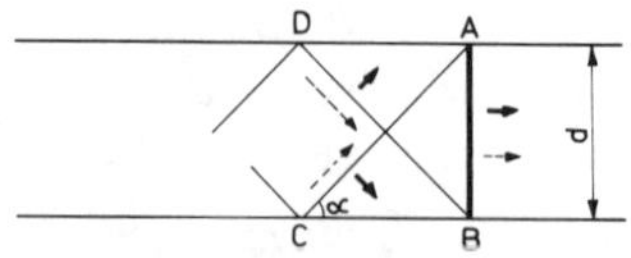

Figure 6.2.1. Fronts of longitudinal and shear waves propagating in a bar. Full arrows are directions of propagation, dotted arrows the directions of particle motion.

Similar procedures are also used in the case when the specimen is subjected to high hydrostatic pressures. It is assumed that the previous formulae are still applicable. Some additional comment are given below.

For other experimental procedures to determine the dynamic elastic moduli, see the papers already mentioned above.

We now describe briefly some experimental results obtained by various authors, results which are of significance for the formulation of the constitutive equation.

First, *all the velocities of propagation generally increase with increasing pressure.* To illustrate this statement, Fig. 6.2.2 shows the variation of the velocities of propagation v_P and v_S with pressure for several *granite* specimens according to the experimental data by Bayuk (1966). The densities and the

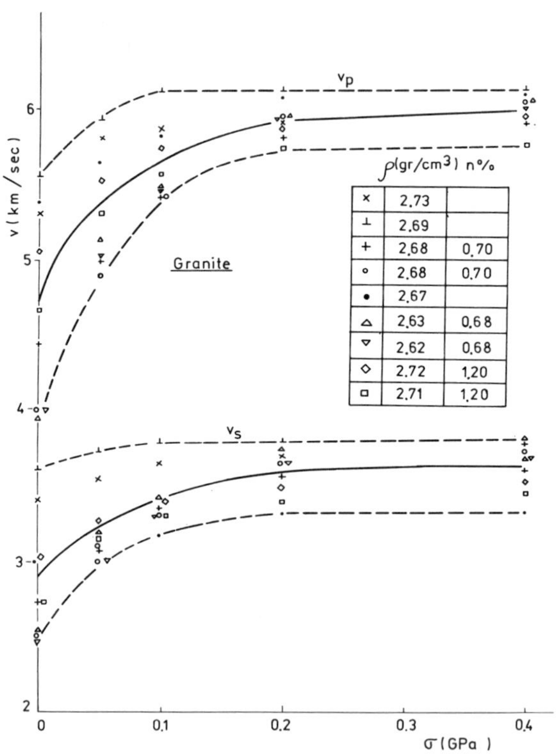

Figure 6.2.2. Variation of longitudinal velocity of propagation v_P and of the shear velocity of propagation v_S with pressure, for granite.

initial porosities are also given. The solid line represents the mean of all shown cases. It is obvious that the increase in the velocities with pressure is more significant for relatively small pressures, while at high pressures this increase is very slow. On the other hand, the increase of v_P with pressure is more significant than that of v_S. The variation of velocities with pressure is also more important for rocks with higher porosities, while for rocks with very small porosity this variation is less pronounced and sometimes even negligible. For example, for *rock salt* this variation is generally quite small. For illustration, in Table 6.1 the variations of v_P in various rocks, when pressures is increased from 1 atm. up to 0.1 GPa are given, according to Volarovich *et al.* (1974). Generally, an increase in temperature first increases and then decreases the velocities of propagation (King and Paulson (1981)).

After the determination of the velocities of propagation, the *dynamic elastic moduli* follow from formulae (6.2.2). The main results which have been obtained concerning the variation of the moduli with pressure are the following. For all rocks the moduli E and G increase with pressure. For instance, in Fig. 6.2.3 the curves showing the variations with pressure of E (solid lines) and of G

Table 6.1. Variation of v_P with pressure for various rocks (Volarovich *et al.* (1974))

Rock	Porosity (%)	Density (g cm^{-3})	v_P at 1 atm. (km s^{-1})	v_P at 0.1 GPa (km s^{-1})
Clay	3.8—3.0	1.77—2.41	1.37—2.70	2.0 —4.2
Diorite	4.9—3.5	1.78—2.38	1.95—5.0	3.0 —5.6
Siltstone	7.4—2.7	2.05—2.46	1.42—5.0	2.94—5.4
Sandstone	0.9—3.0	1.9 —2.78	1.25—6.03	2.88—6.52
Limestone	0.6—1.5	2.40—2.82	3.4 —6.35	4.17—6.68
Rock salt	0.2—2.2	2.02—2.36	3.7 —4.5	4.2 —4.8

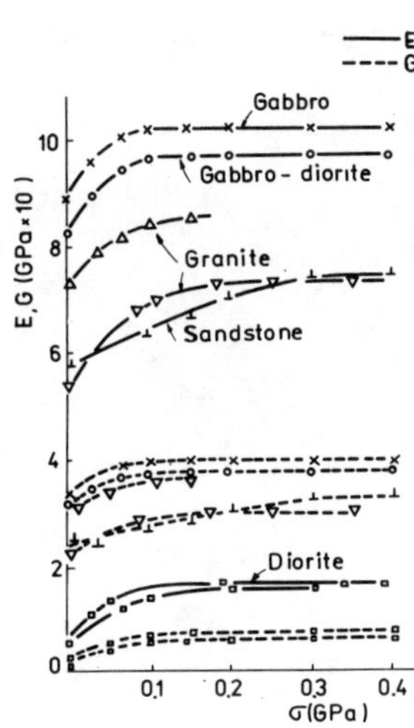

Figure 6.2.3. Variation of elastic parameters with pressure for various rocks.

(broken lines) for various rocks are given after Volarovich and Bayuk (1960). It follows that both E and G increase with pressure, but this increase is more significant at relatively small pressures. At higher pressures these moduli can be considered to be practically constant (see Volarovich *et al.* (1974), and King (1966, 1970)). We recall that above certain pressure, all the cracks of the rock are closed and the rock elasticity is in fact the elasticity of the constituent minerals (without cracks).

The variation of the Poisson ratio with pressure is distinct for various rocks: in some cases it increases and in other cases decreases with increasing pressure. Generally, the variation of the Poisson ratio with pressure is relatively small and at high confining pressure it is constant (King (1970)).

By measuring the ultrasonic velocities one can reveal the effect of porosity on the elastic parameters: the velocities v_P and v_S both decrease when porosity increases (Zimmerman *et al.* (1986)). Using an ultrasonic transmission method, Lo *et al.* (1986) have shown that Chelmsford *granite*, Chicopee *shale*, and Berea *sandstone* are elastically anisotropic up to 0.1 GPa confining pressure; the elastic anisotropy decreases with increasing confining pressure.

Finally, in Fig. 6.2.4 the variation with pressure of the elastic slopes, i.e. the incipient unloading in the unloading diagrams $\sigma_1^R - \varepsilon_1^R$ and $\sigma_1^R - \varepsilon_2^R$, is given for *granite*. For the $\sigma_1^R - \varepsilon_2^R$ slope, it is the absolute value of the elastic slope which is given in the figure. Both slopes have been obtained from the values of E and G for granite given in Fig. 6.2.3. It is obvious that with increasing pressure *both these slopes decrease for relatively small pressures, but further on remain practically constant.* The variation of the slope $(1/3G) + (1/9K)$ is more pronounced than that of the slope $-(1/6G) + (1/9K)$. The significance of these two slopes for the formulation of the constitutive equation will be pointed out in the following two chapters. The meaning of the pressure $\sigma_0 = 0.3445$ GPa shown in the figure is given by the formula (4.3.12) and discussed in §4.3.2, and is the limit pressure at which (i.e. for $\sigma \geqslant \sigma_0$) most of the cracks are already closed. The horizontal arrows mark the mean value of these slopes for the interval $\sigma \in$

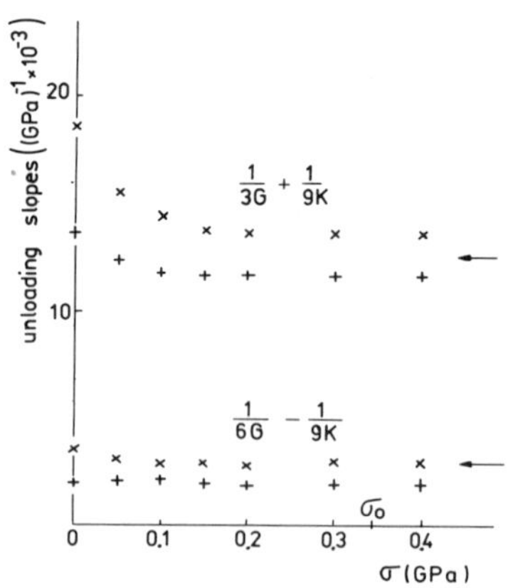

Figure 6.2.4. Variation of elastic unloading slope with pressure, for granite.

(0.1 GPa, 0.4 GPa). This has been considered in the numerical examples given in the following chapters.

A variation of temperature has a marked influence on the velocity of propagation (Kern and Franke (1984)). The variation of wave velocities in various rocks, as function of changes of confining pressure and pore fluid saturants, was studied by King (1966, 1984), and Pandit and King (1979).

Finally, let us observe that the influence of hydrostatic pressure on the elastic parameters has been relatively well studied. The way these parameters may vary with the equivalent stress and perhaps with the third stress invariant is less well known.

6.3. Longitudinal and shear waves in the case of high stresses and finite strains

In the previous section various dynamic methods used to determine the elastic moduli have been described, assuming that the formulae of linear elasticity still hold for high hydrostatic pressures. In the present section we give the main conclusions which follow from a paper by Suliciu (1984) concerning the propagation of acceleration waves in solid bodies subjected to high stresses and finite strains.

The system of equations considered comprises the balance of momentum in the reference configuration and either a rate-type constitutive equation or a physically non-linear elastic constitutive equation. In both kinds of constitutive equations the second Piola—Kirchhoff stress tensor is used (Mihăilescu-Suliciu and Suliciu (1979)). By choosing the concepts of stress and strain in this way, one can use partial derivatives with respect to time in the constitutive equation. The objectivity (frame-independent) of the rate-type constitutive equation is thus ensured. The linear and isotropic elastic case is considered as a special case. General concepts from the mechanics of continua are used (see Truesdell (1972), Malvern (1969), Jaunzemis (1967), Germain (1973), Eringen (1967), Eringen and Şuhubi (1974), etc.).

A *motion* is a mapping

$$\mathbf{x} = \boldsymbol{\chi}(\mathbf{X}, t) \tag{6.3.1}$$

relating the position $\mathbf{X}$ of a particle in the reference configuration to the actual position $\mathbf{x}$. Further,

$$\mathbf{F} = \text{Grad}\, \boldsymbol{\chi} \tag{6.3.2}$$

is the deformation gradient which has the property

$$\mathrm{J} = \det \mathbf{F} > 0. \tag{6.3.3}$$

The particle velocity is defined by

$$\mathbf{v} = \frac{\partial \boldsymbol{\chi}}{\partial t}\,. \tag{6.3.4}$$

With (6.3.2), the concepts of finite strains

$$\mathbf{C} = \mathbf{F}^{\mathrm{T}}\mathbf{F}, \qquad \mathbf{E} = \tfrac{1}{2}(\mathbf{C} - \mathbf{1}) \tag{6.3.5}$$

can be introduced. The superscript T stands for 'transpose'. The first and the second Piola—Kirchhoff stress tensors are

$$\mathbf{S} = \mathrm{J}\boldsymbol{\sigma}(\mathbf{F}^{\mathrm{T}})^{-1}, \qquad \mathbf{T} = \mathrm{J}\mathbf{F}^{-1}\boldsymbol{\sigma}(\mathbf{F}^{\mathrm{T}})^{-1} = \mathbf{F}^{-1}\mathbf{S}, \tag{6.3.6}$$

with $\boldsymbol{\sigma}$ the Cauchy stress.

In the paper by Suliciu (1984) the waves propagating in a body which satisfies the balance of momentum

$$\rho_0 \frac{\partial \mathbf{v}}{\partial t} - \mathrm{Div}\,\mathbf{S} = \rho\mathbf{b} \tag{6.3.7}$$

and a rate-type constitutive equation of the form

$$\frac{\partial \mathbf{T}}{\partial t} - \mathscr{E}(\mathbf{E}, \mathbf{T}) \frac{\partial \mathbf{E}}{\partial t} = \mathbf{G}(\mathbf{E}, \mathbf{T}) \tag{6.3.8}$$

with $\mathscr{E}$ and $\mathbf{G}$ constitutive functions, are studied. In certain particular cases of isotropic bodies, (6.3.8) becomes, in components

$$\frac{\partial T_{kj}}{\partial t} = (\lambda\delta_{kj}\delta_{q\ell} + 2\mu\delta_{kq}\delta_{j\ell}) \frac{\partial E_{q\ell}}{\partial t} + G_{kj}, \tag{6.3.9}$$

where λ and μ are scalar isotropic functions of the symmetric tensors $\mathbf{E}$ and $\mathbf{T}$. It was shown that in an undisturbed medium (all $E_{ij} = 0$ and $T_{ij} = 0$) two kinds of waves with the velocities

$$\rho U_{\mathrm{P}}^2 = \lambda + 2\mu, \qquad \rho U_{\mathrm{S}}^2 = \mu \tag{6.3.10}$$

are propagated, which are the well known expressions of the velocities of propagation from linear elasticity, while the values of λ and μ are those which correspond to $\mathbf{E} = \mathbf{0}$ and $\mathbf{T} = \mathbf{0}$.

Let us consider now the following problem (Suliciu (1984)): do three real acceleration waves exist for any primary state ($\mathbf{E}^{\mathrm{P}}$, $\mathbf{T}^{\mathrm{P}}$), or are there primary states for which there are no real acceleration waves? In other words, is the system (6.3.7), (6.3.9) of partial differential equations hyperbolic for all primary states?

To choose an example, let us consider a primary hydrostatic state and a corresponding volumetric strain (see Chapter 10):

$$\mathbf{T} = -p\mathbf{1}, \qquad \mathbf{C} = \kappa\mathbf{1}, \quad p > 0, \quad \kappa > 0. \tag{6.3.11}$$

We obtain for the velocities of propagation

$$\rho U_{\mathrm{P}}^2 = \kappa(\lambda + 2\mu) - p, \qquad \rho U_{\mathrm{S}}^2 = \kappa\mu - p. \tag{6.3.12}$$

Now λ and μ are considered to be functions of p and κ.

Therefore, if

$$0 \leqslant p \leqslant \min(\kappa\mu,\ \kappa(\lambda + 2\mu)),$$

the system (6.3.7), (6.3.9) is *hyperbolic*, since all the wave speeds are real. If, however,

$$p > \max(\kappa\mu,\ \kappa(\lambda + 2\mu)),$$

the system is *not hyperbolic*.

With

$$\mathbf{E} = -e\mathbf{1}, \qquad e = \frac{1-\kappa}{2} > 0, \tag{6.3.13}$$

the velocities become

$$\rho U_{\mathrm{P}}^2 = (1-2e)(\lambda + 2\mu) - p, \qquad \rho U_{\mathrm{S}}^2 = (1-2e)\mu - p. \tag{6.3.14}$$

These velocities of propagation are computed with respect to the reference configuration. The velocities of propagation $\tilde{U}$ with respect to the actual configuration can be obtained from

$$\tilde{U} = (B_{ij}\tilde{n}_i\tilde{n}_j)^{1/2}U, \tag{6.3.15}$$

where $\tilde{\mathbf{n}}$ is the unit normal to the wave front in the actual configuration, and $\mathbf{B} = \mathbf{F}\mathbf{F}^{\mathrm{T}}$. Since from (6.3.5) and (6.3.11) we have

$$\mathbf{F}^{\mathrm{T}} = \kappa\mathbf{F}^{-1}, \qquad \mathbf{B} = \kappa\mathbf{1},$$

and from (6.3.6)

$$J = \frac{\rho_0}{\rho} = \kappa^{3/2}, \qquad \tilde{p}(\kappa) = \frac{\kappa}{J}\,p(\kappa) = \frac{p(\kappa)}{\sqrt{\kappa}} = \left(\frac{\rho}{\rho_0}\right)^{1/3} p(\kappa), \tag{6.3.16}$$

from (6.3.15) it follows that

$$\tilde{U} = \sqrt{\kappa}\,U = (1-2e)^{1/2}U. \tag{6.3.17}$$

Substituting this in (6.3.15) we get

$$\rho\tilde{U}_{\mathrm{P}}^2 = \left(\frac{\rho_0}{\rho}\right)^{1/3}(\lambda + 2\mu) - \tilde{p}, \qquad \rho\tilde{U}_{\mathrm{S}}^2 = \left(\frac{\rho_0}{\rho}\right)^{1/3}\mu - \tilde{p}, \tag{6.3.18}$$

where λ and μ are constants from the reference configuration, ρ_0 is the initial density, ρ the actual density, and $\tilde{p}$ the pressure in the actual configuration ($\tilde{p} = (\kappa)^{-1/2}p$).

Starting from the experimental data which give ρ_0, λ, and μ in the reference configuration and also ρ and $\tilde{p}$ in the actual configuration, one can calculate $\tilde{U}_{\mathrm{P}}$ and $\tilde{U}_{\mathrm{S}}$. Thus, one can study the influence of the primary hydrostatic pressure on the velocities of propagation and on possible cases when the system (6.3.7), (6.3.9) ceases to be hyperbolic.

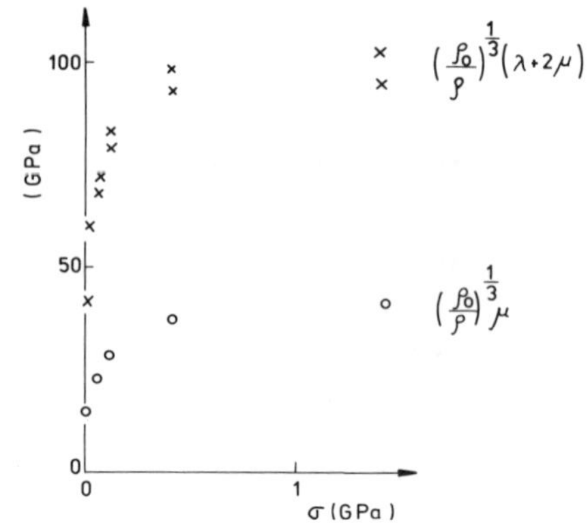

Figure 6.3.1. Increase of the terms with depth for granite, from which it follows that the velocities $\tilde{U}_P$ and $\tilde{U}_S$ have the tendency to decrease with depth.

For instance, if the experimental data for granite given by Balashov and Volarovich (1971) and Bayuk (1966) are used, we obtain the graph shown in Fig. 6.3.1. Here, the variation with pressure of the expressions $(\rho_0/\rho)^{1/3}(\lambda + 2\mu)$ and $(\rho_0/\rho)^{1/3}\mu$ are given. Let us observe that the two pressures practically coincide, i.e. $\tilde{p} \simeq p$. From this figure it follows that both these expressions increase with pressure up to about $p \simeq 0.3$ GPa, while for higher pressures this increase continues, but at a much slower rate. Therefore, there is a tendency of the velocities $\tilde{U}_P$ and $\tilde{U}_S$ to decrease with increasing depth, while at great depth, but not too great to be of interest for mining and oil engineering problems, these velocities of propagation are practically constant. One has taken into account that the pressure increases by about 0.025 GPa for each kilometre of depth.

6.4. Restrictions concerning the elastic parameters

In the following chapters we shall often use the concept of instantaneous response, while in many cases of interest to engineering, the instantaneous response can be approximated by a linearly elastic, and therefore reversible response. Let us examine what the restrictions are which must be satisfied by these elastic parameters (Suliciu (1988)).

We say that the constitutive equation of the form

$$\dot{\boldsymbol{\sigma}} = \mathscr{E}(\boldsymbol{\varepsilon}, \boldsymbol{\sigma})\dot{\boldsymbol{\varepsilon}} + \mathscr{G}(\boldsymbol{\varepsilon}, \boldsymbol{\sigma}) \tag{6.4.1}$$

has an instantaneous response in a domain D if the differential forms

$$\mathrm{d}\boldsymbol{\sigma} = \mathscr{E}(\boldsymbol{\varepsilon}, \boldsymbol{\sigma})\,\mathrm{d}\boldsymbol{\varepsilon} \tag{6.4.2}$$

are closed or completely integrable on a convex and possibly bounded domain D.

The differential forms (6.4.2) are closed in D if for any $(\boldsymbol{\varepsilon}_0, \boldsymbol{\sigma}_0) \in D$ there is a neighbourhood U_0 of $\boldsymbol{\varepsilon}_0$ in D_{ε} and a function $H(\,\cdot\,; \boldsymbol{\varepsilon}_0, \boldsymbol{\sigma}_0)$: $U_0 \to \mathscr{S}$, so that

$$\mathbf{H}(\boldsymbol{\varepsilon}_0;\ \boldsymbol{\varepsilon}_0, \boldsymbol{\sigma}_0) = \boldsymbol{\sigma}_0$$
$$(\boldsymbol{\varepsilon};\ \mathbf{H}(\boldsymbol{\varepsilon};\ \boldsymbol{\varepsilon}_0, \boldsymbol{\sigma}_0)) \in D \quad \text{for any } \boldsymbol{\varepsilon} \in U_0 \tag{6.4.3}$$
$$\frac{\partial \mathbf{H}}{\partial \boldsymbol{\varepsilon}} = \boldsymbol{\mathscr{E}}(\boldsymbol{\varepsilon}, \mathbf{H}) \quad \text{for any } \boldsymbol{\varepsilon} \in U_0.$$

The main results are obtained using the *theorem of Frobenius* according to which, if $\boldsymbol{\mathscr{E}}$ is of class C^1 on D; then (6.4.2) is closed if and only if

$$\frac{\partial \mathscr{E}_{k\ell pq}}{\partial \varepsilon_{mn}} + \mathscr{E}_{rsmn} \frac{\partial \mathscr{E}_{k\ell pq}}{\partial \sigma_{rs}} = \frac{\partial \mathscr{E}_{k\ell mn}}{\partial \varepsilon_{pq}} + \mathscr{E}_{rspq} \frac{\partial \mathscr{E}_{k\ell mn}}{\partial \sigma_{rs}}, \tag{6.4.4}$$

for $(\boldsymbol{\varepsilon}, \boldsymbol{\sigma}) \in D$ and $k, \ell, p, q, m, n = 1, 2, 3$.

We first prove the following corollary. If

$$\mathscr{E}_{ijk\ell}(\boldsymbol{\varepsilon}, \boldsymbol{\sigma}) = \lambda(\boldsymbol{\varepsilon}, \boldsymbol{\sigma})\delta_{ij}\delta_{k\ell} + \mu(\boldsymbol{\varepsilon}, \boldsymbol{\sigma})(\delta_{ik}\delta_{j\ell} + \delta_{i\ell}\delta_{jk}), \tag{6.4.5}$$

with λ and μ scalar functions of class C^1 on D, then (6.4.4) holds if and only if

$$\frac{\partial \mu}{\partial \varepsilon_{mn}} + \lambda \delta_{mn}\delta_{rs} \frac{\partial \mu}{\partial \sigma_{rs}} + 2\mu \frac{\partial \mu}{\partial \sigma_{mn}} = 0, \quad m, n = 1, 2, 3$$
$$\frac{\partial \lambda}{\partial \varepsilon_{mn}} + 2\mu \frac{\partial \lambda}{\partial \sigma_{mn}} = 0, \quad m, n = 1, 2, 3, \quad m \neq n \tag{6.4.6}$$
$$\frac{\partial \lambda}{\partial \varepsilon_{11}} + 2\mu \frac{\partial \lambda}{\partial \sigma_{11}} = \frac{\partial \lambda}{\partial \varepsilon_{22}} + 2\mu \frac{\partial \lambda}{\partial \sigma_{22}} = \frac{\partial \lambda}{\partial \varepsilon_{33}} + 2\mu \frac{\partial \lambda}{\partial \sigma_{33}}.$$

For the proof we introduce (6.4.5) into (6.4.4) to get

$$\begin{aligned}
&\frac{\partial \lambda}{\partial \varepsilon_{mn}} \delta_{k\ell}\delta_{pq} + \frac{\partial \mu}{\partial \varepsilon_{mn}} (\delta_{kp}\delta_{\ell q} + \delta_{kq}\delta_{\ell p}) + \\
&+ \lambda \sum_s \frac{\partial \mu}{\partial \sigma_{ss}} (\delta_{kp}\delta_{\ell q} + \delta_{kq}\delta_{\ell p})\delta_{mn} + 2\mu \frac{\partial \lambda}{\partial \sigma_{mn}} \delta_{k\ell}\delta_{pq} + \\
&+ 2\mu \frac{\partial \mu}{\partial \sigma_{mn}} (\delta_{kp}\delta_{\ell q} + \delta_{kq}\delta_{\ell p}) \\
&\quad = \frac{\partial \lambda}{\partial \varepsilon_{pq}} \delta_{k\ell}\delta_{mn} + \frac{\partial \mu}{\partial \varepsilon_{pq}} (\delta_{km}\delta_{\ell n} + \delta_{kn}\delta_{\ell m}) + \\
&\quad + \lambda \sum_s \frac{\partial \mu}{\partial \sigma_{ss}} (\delta_{km}\delta_{\ell n} + \delta_{kn}\delta_{\ell m})\delta_{pq} + 2\mu \frac{\partial \lambda}{\partial \sigma_{pq}} \delta_{k\ell}\delta_{mn} + \\
&\quad + 2\mu \frac{\partial \mu}{\partial \sigma_{pq}} (\delta_{km}\delta_{\ell n} + \delta_{kn}\delta_{\ell m}).
\end{aligned} \tag{6.4.7}$$

Several cases will be considered.

— If $m = n$, $k = \ell$, and $p = q$, but $k \neq m$ and $p \neq k$, $(6.4.6)_3$ follows straightforwardly from (6.4.7).

— If $m = n$ and $k = p$, but $p \neq q \neq m$ it follows that

$$2\left[\frac{\partial\mu}{\partial\varepsilon_{mn}} + \lambda\sum_s \frac{\partial\mu}{\partial\sigma_{ss}} + 2\mu\frac{\partial\mu}{\partial\sigma_{mn}}\right] = \frac{\partial\lambda}{\partial\varepsilon_{pq}} + 2\mu\frac{\partial\lambda}{\partial\sigma_{pq}}.$$

Also for $m = n$, from $k = q$ and $p \neq q \neq m$ we have

$$\frac{\partial\mu}{\partial\varepsilon_{mm}} + \lambda\sum_s \frac{\partial\mu}{\partial\sigma_{ss}} + 2\mu\frac{\partial\mu}{\partial\sigma_{mm}} = 0,$$

i.e. the relation $(6.4.6)_1$. Combining these last relations, we also get

$$\frac{\partial\lambda}{\partial\varepsilon_{pq}} + 2\mu\frac{\partial\lambda}{\partial\sigma_{pq}} = 0 \quad \text{for } p \neq q$$

which is the relation $(6.4.6)_2$.

Any other combination of indices leads to identities.

A second corollary states that if

1. $\mathscr{E}$ is of the form (6.4.5),

2. the constitutive equation has an instantaneous elastic response (thus (6.4.6) are satisfied),

3. $\partial\mathscr{E}/\partial\boldsymbol{\varepsilon} = 0$ on D,

then

$$\mu = \text{const. on } D,$$
$$\frac{\partial\lambda}{\partial\sigma_{ij}} = 0, \quad i \neq j, \qquad \frac{\partial\lambda}{\partial\sigma_{11}} = \frac{\partial\lambda}{\partial\sigma_{22}} = \frac{\partial\lambda}{\partial\sigma_{33}}. \tag{6.4.8}$$

If further,

4. $\mathscr{E}$ is an isotropic function, then

$$\lambda = \lambda(\text{tr}\,\boldsymbol{\sigma}). \tag{6.4.9}$$

For the proof we observe that from condition 3 and $(6.4.6)_{1,2}$ it follows that

$$\frac{\partial\lambda}{\partial\sigma_{ij}} = \frac{\partial\mu}{\partial\sigma_{ij}} \quad \text{for } i \neq j.$$

On the other hand, from condition 3 and $(6.4.6)_1$ it follows that

$$\lambda\left(\frac{\partial\mu}{\partial\sigma_{11}} + \frac{\partial\mu}{\partial\sigma_{22}} + \frac{\partial\mu}{\partial\sigma_{33}}\right) + 2\mu\frac{\partial\mu}{\partial\sigma_{11}} = 0$$

in the case $m = n = 1$, and two other similar relations for the indices 2 and 3. Since the determinant

$$\begin{vmatrix} \lambda + 2\mu & \lambda & \lambda \\ \lambda & \lambda + 2\mu & \lambda \\ \lambda & \lambda & \lambda + 2\mu \end{vmatrix} = 4\mu^2(3\lambda + 2\mu) \neq 0$$

cannot be zero, it follows that all the derivatives $\partial\mu/\partial\sigma_{ij} = 0$, i.e. μ = const. on D.

The relation $(6.4.8)_3$ follows straightforwardly from $(6.4.6)_3$.

If $\mathscr{E}$ is isotropic, then $\lambda(\boldsymbol{\sigma})$ is an isotropic scalar function of $\boldsymbol{\sigma}$ and will depend on $\boldsymbol{\sigma}$ only through the three stress invariants t_i $(i = 1, 2, 3)$. We have

$$\frac{\partial\lambda}{\partial\sigma_{ij}} = \frac{\partial\lambda}{\partial t_1}\,\delta_{ij} + 2\,\frac{\partial\lambda}{\partial t_2}\,\sigma_{ij} + 3\,\frac{\partial\lambda}{\partial t_3}\,\sigma_{ik}\sigma_{kj}\,.$$

Thus from $(6.4.8)_3$ it follows that

$$2\,\frac{\partial\lambda}{\partial t_2}\,\sigma_{11} + 3\,\frac{\partial\lambda}{\partial t_3}\,\sigma_{1k}\sigma_{k1} = 2\,\frac{\partial\lambda}{\partial t_2}\,\sigma_{22} + 3\,\frac{\partial\lambda}{\partial t_3}\,\sigma_{2k}\sigma_{k2}$$

$$= 2\,\frac{\partial\lambda}{\partial t_2}\,\sigma_{33} + 3\,\frac{\partial\lambda}{\partial t_3}\,\sigma_{3k}\sigma_{k3}.$$

For fixed values of t_1, t_2, and t_3, if we choose $\sigma_{11} = \sigma_{22}$ and $\sigma_{13} \neq \sigma_{23}$, it follows that

$$\frac{\partial\lambda}{\partial t_3}\,(\sigma_{13}^2 - \sigma_{23}^2) = 0,$$

therefore

$$\frac{\partial\lambda}{\partial t_3} = 0.$$

If we choose $\sigma_{11} \neq \sigma_{22}$, it follows that $\partial\lambda/\partial t_2 = 0$.

The conclusion is the following. If in the constitutive equation used (generally of the elastic/viscoplastic type) the instantaneous response is assumed to be elastic, then:

— *the bulk modulus $K(\sigma)$ may depend on the mean stress*;
— *the shear modulus is constant, G = const.*

In the next two chapters these two conclusions will be laid down as constitutive hypotheses.

If we renounce the hypothesis of elastic instantaneous response, then both moduli K and G may be considered to be variable.

Exercises

1. Make uniaxial creep tests of your choice using an arbitrary (but rather high) stress level. Try to make small unloadings, starting at various moments after the initial (unique) loading

moment. Check if the unloading slopes you obtain at various time intervals (but at the same stress level) are all equal. See if the unloading slope you obtain after creep stabilization is constant and if the elastic parameters determined from the $\sigma_1 - \varepsilon_1$ and $\sigma_1 - \varepsilon_2$ unloading slopes are close to the magnitudes determined in dynamic tests.

2. Derive the formula (6.2.5).
3. Give a detailed derivation of the formula (6.4.6).
4. Prove the formulae (6.4.8) and (6.4.9).

7

Rock plasticity

7.1. Historical outline

In order to describe the irreversible deformation of rocks, several authors have formulated variants of incremental constitutive equations, generally inspired by the classical ones. We recall that according to the constitutive equations of classical plasticity no irreversible change of the volume of the material is possible. In what follows we present a short history of the development of the mathematical models so far formulated.

First, we recall that incremental theories often do not describe the entire motion by considering the trajectory $\mathbf{x} = \mathbf{x}(\mathbf{X}, t)$ of a typical particle X, but are rather concerned with the *instantaneous motions* (Hill (1950), Malvern (1969)), i.e. the velocity $\mathbf{v} = \mathbf{v}(\mathbf{x}, t)$ of a typical particle X which at time t is at position $\mathbf{x}$. Thus, instead of following the trajectory of the particles starting from a reference configuration (possibly the stress-free and strain-free configuration, or maybe the relaxed configuration) and moving to the current one, the motion is determined by considering at the moment t the velocities $\mathbf{v}(\mathbf{x}, t)$, since even the reference configuration is generally unknown. Following this procedure, from the velocities of neighbouring points we can determine the rate-of-deformation tensor and the spin tensor. The relative velocity of the particle q with respect to the particle p (Fig. 7.1.1) is

$$d\mathbf{v} = \mathbf{L}\, d\mathbf{x} \quad \text{or} \quad dv_i = \frac{\partial v_i}{\partial x_j}\, dx_j,$$

where $\mathbf{L}$ is the *velocity gradient*. The symmetric and skew-symmetric tensors

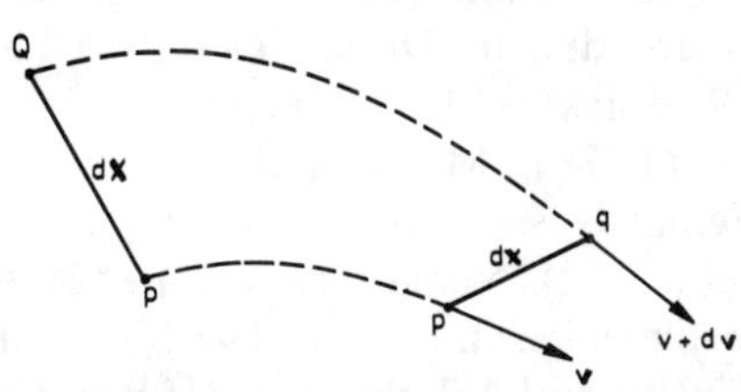

Figure 7.1.1. Notation used in defining 'instantaneous' motion.

$$\mathbf{D} = \tfrac{1}{2}(\mathbf{L} + \mathbf{L}^{\mathrm{T}}) \quad \text{and} \quad \mathbf{W} = \tfrac{1}{2}(\mathbf{L} - \mathbf{L}^{\mathrm{T}})$$

are called the *rate-of-deformation tensor* and the *spin tensor*.

If we denote by $\mathrm{d}u_i$ the displacement in the time interval $\mathrm{d}t$ of the particle which at time t is at $\mathbf{x}$, then

$$\mathrm{d}u_i = v_i \, \mathrm{d}t.$$

With these displacements, an *increment of the natural strain* is introduced as

$$\mathrm{d}\boldsymbol{\varepsilon} = \mathbf{D}\,\mathrm{d}t \quad \text{or} \quad \mathrm{d}\varepsilon_{ij} = \frac{1}{2}\left[\frac{\partial(\mathrm{d}u_i)}{\partial x_j} + \frac{\partial(\mathrm{d}u_j)}{\partial x_i}\right].$$

Further, $\mathrm{d}\boldsymbol{\sigma}$ or $\mathring{\boldsymbol{\sigma}}$ will denote a *stress flux*. For rocks, particularly hard rocks, the displacements and material rotations are generally small, so $\mathrm{d}\boldsymbol{\sigma}$ is in fact $\dot{\boldsymbol{\sigma}}$, i.e. the partial derivative with respect to time.

The notation and concepts introduced above will be used in the following. They are appropriate for rock mechanics since for rocks an initial reference configuration (stress-free and strain-free) does not exist (not even a relaxed one). The state actually existing *in situ* before disturbances produced by human intervention will be considered to be the natural reference configuration for rocks.

Various generalizations for finite strains of the incremental constitutive equations of classical plasticity theory have been proposed. The thermodynamic restrictions imposed on the constitutive equations have also have considered. The main contributions are due to Pipkin and Rivlin (1965), Green and Naghdi (1965), Lee (1969, 1970), Mihăilescu-Suliciu and Suliciu (1979, 1981), Eringen (1981), and others. Most of these general formulations can also be applied to porous or rock-like materials. However, as pointed out in previous chapters, rocks have some peculiar mechanical properties. For instance, depending on their mechanical properties and on the loading history, their volume may be either irreversibly compressible or irreversibly dilatant. On the other hand, for instance, rocks exhibit linear elastic behaviour only at high hydrostatic pressures and not in the neighbourhood of the stress-free state, as is the case for other materials. Therefore, mathematical models developed essentially for metal-like materials are not suitable for the description of the mechanical properties of rocks. Only constitutive equations which can describe phenomena such as irreversible compressibility or irreversible dilatancy, as well as volumetric elasticity at high pressures only, etc., are appropriate for rock-like materials.

Several approaches to the formulation of elastic/plastic constitutive equations for rock-like materials are due to Dreyer (1973), Cherry *et al.* (1975), Nur (1975), Rice (1975), Rudnicki and Rice (1975), Johnson and Green (1976), Bažant (1978), Krenk (1978), Maier and Hueckel (1979), Stavrogin and Protosenya (1979), Nemat-Nasser and Shokooh (1980), Dorris and Nemat-Nasser (1982), Hueckel (1982), Nemat-Nasser (1982), Rudnicki (1982, 1984), Vardoulakis (1984), Zienkiewicz and Mróz (1984), Desai *et al.* (1985), Ichikawa *et al.* (1985), Kapustyanskii and Nikolaevskii (1985), Oka and Adachi (1985), Okamoto (1985), Xiong (1985), and Desai *et al.* (1986).

Various authors have discussed the formulation of hardening rules for rocks, following the approach from metal plasticity (Celle and Cheatham (1981), Cheatham *et al.* (1984)). An 'elastic-crack-plastic' constitutive equation was developed by Lippmann (1984), while elastic/plastic saturated porous rocks were considered by Yamada *et al.* (1981). An experimental study and development of a constitutive equation for anisotropic rocks is due to Allirot and Boehler (1979) (see also Isobe and Tominga (1979), Rodrigues (1979), Nova (1980), Smith and Cheatham (1980) and the references cited there). For uniaxial stress-states, a simplified version of the elastic/plastic constitutive equation proposed for cement concrete was given by Constantinescu and Cristescu (1982).

In the present chapter an incremental elastic/plastic constitutive equation of classical structure for rock-like materials will be formulated, following the paper by Cristescu (1985*d*). What distinguishes this constitutive equation from those previously mentioned is not only a more general formulation (in the sense that no restrictions are *a priori* accepted either for the constitutive coefficients, or for the existence of linear elastic behaviour at small stresses, or for the shape of the yield surface, etc.), but also the fact that all concepts and coefficients involved in the constitutive equation are precisely defined by mathematical formulation. Moreover, the constitutive coefficients are not left vaguely defined; it will be shown how these coefficients can be explicitly obtained from the experimental data. Finally, the concepts of dilatancy, compressibility, and the compressibility/dilatancy boundary will be defined by precise, mathematical definitions.

Since the main applications of a constitutive equation written for rocks are designed for use in studying the deformation of rocks around underground openings, where there is already a certain stress state before excavation, the natural reference configuration is the actual one. However, in order to determine the constitutive equation in the laboratory we will choose the relaxed configuration (i.e. the stress-free configuration) as reference configuration. The strains will be related to this configuration too.

The mathematical model given below is a generalization for compressible/dilatant materials of the classical Prandtl—Reuss constitutive equation. For the moment time effects are disregarded, but they are considered in the next chapter.

7.2. Constitutive hypotheses

Five constitutive hypotheses are given (Cristescu (1985*d*)).

H1. The rock-like material is assumed to be a homogeneous and isotropic continuous body.

H2. The displacements and material rotations are always small during rock deformation so that the elastic (reversible in an incremental loading/unloading cycle) and plastic (or inelastic or irreversible) components of the rate of deformation tensor are additive:

$$\dot{\boldsymbol{\varepsilon}} = \dot{\boldsymbol{\varepsilon}}^{\mathrm{E}} + \dot{\boldsymbol{\varepsilon}}^{\mathrm{I}}. \tag{7.2.1}$$

The elastic component is determined from the constitutive law

$$\dot{\boldsymbol{\varepsilon}}^{\mathrm{E}} = \frac{\dot{\boldsymbol{\sigma}}}{2G} + \left(\frac{1}{3K} - \frac{1}{2G} \right) \dot{\sigma}\mathbf{1}, \tag{7.2.2}$$

where the elastic parameters are determined either from the slopes of the stress—strain curves at the beginning of the unloading processes or using dynamic procedures. $K(\sigma)$ may vary smoothly with the mean stress σ, while G is assumed to be constant. In some cases the 'secant' values for G and K determined during unloading processes are reasonable values for a simplified version of the model. Finally, in some engineering applications, where σ does not vary too much, constant values for both these coefficients are reasonable.

H3. The inelastic component of the rate of deformation is obtained from

$$\dot{\boldsymbol{\varepsilon}}^{\mathrm{I}} = \langle \dot{\lambda} \rangle \frac{\partial F(\sigma, \bar{\sigma}, \Delta)}{\partial \boldsymbol{\sigma}}, \tag{7.2.3}$$

where $\dot{\lambda} \geqslant 0$ is a non-negative variable scalar which will be defined below, and $\langle A \rangle = \frac{1}{2}(A + |A|)$ is the positive part of A. $F(\sigma, \bar{\sigma}, \Delta)$ is a *plastic potential.* It will be assumed that for any stress-states for which $\boldsymbol{\sigma} \cdot \dot{\boldsymbol{\varepsilon}}^{\mathrm{I}} > 0$, we have

$$\frac{\partial F}{\partial \boldsymbol{\sigma}} \cdot \boldsymbol{\sigma} > 0. \tag{7.2.4}$$

In other words, the stress-states for which possibly $(\partial F/\partial \boldsymbol{\sigma}) \cdot \boldsymbol{\sigma} = 0$ do not belong to the inelastic domain.

H4. Generally, rocks and rock-like materials have a zero yield stress. The *work-hardening* condition for such materials will be defined using the classical concept of irreversible stress power

$$P^{\mathrm{I}}(s) = \boldsymbol{\sigma}(s) \cdot \dot{\boldsymbol{\varepsilon}}^{\mathrm{I}}(s), \tag{7.2.5}$$

where the loading parameter s has the meaning of 'time' from classical plasticity theory and P^{I} is non-negative. The expression (7.2.5) is integrated over the time interval $s \in [0, S]$, where $s = 0$ corresponds to the beginning of the deformation process and S to the actual time, to get the irreversible stress-work as

$$W^{\mathrm{I}}(S) = \int_0^S \boldsymbol{\sigma}(s) \cdot \dot{\boldsymbol{\varepsilon}}^{\mathrm{I}}(s)\, \mathrm{d}s. \tag{7.2.6}$$

The *work-hardening condition* is written in the form

$$H(\sigma, \bar{\sigma}, \Delta) = W^{\mathrm{I}}(S), \tag{7.2.7}$$

where $H(\sigma, \bar{\sigma}, \Delta)$ is the yield function assumed to depend on the three stress invariants shown. This function can be determined from laboratory tests on rock specimens; this will be described in §7.4. Starting from a plastic stress-state (i.e. satisfying (7.2.7)) any additional loading (i.e. $\dot{\lambda} > 0$) produces work-hardening governed by the evolution law

$$\dot{H}(\sigma, \bar{\sigma}, \Delta) = P^{\mathrm{I}}(S), \tag{7.2.8}$$

with

$$\dot{H}(\sigma, \bar{\sigma}, \Delta) = \frac{\partial H}{\partial \sigma}\dot{\sigma} + \frac{\partial H}{\partial \bar{\sigma}}\dot{\bar{\sigma}} + \frac{\partial H}{\partial \Delta}\dot{\Delta}. \tag{7.2.9}$$

H5. The domain in the stress space $0\sigma_1\sigma_2\sigma_3$, where the constitutive equation is defined as

$$\begin{aligned} &\sigma_i \geqslant 0 \quad (i = 1, 2, 3) \\ &\bar{\sigma} < R(\sigma, \Delta, d), \end{aligned} \tag{7.2.10}$$

i.e. only compressive stress states will be considered. Here $\bar{\sigma} = R(\sigma, \Delta, d)$ is the equation of the failure surface determined by experiment, with d a damage parameter (see Chapter 9). The failure surface is not uniquely determined, since it depends on the loading history as well. An extension of the constitutive domain for tensile stresses is possible, but has not been considered here due to the lack of certain experimental data.

An example of a constitutive domain containing compressive stresses only is shown in Fig. 7.2.1 (the possible influence of Δ has been disregarded). For relatively small mean stresses in the neighbourhood of the 0σ-axis ($\sigma_1 = \sigma_2 = \sigma_3$) most rocks are compressible. The domain of partial inelastic volumetric compressibility is denoted by D_c. For higher values of $\bar{\sigma}$, i.e. for stress states

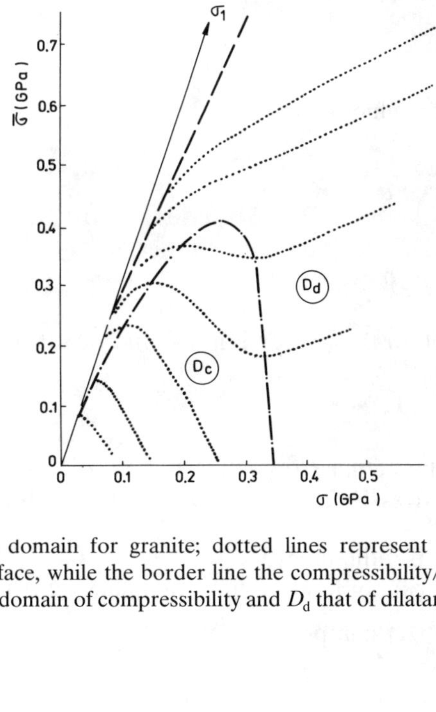

Figure 7.2.1. Constitutive domain for granite; dotted lines represent the yield surfaces, interrupted line the failure surface, while the border line the compressibility/dilatancy boundary. D_c is the domain of compressibility and D_d that of dilatancy.

which are relatively far from the 0σ-axis, most rocks are dilatant. The stress-states where the rock is dilatant (more precisely, a dilatancy of the volume of the rock takes place and a part of the volumetric deformation is irreversible) belong to the dilatancy domain D_d. It is also known (see Chapter 4) that at high hydrostatic stresses the behaviour of the volume is nearly linearly elastic for relatively small values of $\bar{\sigma}$, but dilatant at very high values of $\bar{\sigma}$. Therefore, the domain D_c of irreversible compressibility does not extend beyond a certain value of σ. The boundary between the domains D_c and D_d, (— · — · — in Fig. 7.2.1), is the dilatancy threshold boundary or compressibility/dilatancy boundary. The dotted lines in Fig. 7.2.1 represent several yield surfaces (7.2.7) corresponding to several values of W^I; the details shown in this figure correspond to an example for *granite* given below.

In the present monograph mainly compressive stresses are considered, since the inelastic behaviour of rocks subjected to tensile stresses has received less attention. However, from the few experimental data known to the author (Schock and Louis (1982)) it seems that the mathematical models can be smoothly extended to the domain of tensile stresses.

7.3. Constitutive equation

Following a classical procedure from H3 and H4 yields

$$\dot{\boldsymbol{\varepsilon}}^I = \frac{\left\langle \dfrac{\partial H}{\partial \boldsymbol{\sigma}} \cdot \dot{\boldsymbol{\sigma}} \right\rangle}{\dfrac{\partial F}{\partial \boldsymbol{\sigma}} \cdot \boldsymbol{\sigma}} \frac{\partial F}{\partial \boldsymbol{\sigma}}, \tag{7.3.1}$$

and therefore the loading factor is

$$\langle \dot{\lambda} \rangle = \frac{\left\langle \dfrac{\partial H}{\partial \boldsymbol{\sigma}} \cdot \dot{\boldsymbol{\sigma}} \right\rangle}{\dfrac{\partial F}{\partial \boldsymbol{\sigma}} \cdot \boldsymbol{\sigma}} \geqslant 0. \tag{7.3.2}$$

From (7.3.1) we get for the irreversible volumetric rate of deformation

$$\dot{\varepsilon}_v^I = \langle \dot{\lambda} \rangle \frac{\partial F}{\partial \boldsymbol{\sigma}} \cdot \mathbf{1}. \tag{7.3.3}$$

Let us assume that the actual stress-state is a plastic one, i.e. satisfying (7.2.7). For a variation of the stresses for which $\dot{\lambda} > 0$, three cases are possible:

$$\frac{\partial F}{\partial \boldsymbol{\sigma}} \cdot \mathbf{1} \begin{cases} < 0 & \text{dilatancy,} \\ = 0 & \text{compressibility/dilatancy boundary,} \\ > 0 & \text{compressibility.} \end{cases} \tag{7.3.4}$$

The relation $(7.3.4)_1$ is the mathematical definition of volume dilatancy; relation $(7.3.4)_2$ corresponds to an elastic behaviour of the volume, similar to classical plasticity theory; relation $(7.3.4)_3$ is the definition of irreversible compressibility of the volume. Since $(\partial F/\partial \boldsymbol{\sigma}) \cdot \mathbf{1}$ can be interpreted as the projection on the hydrostatic line $\sigma_1 = \sigma_2 = \sigma_3$ of the external normal to the surface $F =$ const., the geometrical meaning of the relations (7.3.4) is obvious and is shown in Fig. 7.3.1. If this projection points towards σ positive, the rock-like material is in a compressible process (Fig. 7.3.1*c*). If the projection of the normal on the line $\sigma_1 = \sigma_2 = \sigma_3$ points in the direction of decreasing σ, the rock-like material is in a dilatant process (Fig. 7.3.1*a*). Finally, the case of zero projection corresponds to a behaviour reminiscent of classical plasticity theory when the yield surface is a cylinder with the generatrices parallel to the hydrostatic line $\sigma_1 = \sigma_2 = \sigma_3$, and the volume compressibility is linearly elastic (reversible). If we further assume that F depends on σ and $\bar{\sigma}$ alone, it is easy to see that $(\partial F/\partial \boldsymbol{\sigma}) \cdot \mathbf{1} = \partial F/\partial \sigma$; therefore in this particular case the conditions (7.3.4) can be written as

$$\frac{\partial F}{\partial \sigma} \begin{cases} < 0 & \text{dilatancy,} \\ = 0 & \text{compressibility/dilatancy boundary,} \\ > 0 & \text{compressibility.} \end{cases} \qquad (7.3.4)'$$

Depending on the stress state, all three cases described above are possible for most rock-like materials. The equation $(7.3.4)_2'$ is by definition the equation of the dilatancy threshold surface or of the compressibility/dilatancy boundary.

Therefore, the constitutive equation proposed for rocks or rock-like materials for any stress state, dilatant or compressible, can be written as

$$\dot{\boldsymbol{\varepsilon}} = \frac{\dot{\boldsymbol{\sigma}}}{2G} + \left(\frac{1}{3K} - \frac{1}{2G}\right)\dot{\sigma}\mathbf{1} + \frac{\left\langle \dfrac{\partial H}{\partial \boldsymbol{\sigma}} \cdot \dot{\boldsymbol{\sigma}} \right\rangle}{\dfrac{\partial F}{\partial \boldsymbol{\sigma}} \cdot \boldsymbol{\sigma}} \frac{\partial F}{\partial \boldsymbol{\sigma}} \quad \text{if} \quad H = W^{\mathrm{I}}$$

$$\dot{\boldsymbol{\varepsilon}} = \frac{\dot{\boldsymbol{\sigma}}}{2G} + \left(\frac{1}{3K} - \frac{1}{2G}\right)\dot{\sigma}\mathbf{1} \quad \text{if} \quad H < W^{\mathrm{I}}, \qquad (7.3.5)$$

with K varying gently with σ, and H and F depending on stress invariants. This constitutive equation holds in the constitutive domain defined by H5.

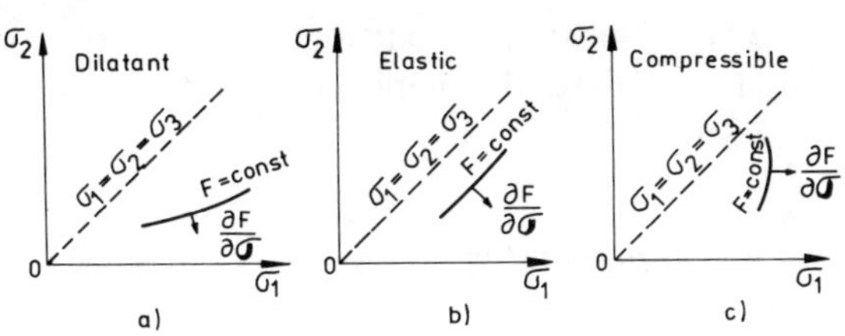

Figure 7.3.1. Three possible local properties of the yield surface.

In the case of saturated rock-like materials, if the pore pressure is denoted by p, the stress $\boldsymbol{\sigma}$ involved in the constitutive equation is to be replaced by the 'equivalent' stress $\boldsymbol{\sigma} - p\mathbf{1}$ (see Biot (1973), Rice (1975), Serafim (1968), etc.). If the constitutive equation (7.3.5) has already been formulated for the case of the absence of pore-pressure variation, then the influence of a pore-pressure variation on the response of rock-like materials can be predicted very accurately by checking the conditions (7.3.4) for the existing stress-state and for the stress variation considered. Thus, an increase in pore pressure will decrease the mean stress in the solid skeleton and this decrease may in turn increase the irreversible strains, or may produce an unloading, or may modify the rock from a dilatant to a compressible state or *vice versa* (which in turn will influence the pressure), depending on the location of the stress-state in one of the domains shown in Fig. 7.2.1 and on the migration of this location due to the change in the pore pressure. With a constant volume of the fluid in the rock, dilatancy decreases the pore pressure, while compressibility increases the pore pressure.

7.4. Yield function and plastic potential

The work-hardening condition (7.2.7) can be determined from experimental data, if the integral (7.2.6) is calculated along various loading paths. Generally, these loading paths can be chosen in an arbitrary manner, but they must cover the whole stress range that may be involved in engineering problems. In current experimental practice, these loading paths are as a rule obtained in a 'triaxial' apparatus in which the variation of only two principal components of the stresses can be prescribed: the axial component σ_1 and the circumferential one $\sigma_2 = \sigma_3$. That is why the function H is assumed to depend on two invariants σ and $\bar{\sigma}$ only. The dependence on Δ is assumed to be less significant, and is disregarded from now on.

Generally, the loading paths most used are those in which in the first stage of the experiment a hydrostatic loading up to $\sigma_1^{\mathrm{H}} = \sigma_2^{\mathrm{H}} = \sigma_3^{\mathrm{H}} = \sigma^{\mathrm{H}}$ is applied to the specimen. In the second stage of the experiment the lateral pressure is kept constant $\sigma_2 = \sigma_3 = \sigma^{\mathrm{H}}$, and the only component which is increased is the axial one, σ_1. The irreversible stress power now becomes

$$P^{\mathrm{I}} = \sigma_1 \dot{\varepsilon}_1^{\mathrm{I}} + 2\sigma_2 \dot{\varepsilon}_2^{\mathrm{I}} = (\sigma_1^{\mathrm{H}} + \sigma_1^{\mathrm{R}})\dot{\varepsilon}_1^{\mathrm{I}} + 2(\sigma_2^{\mathrm{H}} + \sigma_2^{\mathrm{R}})\dot{\varepsilon}_2^{\mathrm{I}}. \qquad (7.4.1)$$

In the following, we first consider a particular model of volume compressibility, and subsequently analyse the general case (see §8.2, for additional explanations of the notation used).

7.4.1. *Elastic hydrostatic compressibility*

For those rocks (generally, hard rocks with very small porosity) for which the hydrostatic loadings produce irreversible volumetric compressibilities which are negligible (at least for certain problems), a particular model can be imagined according to which *hydrostatic loadings produce elastic deformations of the volume only*. Therefore all along the hydrostatic line the rock behaviour is linearly elastic. This model can be called *hydrostatic elasticity*. In the second stage of the loading, $\sigma_2^R = 0$ and $\dot{\boldsymbol{\varepsilon}}^I = \dot{\boldsymbol{\varepsilon}} - \dot{\boldsymbol{\varepsilon}}^H - \dot{\boldsymbol{\varepsilon}}^{ER} = \dot{\boldsymbol{\varepsilon}}^R - \dot{\boldsymbol{\varepsilon}}^{ER}$, etc. We recall that $\boldsymbol{\varepsilon}^H$ are elastic deformations. Therefore we have

$$\begin{aligned} P^I &= \sigma_1^H(\dot{\varepsilon}_1^R - \dot{\varepsilon}_1^{ER}) + 2\sigma_2^H(\dot{\varepsilon}_2^R - \dot{\varepsilon}_2^{ER}) + \sigma_1^R(\dot{\varepsilon}_1^R - \dot{\varepsilon}_1^{ER}) \\ &= \sigma_1^H\left[\dot{\varepsilon}_1^R - \left(\frac{1}{3G} + \frac{1}{9K}\right)\dot{\sigma}_1^R\right] + \\ &\quad + 2\sigma_2^H\left[\dot{\varepsilon}_2^R - \left(-\frac{1}{6G} + \frac{1}{9K}\right)\dot{\sigma}_1^R\right] + \\ &\quad + \sigma_1^R\left[\dot{\varepsilon}_1^R - \left(\frac{1}{3G} + \frac{1}{9K}\right)\dot{\sigma}_1^R\right]. \end{aligned} \tag{7.4.2}$$

By integration of this expression in the interval $s \in [0, S]$, where $s = 0$ corresponds to the beginning of the second stage of the experiment, and taking into account that during this interval all components with an superscript H are constant, we get

$$\begin{aligned} W^I(S) &= \sigma_1^H\varepsilon_1^R(S) + 2\sigma_2^H\varepsilon_2^R(S) - \left(\frac{1}{3G} + \frac{1}{9K}\right)\sigma_1^H\sigma_1^R(S) - \\ &\quad - 2\left(-\frac{1}{6G} + \frac{1}{9K}\right)\sigma_2^H\sigma_1^R(S) + \int_0^S \sigma_1^R(s)\dot{\varepsilon}_1^R(s)\,ds - \\ &\quad - \frac{1}{2}\left(\frac{1}{3G} + \frac{1}{9K}\right)(\sigma_1^R(S))^2. \end{aligned} \tag{7.4.3}$$

The experimental data obtained in triaxial tests provide enough information for the calculation of this expression.

Further, the work-hardening condition (7.2.7) can be determined by choosing for the yield function, $H(\sigma, \bar{\sigma}) = \overline{H}(\sigma_1^R, \sigma_2)$, the simplest possible expression with the property $H(\sigma, 0) = 0$, imposed by the hypothesis of hydrostatic elasticity. Let us observe that function $\overline{H}(\sigma_1^R, 0)$ can be determined from the data obtained in uniaxial compressive tests. To pass from the expression $\bar{H}$ to that of H, we recall that in 'triaxial' tests, when $\sigma_1 \geqslant \sigma_2$, we have

$$\sigma_1^{\mathrm{H}} = \sigma_1 - \sigma_1^{\mathrm{R}} = \sigma - \frac{\bar{\sigma}}{3}$$

$$\sigma_1^{\mathrm{R}} = \sigma_1 - \sigma_1^{\mathrm{H}} = \bar{\sigma} = \bar{\sigma}^{\mathrm{R}}$$

$$\sigma_2^{\mathrm{H}} = \sigma_2 - \sigma_2^{\mathrm{R}} = \sigma_2 = \sigma - \frac{\bar{\sigma}}{3}$$

$$\sigma_2^{\mathrm{R}} = 0.$$

These relations allow us to write the work-hardening conditions in terms of invariants.

7.4.2. *The general case of irreversible hydrostatic compressibility*

Let us now consider the case when a hydrostatic loading can produce an irreversible compressibility of the volume, a part of the volumetric strain being reversible, however. Generally, this is the property exhibited by most rocks. We decompose the rate of deformation as follows:

$$\dot{\varepsilon}^{\mathrm{H}} = \dot{\varepsilon}^{\mathrm{EH}} + \dot{\varepsilon}^{\mathrm{IH}}, \qquad \dot{\varepsilon}^{\mathrm{R}} = \dot{\varepsilon}^{\mathrm{ER}} + \dot{\varepsilon}^{\mathrm{IR}}, \qquad \dot{\varepsilon} = \dot{\varepsilon}^{\mathrm{R}} + \dot{\varepsilon}^{\mathrm{H}}. \tag{7.4.4}$$

Thus, the irreversible stress power becomes

$$\begin{aligned} P^{\mathrm{I}} &= (\sigma_1^{\mathrm{H}} + \sigma_1^{\mathrm{R}})(\dot{\varepsilon}_1^{\mathrm{R}} + \dot{\varepsilon}_1^{\mathrm{IH}} - \dot{\varepsilon}_1^{\mathrm{ER}}) + 2(\sigma_2^{\mathrm{H}} + \sigma_2^{\mathrm{R}})(\dot{\varepsilon}_2^{\mathrm{R}} + \dot{\varepsilon}_2^{\mathrm{IH}} - \dot{\varepsilon}_2^{\mathrm{ER}}) \\ &= \sigma_1^{\mathrm{H}}\left[\dot{\varepsilon}_1^{\mathrm{R}} - \left(\frac{1}{3G} + \frac{1}{9K}\right)\dot{\sigma}_1^{\mathrm{R}}\right] + \\ &\quad + 2\sigma_2^{\mathrm{H}}\left[\dot{\varepsilon}_2^{\mathrm{R}} - \left(-\frac{1}{6G} + \frac{1}{9K}\right)\dot{\sigma}_1^{\mathrm{R}}\right] + \\ &\quad + \sigma_1^{\mathrm{R}}\left[\dot{\varepsilon}_1^{\mathrm{R}} - \left(\frac{1}{3G} + \frac{1}{9K}\right)\dot{\sigma}_1^{\mathrm{R}}\right] + \\ &\quad + (\sigma_1^{\mathrm{H}} + \sigma_1^{\mathrm{R}} + 2\sigma_2^{\mathrm{H}})\frac{1}{3}\,\dot{\varepsilon}_{\mathrm{v}}^{\mathrm{IH}}, \end{aligned} \tag{7.4.5}$$

with $\dot{\varepsilon}_1^{\mathrm{IH}} = \dot{\varepsilon}_2^{\mathrm{IH}} = \frac{1}{3}\dot{\varepsilon}_{\mathrm{v}}^{\mathrm{IH}}$. Integrating this expression in the interval $s \in [0, S]$, where $s = 0$ is the moment of the beginning of the hydrostatic loading, and denoting by $S_{\mathrm{H}} \leqslant S$ the moment marking the end of the hydrostatic loading stage, then (we recall that in the interval $s \in [0, S_{\mathrm{H}}]$ all components with the superscript R are zero, while in the interval $s \in [S_H, S]$ all components with the superscript H are constant) we have at time S

$$W^I(S) = \sigma_1^H \varepsilon_1^R(S) + 2\sigma_2^H \varepsilon_2^R(S) - \sigma_1^H \left(\frac{1}{3G} + \frac{1}{9K} \right) \sigma_1^R(S) -$$

$$- 2\sigma_2^H \left(-\frac{1}{6G} + \frac{1}{9K} \right) \sigma_1^R(S) + \int_{S_H}^{S} \sigma_1^R(s) \dot{\varepsilon}_1^R(s)\, \mathrm{d}s -$$

$$- \frac{1}{2} \left(\frac{1}{3G} + \frac{1}{9K} \right) (\sigma_1^R(S))^2 + \int_0^{S_H} \sigma(s) \dot{\varepsilon}_v^I(s)\, \mathrm{d}s. \tag{7.4.6}$$

For the evaluation of the latter integral from (7.4.6), the experimental data obtained in hydrostatic tests are necessary; in these tests, W^I reduces just to this latter term. As a rule, the hydrostatic compressibility law during loading is expressed as

$$\varepsilon_v^H = f(\sigma), \tag{7.4.7}$$

with $f > 0$, $f' > 0$ and $f'' \leqslant 0$ for any σ. Also during hydrostatic unloading, a linear relationship is assumed (see Chapter 4). With these simplifying relationships, the computation of the last integral, which will be denoted by W^{IH}, is somewhat facilitated. Generally, however, the term W^{IH} can be obtained directly from experimental data, without making use of empirical formulae of the type (7.4.7). The remaining terms from (7.4.6) are evaluated from the data obtained in the second stage of the triaxial tests.

The determination of the plastic potential F from experimental data is a much more difficult task. Following an idea from classical plasticity theory, however, in a first variant of the model it can be assumed that the plastic potential coincides with the yield function

$$F(\sigma, \bar{\sigma}, \Delta) \equiv H(\sigma, \bar{\sigma}, \Delta), \tag{7.4.8}$$

while this latter function can be determined much more easily from the experimental data, using, for instance, the procedure just described. In order to test if this simplifying assumption is reasonable, one can check whether the relationship

$$\dot{\varepsilon}_1^I / \dot{\varepsilon}_2^I = \left(\frac{\partial H}{\partial \sigma_1} \right) \Big/ \left(\frac{\partial H}{\partial \sigma_2} \right) \tag{7.4.9}$$

which results from the constitutive equation and the assumption (7.4.8), is in agreement with the experimental data. Here, the left-hand side of (7.4.9) can be estimated from the experimental data and the right-hand side can be obtained from the previously determined expression for $H(\sigma, \bar{\sigma})$. Further, if the overall prediction of the constitutive equation, written with the assumption (7.4.8), compares favourably with the experimental data, then this assumption can be accepted as a reasonable one.

Generally, the properties of compressibility, dilatancy, and the dilatancy threshold may be expected to be loading-path-dependent. The classification (7.3.4) is to be analysed, depending on the loading paths considered in the $\bar{\sigma}0\sigma$-plane. In the example shown in Fig. 7.2.1, the loading paths used were, in the first stage, hydrostatic and, in the second stage, $\dot{\sigma}_2 = \dot{\sigma}_3 = 0$ and $\dot{\sigma}_1 > 0$. These loading paths are traditionally used in triaxial tests and are nowhere neutral loading paths, as one can see from Fig. 7.2.1. Finally, let us observe that Schock *et al.* (1973) have found experimentally for *granodiorite* that the curve establishing the onset of dilatancy is stress-path-independent (see also Schock and Heard (1974), Schock (1976)).

7.5. Example for a dilatant rock

In order to give an initial example for a rock which is dilatant only (negligible compressibility), the *hydrostatic elasticity model* will be assumed. We will use the curves σ_1^R—ε_1^R and σ_1^R—ε_2^R from the experimental data by Brace *et al.* (1966) for *granite* (we do not claim, however, to reproduce these experimental data exactly).

Using the method described above, we get for the function $H(\sigma, \bar{\sigma})$

$$H(\sigma, \bar{\sigma}) := a\left(\sigma - \frac{\bar{\sigma}}{3}\right)\left(\frac{\bar{\sigma}}{\sigma_*}\right)^n + b\,\frac{\bar{\sigma}}{\sigma_*}, \tag{7.5.1}$$

with the constitutive coefficient function

$$a\left(\sigma - \frac{\bar{\sigma}}{3}\right) = \frac{a_0}{\left(\sigma - \dfrac{\bar{\sigma}}{3}\right)\dfrac{1}{\sigma_*} + a_1}, \tag{7.5.2}$$

and $a_0 = 0.417 \times 10^{-3}$ GPa, $a_1 = 0.01082$, $b = 0.5 \times 10^{-3}$ GPa, $n = 4$, and $\sigma_* = 1$ GPa.

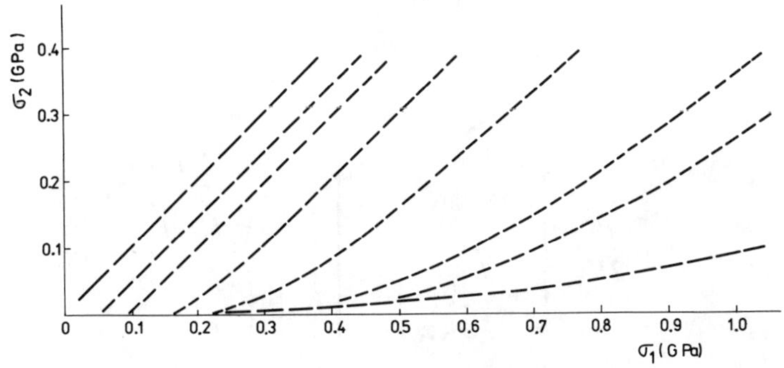

Figure 7.5.1. Yield surfaces (broken lines) for the hydrostatic elasticity model.

For the irreversible volumetric rate of deformation we have

$$\dot{\varepsilon}_v^I = \langle \dot{\lambda} \rangle \frac{-a_0}{\left[\dfrac{1}{\sigma_*}\left(\sigma - \dfrac{\bar{\sigma}}{3}\right) + a_1\right]^2} \left(\frac{\bar{\sigma}}{\sigma_*}\right)^4, \tag{7.5.3}$$

and, since $a_0 > 0$, the potential (7.5.1) will describe dilatancy for all stress-states for which $\bar{\sigma} \neq 0$. It is obvious from the expression of the coefficient a that for $\bar{\sigma} =$ const., if σ increases very much, the surfaces $H =$ const. tend to become circular cylinders, i.e. a Mises kind of yield function. This results also from Fig. 7.5.1, where several surfaces $H(\sigma, \bar{\sigma}) = W^I =$ const. are plotted, for the following values of W^I given in MPa: 0.025; 0.05; 0.1; 0.2; 0.5; 0.8. The broken line is an approximate position of the failure surface.

7.6. Example of compressible/dilatant rock

Let us consider again the same experimental data for *granite*, this time without disregarding the irreversible compressibility of the volume. A mathematical model for this compressibility was given in §4.3.2. Using the formulae given there, the last integral from (7.4.6) becomes

$$W^{IH}(S_H) = \int_0^{S_H} \sigma(s) \left[q \frac{\dot{\sigma}(s)}{2\sqrt{\sigma(s)} + r} - \frac{\dot{\sigma}(s)}{K(\sigma(s))} \right] ds. \tag{7.6.1}$$

For $\sigma^{\mathrm{H}} < \sigma_0$, we get for W^{IH} at time S_{H}:

$$W^{\mathrm{IH}}(S_{\mathrm{H}}) = \frac{q(\sigma^{\mathrm{H}} - 2r)}{3}\sqrt{\sigma^{\mathrm{H}} + r} + q\,\frac{2r}{3}\sqrt{r} +$$

$$+ \frac{\sigma_0^2}{2K_1}\left\{\frac{1}{2\left(\dfrac{K_0}{K_1}\right)^{3/4}}\left[\ln\left|\frac{\left(\dfrac{K_0}{K_1}\right)^{1/4} + 1 - \dfrac{\sigma^{\mathrm{H}}}{\sigma_0}}{\left(\dfrac{K_0}{K_1}\right)^{1/4} - 1 + \dfrac{\sigma^{\mathrm{H}}}{\sigma_0}}\right| +\right.\right.$$

$$\left. + 2\operatorname{arctg}\frac{1 - \dfrac{\sigma^{\mathrm{H}}}{\sigma_0}}{\left(\dfrac{K_0}{K_1}\right)^{1/4}}\right] -$$

$$- \frac{1}{2\left(\dfrac{K_0}{K_1}\right)^{1/2}}\ln\left|\frac{\left(\dfrac{K_0}{K_1}\right)^{1/2} + \left(1 - \dfrac{\sigma^{\mathrm{H}}}{\sigma_0}\right)^2}{\left(\dfrac{K_0}{K_1}\right)^{1/2} - \left(1 - \dfrac{\sigma^{\mathrm{H}}}{\sigma_0}\right)^2}\right| -$$

$$- \frac{1}{2\left(\dfrac{K_0}{K_1}\right)^{3/4}}\left[\ln\left|\frac{\left(\dfrac{K_0}{K_1}\right)^{1/4} + 1}{\left(\dfrac{K_0}{K_1}\right)^{1/4} - 1}\right| +\right.$$

$$\left. + 2\operatorname{arctg}\frac{1}{\left(\dfrac{K_0}{K_1}\right)^{1/4}}\right] +$$

$$\left. + \frac{1}{2\left(\dfrac{K_0}{K_1}\right)^{1/2}}\ln\left|\frac{\left(\dfrac{K_0}{K_1}\right)^{1/2} + 1}{\left(\dfrac{K_0}{K_1}\right)^{1/2} - 1}\right|\right\}. \tag{7.6.2}$$

For $\sigma^{\mathrm{H}} \geqslant \sigma_0$ the term W^{IH} is constant and equal to its value taken at $\sigma^{\mathrm{H}} = \sigma_0$.

Using the experimental data given in Table 4.2 and formula (7.6.2), we get for W^{IH} the numerical values shown by crosses in Fig. 7.6.1. For convenience we can approximate the relationship $H^{H}(\sigma(s)) = W^{IH}(s)$ by a simple empirical expression

$$H^{H}(\sigma) := c_0 \sin\left(\omega \frac{\sigma}{\sigma_*} + \varphi\right) + c_1, \tag{7.6.3}$$

with $\omega = 493.14748°$, $\varphi = -79.889307°$, $c_0 = 1.16454 \times 10^{-4}$ GPa, $c_1 = 1.14645 \times 10^{-4}$ GPa. Since for $\sigma \geqslant \sigma_0$ we must have $H^{H} = \text{const.}$, we choose the following expression for this function:

$$H^{H}(\sigma) := \begin{cases} c_0 \sin\left(\omega \dfrac{\sigma}{\sigma_*} + \varphi\right) + c_1 & \text{if } 0 \leqslant \sigma \leqslant \sigma_0 \\ c_0 + c_1 \quad \text{if } \sigma_0 \leqslant \sigma. \end{cases} \tag{7.6.4}$$

The variation of this function with σ is shown by the dotted line in Fig. 7.6.1. In order to simplify the notation, we write $\sigma = \sigma^{H} = \sigma(S_H)$.

If simple empirical formulae describing volume compressibility (those given in §4.3.2, for instance) are not available, then after numerical integration of the latter integral from (7.4.6), we can approximate the numerical set of values obtained by formulae of the form (7.6.3), etc.

Let us estimate the group of remaining terms involved in (7.4.6), which will be denoted by W^{IR} or W^{ID}. In order to simplify the computations, and also because of the lack of very accurate experimental data, it is assumed that in the interval 0.1 GPa $\leqslant \sigma^{H} \leqslant$ 0.4 GPa, the slopes of the curves $\sigma_1^R - \varepsilon_1^R$ and $\sigma_1^R - \varepsilon_2^R$ during unloading are approximately constant and equal to

$$\frac{1}{3G} + \frac{1}{9K} = 0.0125\ (\text{GPa})^{-1}, \qquad -\frac{1}{6G} + \frac{1}{9K} = -0.0030\ (\text{GPa})^{-1}.$$

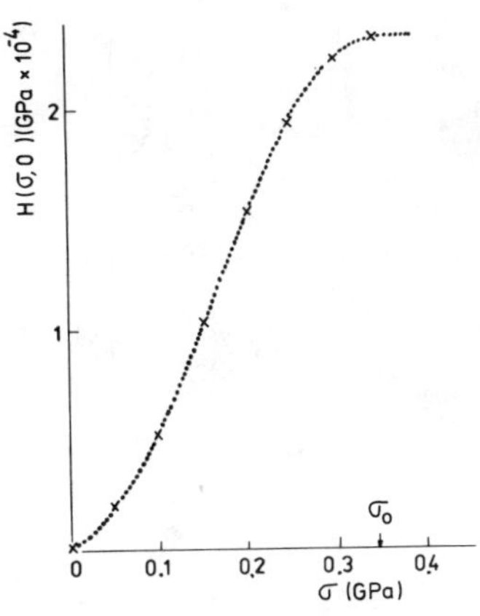

Figure 7.6.1. The plastic potential for the hydrostatic case. Crosses are the values of W^{IH} obtained from experimental data for granite; the dotted line represents the function (7.6.3).

With the experimental data mentioned, we first compute W^{IR}. The results are shown by crosses in Fig. 7.6.2 and can be approximated by the following expression for $H(\sigma, \bar{\sigma})$:

$$H(\sigma, \bar{\sigma}) := \left(\frac{a_0}{\frac{1}{\sigma_*}\left(\sigma - \frac{\bar{\sigma}}{3}\right) + a_1} + a_2 \right) \left(\frac{\bar{\sigma}}{\sigma_*} \right)^4 + $$

$$+ \left(\frac{\left(\frac{1}{\sigma_*}\left(\sigma - \frac{\bar{\sigma}}{3}\right) - b_0 \right)^2}{b_1} + b_2 \right) \frac{\bar{\sigma}}{\sigma_*} +$$

$$+ \begin{cases} c_0 \sin\left(\omega \frac{\sigma}{\sigma_*} + \varphi\right) + c_1 & \text{if } 0 \leqslant \sigma \leqslant \sigma_0 \\ c_0 + c_1 & \text{if } \sigma_0 \leqslant \sigma, \end{cases} \tag{7.6.5}$$

with $a_0 = 3.8327 \times 10^{-4}$ GPa, $a_1 = 9.2245 \times 10^{-3}$, $a_2 = 2.622 \times 10^{-4}$ GPa, $b_0 = 6.0045 \times 10^{-3}$, $b_1 = -969.4893\ (\text{GPa})^{-1}$, $b_2 = 4.5087 \times 10^{-4}$ GPa.

Several surfaces $H(\sigma, \bar{\sigma}) = W^{I} =$ const., for various values of this constant are shown in Fig. 7.2.1 by dotted lines. The following values of W^{I} given in MPa, in the order of increasing $\bar{\sigma}$, have been used: 0.05; 0.1; 0.2; 0.3; 0.4; 0.6; 0.8.

The compressibility/dilatancy boundary appears in this figure as a border line (— · — ·), while the failure surface is plotted from the experimental data by Brace (1964) as a broken line. For $\sigma > \sigma_0$, the straight line $\sigma_1 = \sigma_2 = \sigma_3$ can be thought of as an extension of the compressibility/dilatancy boundary (along this line, $H(\sigma, 0) = H^{H}(\sigma) = H^{H}(\sigma_0) =$ const.).

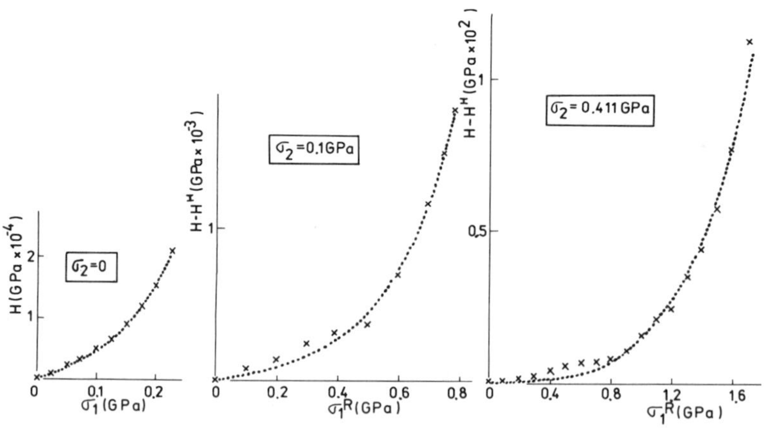

Figure 7.6.2. Variation of function H with stress (dotted lines); crosses are the numerical values of W^{I} for granite.

In a preliminary check, relation (7.4.9) is in reasonable agreement with the experimental data, though an accurate estimate of $\dot{\varepsilon}_1^{\mathrm{I}}$ and $\dot{\varepsilon}_2^{\mathrm{I}}$ from experimental data is quite difficult.

Let us make several remarks in conjunction with the typical example given above.

For relatively small stresses the 'yield surfaces' are conical with the vertices pointing towards the positive sense of σ. Thus, for relatively small hydrostatic stresses (i.e. $\sigma < \sigma_0$) and relatively small $\bar{\sigma}$, the rock is compressible only. For higher values of $\bar{\sigma}$ however, the rock is dilatant. At higher mean stresses, (for $\sigma > \sigma_0$) the rock is dilatant only, though this dilatancy is negligibly small for a very wide domain around the 0σ-axis, and is significant again only for high values of $\bar{\sigma}$. The smooth departure of the yield surfaces from the 0σ-axis at $\sigma > \sigma_0$, as σ increases, and the increased distances between these surfaces, compared to the distances between the same surfaces along the $0\sigma_1$-axis, means that at high hydrostatic stresses there is a wide domain around the 0σ-axis where the behaviour of the volume of the rock is nearly linearly elastic, and this domain enlarges when σ increases. Thus, although at high hydrostatic stresses the rock is dilatant only, this dilatancy is very pronounced for high values of $\bar{\sigma}$ only.

The 'associated law' (i.e. relation (7.2.3)) seems to give reasonable results for rocks too, if the plastic potential is correctly determined. Along the hydrostatic line $\sigma_1 = \sigma_2 = \sigma_3$ for $0 \leqslant \sigma \leqslant \sigma_0$, as expected, the normal to the yield surface is not uniquely determined, as can be seen from the expression of the derivatives $(\partial H/\partial \sigma_i)$.

A *second variant* of the model can be obtained starting from the idea suggested by Fig. 7.2.1, that the *failure surface is the envelope of the yield surfaces.* Let us assume that the equation of the failure surface can be written in the form

$$\frac{\bar{\sigma}}{\sigma_*} = \left[\frac{A_0}{\sigma_*^2}\left(\sigma - \frac{\bar{\sigma}}{3}\right)^2 + \frac{A_1}{\sigma_*}\left(\sigma - \frac{\bar{\sigma}}{3}\right) + A^2\right]^{1/2}, \tag{7.6.6}$$

where A_0, A_1, and A_2 are material constants, and that it is the envelope of the yield surfaces. This equation can be determined either directly from experimental data concerning the failure, or from the asymptotes of the W^{I}—σ_1^{R} curves obtained for various values $\sigma_2 =$ const. in triaxial tests. It is more convenient to determine equation (7.6.6) before determining the function $H(\sigma, \bar{\sigma})$, but one can also determine (7.6.6) simultaneously with the other coefficients involved in $H(\sigma, \bar{\sigma})$. We will choose for the function $H(\sigma, \bar{\sigma})$ the expression

$$H(\sigma, \bar{\sigma}) := \left[\frac{a}{\left[\frac{A_0}{\sigma_*^2}\left(\sigma - \frac{\bar{\sigma}}{3}\right)^2 + \frac{A_1}{\sigma_*}\left(\sigma - \frac{\bar{\sigma}}{3}\right) + A_2\right]^{1/2} - \frac{\bar{\sigma}}{\sigma_*}} + b\right]\frac{\bar{\sigma}}{\sigma_*} + H^{\mathrm{H}}(\sigma), \tag{7.6.7}$$

where $H^{\mathrm{H}}(\sigma)$ is determined in the same way as above. We determine also the coefficient functions as

$$a\left(\sigma-\frac{\bar{\sigma}}{3}\right)=\frac{a_0}{\sigma_*^2}\left(\sigma-\frac{\bar{\sigma}}{3}\right)^2+\frac{a_1}{\sigma_*}\left(\sigma-\frac{\bar{\sigma}}{3}\right)+a_2$$

$$b\left(\sigma-\frac{\bar{\sigma}}{3}\right)=\frac{1}{b_1}\left[\frac{1}{\sigma_*}\left(\sigma-\frac{\bar{\sigma}}{3}\right)-b_0\right]^2+b_2, \tag{7.6.8}$$

with $a_0 = 6.4052 \times 10^{-3}$ GPa, $a_1 = 5.2157 \times 10^{-4}$ GPa, $a_2 = 1.822 \times 10^{-5}$ GPa, $b_0 = 0.02$, $b_1 = -154.0859$ (GPa)$^{-1}$, $b_2 = 6.391 \times 10^{-4}$ GPa, $A_0 = 6.908888$, $A_1 = 5.693111$, $A_2 = 0.0841$. Once function H is obtained from experimental data, the constitutive equation is fully determined. Using expression (7.6.7), the domains D_c and D_d as well as the compressibility/dilatancy boundary have similar shapes to those obtained for the first variant given above.

Finally, we observe that failure of rocks can be only poorly described by a failure condition written in stresses alone. This subject is tackled in Chapter 9.

7.7. Generalization of the model for finite rotations

The constitutive equation (7.3.5) has been formulated with respect to the actual configuration and can certainly be applied for the cases of finite strains but small material rotations. However, since in these constitutive equations derivatives with respect to time of the stress tensor are also involved, and since these derivatives are not objective, some comment is necessary (see also the models proposed by Rudnicki and Rice (1975), and Dorris and Nemat-Nasser (1982)). In a generalized version of the model the time derivatives of the stress tensor must be replaced by a stress flux, both in the terms of the constitutive equation referring to the elastic rate of deformation and in the evolution law (7.2.8), if work-hardening is defined using the concept of irreversible stress power.

Therefore, in the generalized version of the constitutive equation (7.3.5), the rate of deformation tensor $\mathbf{D}$, the Cauchy stress tensor $\boldsymbol{\sigma}$ and the stress flux $\mathring{\boldsymbol{\sigma}}$ are involved, i.e. the constitutive equation is written as

$$\mathbf{D}=\frac{\mathring{\boldsymbol{\sigma}}}{2G}+\left(\frac{1}{3K}-\frac{1}{2G}\right)\mathring{\sigma}\mathbf{1}+\frac{\left\langle\dfrac{\partial H}{\partial\boldsymbol{\sigma}}\cdot\mathring{\boldsymbol{\sigma}}\right\rangle}{\dfrac{\partial F}{\partial\boldsymbol{\sigma}}\cdot\boldsymbol{\sigma}}\frac{\partial F}{\partial\boldsymbol{\sigma}} \quad \text{if } H = W^{\mathrm{I}}$$

$$\mathbf{D}=\frac{\mathring{\boldsymbol{\sigma}}}{2G}+\left(\frac{1}{3K}-\frac{1}{2G}\right)\mathring{\sigma}\mathbf{1} \quad \text{if } H < W^{\mathrm{I}}. \tag{7.7.1}$$

Many other variants may be formulated for a generalized constitutive equation. It is obvious that the determination of the function $H(\sigma, \bar{\sigma}, \Delta)$, from experimental data, is a most difficult task. Generally, without giving explicit expressions for all the constitutive parameters involved, the constitutive equation is of little value for rock mechanics.

Exercises

1. Give a detailed derivation of formula (7.3.1).
2. Starting from (7.3.3) and (7.3.4), discuss the details presented in Fig. 7.3.1.
3. Give detailed derivations of formulae (7.4.3) and (7.4.6).
4. Prove that the constitutive equation obtained using the plastic potential (7.5.1) describes dilatancy only, and that the surfaces $H = \text{const.}$ are of the form shown in Fig. 7.5.1.
5. Derive the formula (7.6.2).
6. Find the domain in the $\sigma 0 \bar{\sigma}$-plane where the constitutive equation obtained using the plastic potential (7.6.5) expresses dilatancy, and where compressibility. Find the equation of the compressibility/dilatancy boundary.
7. From triaxial test data available to you, determine a plastic potential following the procedure from §7.6.

8

Elastic/viscoplastic constitutive equations

8.1. General considerations

In previous chapters it has been shown that the mechanical properties of rocks are quite complex, and that they cannot be described satisfactorily by a classical constitutive equation of the Hooke type or of linear viscoelastic type, for example, with the possible exception of restricted constitutive domains. The experimental data reveal some specific mechanical properties of rocks which cannot be described by either of the constitutive equations presented in the previous chapters. Some of these properties are not even described qualitatively, while others are poorly described quantitatively.

In order to describe more accurately the mechanical properties of rocks and rock-like materials, in the present chapter we show how an elastic/viscoplastic constitutive equation for rocks can be formulated for isothermal processes. Various variants have been tested. One starts from *quasi-linear* isotropic constitutive equations. Since the number of invariants involved in the coefficient functions is too large to obtain reasonably simple models, we have restricted ourselves to a reduced number of invariants on which the constitutive equations may depend. These are thought to be the ones most significant in the mechanics of rocks. Generally, the following invariants will be involved in the constitutive coefficients:

— the mean strain

$$\varepsilon = \tfrac{1}{3}\,(\varepsilon_1 + \varepsilon_2 + \varepsilon_3); \tag{8.1.1}$$

— the equivalent strain

$$(\tfrac{3}{2}\,\bar{\varepsilon})^2 = \varepsilon_1^2 + \varepsilon_2^2 + \varepsilon_3^2 - \varepsilon_1\,\varepsilon_2 - \varepsilon_2\,\varepsilon_3 - \varepsilon_3\,\varepsilon_1; \tag{8.1.2}$$

— the third invariant of the strain tensor det $\boldsymbol{\varepsilon}$;
— the mean stress (see Chapter 4, exercise 7)

$$\sigma = \tfrac{1}{3}\,(\sigma_1 + \sigma_2 + \sigma_3); \tag{8.1.3}$$

— the equivalent stress

$$\bar{\sigma}^2 = \sigma_1^2 + \sigma_2^2 + \sigma_3^2 - \sigma_1\,\sigma_2 - \sigma_2\,\sigma_3 - \sigma_3\,\sigma_1; \tag{8.1.4}$$

— the third invariant of the stress tensor $\Delta = \det \boldsymbol{\sigma}$.

As will be shown below, the concept of *viscoplastic potential*, $F(\sigma, \bar{\sigma}, \Delta)$, is very useful. According to this concept the irreversible part of the rate of deformation is obtained from

$$\dot{\boldsymbol{\varepsilon}}^{\mathrm{I}} = \lambda \frac{\partial F}{\partial \boldsymbol{\sigma}}, \qquad (8.1.5)$$

with λ a scalar variable which may depend on the previously mentioned invariants and on the irreversible stress work

$$W^{\mathrm{I}}(T) = \int_0^T \boldsymbol{\sigma}(t) \cdot \dot{\boldsymbol{\varepsilon}}^{\mathrm{I}}(t)\,\mathrm{d}t, \qquad (8.1.6)$$

and, possibly independently, on the volumetric irreversible stress work

$$W_{\mathrm{v}}^{\mathrm{I}}(T) = \int_0^T \sigma(t)\,\dot{\varepsilon}_{\mathrm{v}}^{\mathrm{I}}(t)\,\mathrm{d}t. \qquad (8.1.7)$$

As will be shown below, using the concept of viscoplastic potential in the framework of an elastic/viscoplastic constitutive equation, both dilatancy and compressibility of rocks during creep can be described.

Several elastic/viscoplastic constitutive equations have been formulated for rocks by various authors. The most important approaches are mentioned below.

The Dragon and Mróz (1979) approach was aimed at describing the secondary and tertiary creep of rocks; primary creep is disregarded. These authors have considered only those processes of deformation by creep which occur under relatively high stresses (over, say, 0.4 to 0.6 times the compressive strength). These limit stresses are considered to be 'yield stresses' for that particular rock. Relation (8.1.5) is assumed, together with the yield condition

$$F(\boldsymbol{\sigma}, W, \beta) := \boldsymbol{\sigma}' \cdot \boldsymbol{\sigma}' - 2p(W, \beta)(I_1^0 - \operatorname{tr} \boldsymbol{\sigma}) = 0$$
$$p(W, \beta) = p_0 + W - \beta^2,$$

where I_1^0 and p_0 are constants, while β is the scalar fracture parameter, for which an evolution law is postulated in the form

$$\dot{\beta} = K_1\,\dot{\varepsilon}_{\mathrm{v}}^{\mathrm{I}} + K_2(\beta)\,g(\sigma, \beta),$$

where K_1 is a constant which is thought to describe the softening of the rock. The factor W is designed to describe the opposite effect, i.e. the hardening. In fact, β and W are not independent. Generally, a potential of this form describes dilatancy only.

The Gioda (1981) approach is aimed also at describing secondary and tertiary creep. The model can be termed viscoelastic/viscoplastic; it uses for the potential and yield surfaces

$$F(\sigma, \bar{\sigma}) := 3\alpha\sigma - \frac{\bar{\sigma}}{3} - k = 0,$$

with α and k constants. This potential too can describe dilatancy only.

An elastic/viscoplastic constitutive equation for rock salt was developed by Langer (1982), starting from the assumption

$$\dot{\boldsymbol{\varepsilon}} = \dot{\boldsymbol{\varepsilon}}^{\mathrm{E}} + \dot{\boldsymbol{\varepsilon}}^{\mathrm{cr}} + \dot{\boldsymbol{\varepsilon}}^{\mathrm{f}},$$

where $\dot{\boldsymbol{\varepsilon}}^{\mathrm{E}}$ is the elastic, reversible, component, $\dot{\boldsymbol{\varepsilon}}^{\mathrm{cr}}$ is the viscoplastic component, and $\dot{\boldsymbol{\varepsilon}}^{\mathrm{f}}$ is the irreversible, plastic, fracture component. Assuming incompressibility, $\dot{\boldsymbol{\varepsilon}}^{\mathrm{cr}}$ is obtained from a Saint-Venant kind of flow rule (i.e. $\dot{\boldsymbol{\varepsilon}}^{\mathrm{cr}} \sim \boldsymbol{\sigma}'$, but with an additional time-dependent relationship between the second invariants of the two deviators), while $\dot{\boldsymbol{\varepsilon}}^{\mathrm{f}}$ is derived from a fracture (flow) condition and an associated flow law.

In order to describe creep of *rock salt*, Wawersik *et al.* (1982) propose the constitutive equation

$$\dot{\boldsymbol{\varepsilon}} = \dot{\boldsymbol{\varepsilon}}^{\mathrm{E}} + \frac{3}{2}\,\frac{\dot{\bar{\varepsilon}}^{\mathrm{cr}}}{\bar{\sigma}}\,\boldsymbol{\sigma}',$$

where $\dot{\boldsymbol{\varepsilon}}^{\mathrm{cr}}$ are the components of the creep strain-rate tensor, with an additional relationship between $\dot{\bar{\varepsilon}}^{\mathrm{cr}}$, $\bar{\sigma}$ and time obtained from triaxial creep tests. Transient and steady-state creep are both considered, but not volumetric creep.

A kinematic hardening model for *halite* was developed by Krieg (1982), assuming, however, a linear elastic volumetric behaviour. The model is aimed at describing primary and secondary creep behaviour of halite.

A microphysical model for inelastic material response formulated in terms of the evolution of some microstructures (growth and coalescence of microcracks) is due to Margolin (1984).

Another constitutive equation under development is aimed at describing the constant (creep) and cyclic compressive loading of rocks, starting from the mechanics of microcracks (Costin and Holcomb (1981, 1983)). It is assumed that time- and stress-dependent microcrack growth is responsible for the plastic deformation of rocks. A specific form of elastic/viscoplastic constitutive equation for *mudstone* was developed by Sekiguchi (see Sekiguchi *et al.* (1985)).

In the models formulated below (Cristescu (1985*a, c*, 1987)) we start from the experimental evidence presented in Chapter 3, according to which *primary creep also produces irreversible deformations.* The irreversible deformations due to creep occur for stresses much smaller than $0.4\sigma_c$ (say) (where σ_c is the short-term compressive strength of the rock determined by standard procedures). On the other hand, any constitutive equation intended to be used for rocks must describe the peculiar mechanical properties of those materials as *dilatancy* and/or *compressibility* of the volume. Since *time effects* are also always present during the deformation of such materials, at any stress level these effects must also be described by the constitutive equation (see also Hansen and Carter (1984), Hansen *et al.* (1984)).

Experimental evidence shows that any loading/unloading cycle applied to a rock reveals an irreversible part of the strain, even if the applied stresses are relatively small. For this reason and taking into account that at small stresses time effects are also present, we assume that rock-like materials possess a *zero yield stress* (see also Itô (1979), Hansen and Carter (1980), Wawersik and Hannum (1980), Mellegard *et al.* (1981), Akagi (1985)).

On the other hand, we recall that if a rock is subjected to relatively small compressive stresses, the volume of the rock is always compressible with a part of the volumetric strain irreversible. It is only at higher compressive stresses that the rock has a behaviour reminiscent of *linear elastic behaviour*, in the sense that small loading/unloading cycles essentially exhibit reversible strains only. Let us remember also that the stress—strain curves are non-linear in the neighbourhood of the zero stress-states in both tensile and compressive tests (see Schock and Louis (1982)). Therefore, the elastic domain is not to be found in the neighbourhood of the zero stress-state. It follows from this that the concept of 'elasticity' has a different meaning for rocks and rock-like materials than for metals, for instance, even if the time effects are negligible. Extremely small stresses of no significance for engineering problems are disregarded here.

8.2. Experimental foundation

In order to formulate a constitutive equation, two kinds of tests are necessary: confined tests in which hydrostatic loadings are applied to the specimen in order to get the compressibility properties of the volume, and 'triaxial' tests which reveal the properties of the rock subjected to a deviatoric stress superimposed on a hydrostatic one. The results obtained with the first type of test are reviewed only briefly here, because they have already been presented in detail in Chapter 4.

If a cylindrical rock specimen is subjected to a hydrostatic stress σ which is applied incrementally, after each increment $\Delta\sigma$ of the stress an instantaneous increase of the volumetric strain takes place which, in the first approximation, can be obtained from a relationship of the form

$$\dot{\sigma} = K(\sigma)\, \dot{\varepsilon}_{\mathrm{v}}, \tag{8.2.1}$$

where $K(\sigma)$ depends on the mean stress. Furthermore, between successive loadings, during the long time-interval in which σ is kept constant, a volumetric creep is recorded (see also Wawersik and Hannum (1980)). When the deformation by creep of the volume comes to a halt, the stress is again increased by another step, etc. The geometrical locus of points in the plane σ—ε_{v}, where the deformation by creep has stopped, is called the '*stabilization boundary*' for volumetric creep and is shown as a broken line in Fig. 8.2.1. An unloading

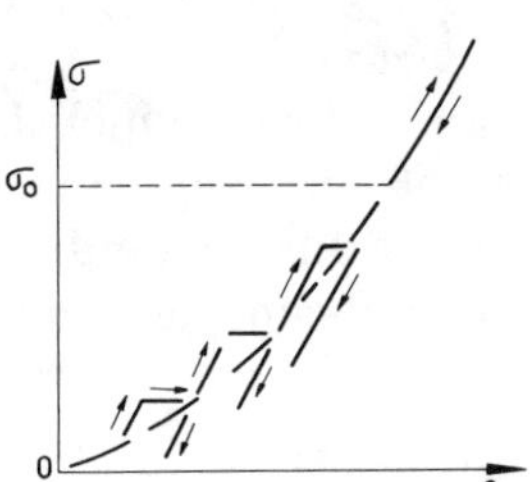

Figure 8.2.1. Hydrostatic volumetric creep test: incremental loadings followed by unloadings. σ_0 is the pressure which closes most cracks.

starting from one of the points situated on this boundary reveals at the beginning of the process a behaviour which can be approximated by relationship (8.2.1). K is an increasing function of σ in the interval $0 \leqslant \sigma \leqslant \sigma_0$ only. For $\sigma > \sigma_0$ volumetric creep is no longer observed and the relationship σ—ε_v becomes nearly linear, for both loading and unloading during cyclic processes. Thus

$$K(\sigma) := \begin{cases} \hat{K}(\sigma) & \text{if } 0 \leqslant \sigma \leqslant \sigma_0, \\ K_0 & \text{if } \sigma_0 \leqslant \sigma. \end{cases} \tag{8.2.2}$$

Experimental data show that the end of the unloading process is sometimes non-linear. However, in the model which follows, for the sake of simplicity, we approximate the whole unloading period with a relationship of the form (8.2.1), choosing for K a 'secant' value. For some rocks, such as *salt*, the unloading process is quite accurately linearly elastic (Wawersik and Hannum (1980)). We recall that σ_0 is the smallest mean stress for which all the cracks are already closed, and that is why for $\sigma > \sigma_0$ the behaviour of the volume of the rock is nearly linearly elastic and reversible — the hydrostatic elasticity of the constituent minerals.

The volumetric deformation is generally small, and

$$\dot{\varepsilon}_v = \dot{\varepsilon}_v^E + \dot{\varepsilon}_v^I. \tag{8.2.3}$$

The volumetric work-hardening during compressibility will be expressed by

$$H^H(\sigma(t)) = W_v^I(t) \quad \text{with} \quad W_v^I(T) = \int_0^T \sigma(t)\, \dot{\varepsilon}_v^I(t)\, dt \tag{8.2.4}$$

and therefore the stabilization boundary is generally not unique, since it depends on the loading history. From the peculiarities of the volume compressibility results that during loading, W_v^I is zero at $\sigma = 0$, then it increases with σ, and reaches its maximal value at $\sigma = \sigma_0$. For $\sigma > \sigma_0$, W_v^I remains constant.

The second set of experimental data necessary for the formulation of a mathematical model is obtained in triaxial tests. As a rule, in the first stage of such tests, all the stress components are kept equal and simultaneously increased from zero up to a certain value

$$\sigma_1^H = \sigma_2^H = \sigma_3^H = \sigma^H, \tag{8.2.5}$$

where the superscript H denotes 'hydrostatic'. In the second stage of the experiment, $\dot{\sigma}_2 = \dot{\sigma}_3 = 0$, and only the component σ_1 is further increased; the variation in time of all strain components is recorded in both stages.

Thus both stress and strain components are increased in two stages, denoted by superscripts H and R respectively (R denoting 'relative'), i.e.

$$\begin{aligned} &\sigma_1(t) = \sigma_1^H(t) + \sigma_1^R(t), \text{ etc.,} \\ &\dot{\varepsilon}_1^R(t) = \dot{\varepsilon}_1(t) - \dot{\varepsilon}_1^H(t) = \dot{\varepsilon}_1(t) - \dot{\varepsilon}_1^{EH}(t) - \dot{\varepsilon}_1^{IH}(t), \text{ etc.} \end{aligned} \tag{8.2.6}$$

Sometimes the components $\boldsymbol{\sigma}^{\mathrm{R}}$ are also called 'stress differences' or 'differential stresses'. In each stage of the experiment the rate of deformation tensor possesses a reversible component (elastic) and an irreversible one. Typical results obtained in the second stage of the experiment are shown in Fig. 8.2.2, where the stress—strain curves have a stepped appearance (see also Cristescu (1982)). A sudden increase $\Delta\sigma_1^{\mathrm{R}}$ of the component σ_1^{R} produces a sudden variation of the strain components, the relationship between $\Delta\sigma_1^{\mathrm{R}}$ and $\Delta\varepsilon_1^{\mathrm{ER}}$ and between $\Delta\sigma_1^{\mathrm{R}}$ and $\Delta\varepsilon_2^{\mathrm{ER}}$ being nearly linear. The corresponding slopes

$$\left(\frac{1}{3G}+\frac{1}{9K}\right)^{-1}, \qquad \left(-\frac{1}{6G}+\frac{1}{9K}\right)^{-1}, \tag{8.2.7}$$

may possibly vary slowly with σ. We recall that, in fact, the slopes (8.2.7) are determined in unloading/reloading cycles, where the stress—strain curves are quite accurately straight lines. In many mining engineering applications, K and G can be assumed to be constant and to take specific values for the particular depth under consideration.

After the sudden increase of the strain with the elastic components $\Delta\varepsilon_1^{\mathrm{ER}}$ and $\Delta\varepsilon_2^{\mathrm{ER}}$, there follows a very long time interval during which the component σ_1^{R} is kept constant as well, and the variation in time of the deformation by creep is recorded. The end of the deformation process, marked by a small circle in Fig. 8.2.2, corresponds to the stabilization of the creep. Next, the stress is increased again by another increment $\Delta\sigma_1^{\mathrm{R}}$, etc. Such experiments have been described in Chapter 4. We recall here that stabilization of the creep takes place if the stress does not exceed a certain limit; if this limit is surpassed, a deformation by secondary creep takes place, followed by a tertiary one, and eventually by failure. The volumetric creep is obtained from ε_1 and ε_2 by using the formula

$$\varepsilon_{\mathrm{v}} = \varepsilon_1 + 2\varepsilon_2. \tag{8.2.8}$$

Typical properties of volumetric creep are shown in Fig. 8.2.2. Generally, at relatively small stresses volumetric creep is not very significant, and in most

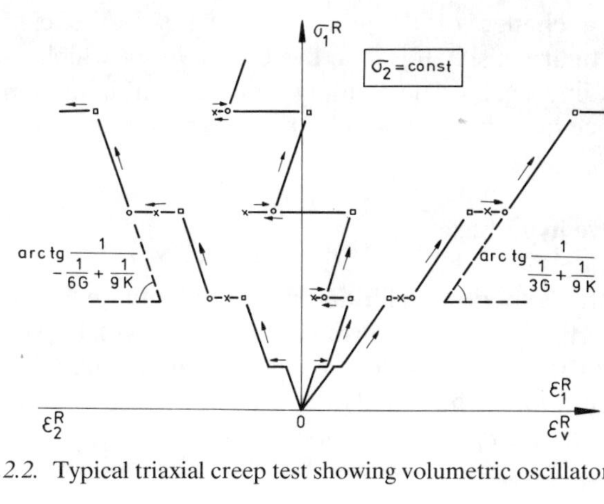

Figure 8.2.2. Typical triaxial creep test showing volumetric oscillatory creep.

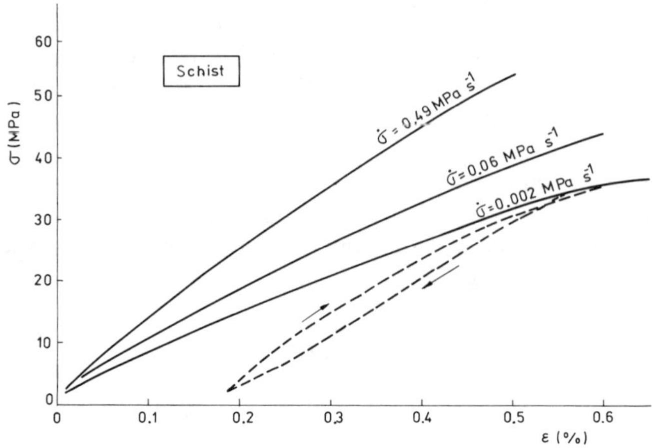

Figure 8.2.3. Uniaxial stress—strain curves for schist showing that the loading rate influences these curves starting from the origin.

cases produces compressibility. At somewhat higher stresses the behaviour of the volume is quite close to the linearly elastic one. At even higher stresses the volume is generally dilatant. Sometimes oscillatory volumetric creep can be observed (Cristescu (1982)). This oscillatory feature is shown in Fig. 8.2.2: squares mark the beginning of the deformation by creep, crosses mark the end of volume dilatancy during creep, and circles correspond to the stabilization of the creep process.

Other experiments showing *important time effects* during the deformation of rocks are the uniaxial compression tests performed with various loading rates. These experiments have been discussed already in Chapter 2. To recall some of the conclusions, Fig. 8.2.3 shows the stress—strain curves for *schist* obtained with several constant loading rates. The general feature is the raising of the entire stress—strain curve, if a higher constant rate of loading is used. There is no initial elastic portion on these stress—strain curves, and the whole stress—strain curve is influenced by the loading rate; therefore, no portion of this curve is invariant to a change of the loading rate (as is the case for metals, for instance). The figure also shows a hysteresis loop which is obtained in an unloading/reloading cycle. The loading rate also influences rock dilatancy, as already mentioned in Chapter 4. Similar results are obtained in triaxial tests.

8.3. Constitutive hypotheses

Starting from the experimental data presented above, a certain number of constitutive hypotheses will be formulated. These will be simplifying assumptions made in order to obtain a reasonably simple mathematical model. The reference configuration is the actual one.

H1. The rock is *isotropic* and *homogeneous*.

H2. The *displacements and material rotations are assumed to be small.* Thus the elastic (reversible) component of the rate of deformation and the irreversible (plastic) one are additive

$$\dot{\boldsymbol{\varepsilon}} = \dot{\boldsymbol{\varepsilon}}^{\mathrm{E}} + \dot{\boldsymbol{\varepsilon}}^{\mathrm{I}}, \tag{8.3.1}$$

while the elastic component of the rate of deformation is obtained from the general constitutive equation

$$\dot{\boldsymbol{\varepsilon}}^{\mathrm{E}} = \left(\frac{1}{3K} - \frac{1}{2G}\right)\dot{\sigma}\mathbf{1} + \frac{1}{2G}\,\dot{\boldsymbol{\sigma}} \tag{8.3.2}$$

in which the elastic parameter $K(\sigma)$ may depend on the mean stress.

H3. The equation of the *stabilization boundary* will be written in the form

$$H(\sigma, \bar{\sigma}, \Delta) = W^{\mathrm{I}}(t), \tag{8.3.3}$$

where the *work-hardening parameter* $W^{\mathrm{I}}(t)$ is the irreversible stress work

$$W^{\mathrm{I}}(T) = \int_0^T \boldsymbol{\sigma}(t) \cdot \dot{\boldsymbol{\varepsilon}}^{\mathrm{I}}(t)\,\mathrm{d}t, \tag{8.3.4}$$

with T the actual time, while the non-negative expression

$$P^{\mathrm{I}}(t) = \boldsymbol{\sigma}(t) \cdot \dot{\boldsymbol{\varepsilon}}^{\mathrm{I}}(t) \tag{8.3.5}$$

is the irreversible stress power. Since the yield stress is assumed to be zero, irreversible deformation is possible starting from the smallest values of the stress (either hydrostatic and/or deviatoric).

Sometimes it is useful to decompose expression (8.3.4) into two terms

$$W^{\mathrm{I}}(T) = W^{\mathrm{I}}_{\mathrm{V}}(T) + W^{\mathrm{I}}_{\mathrm{D}}(T) = \int_0^T \sigma(t)\,\dot{\varepsilon}^{\mathrm{I}}_{\mathrm{V}}(t)\,\mathrm{d}t + \int_0^T \boldsymbol{\sigma}'(t) \cdot \dot{\boldsymbol{\varepsilon}}'^{\mathrm{I}}(t)\,\mathrm{d}t \tag{8.3.6}$$

corresponding to the irreversible work for the change of the volume and the irreversible work for the change in shape.

H4. The irreversible component $\dot{\boldsymbol{\varepsilon}}^{\mathrm{I}}$ of the rate of deformation tensor is obtained from

$$\dot{\boldsymbol{\varepsilon}}^{\mathrm{I}} = k(\sigma, \bar{\sigma}, \Delta, d)\left\langle 1 - \frac{W^{\mathrm{I}}(t)}{H(\sigma, \bar{\sigma}, \Delta)}\right\rangle \frac{\partial F(\sigma, \bar{\sigma}, \Delta)}{\partial \boldsymbol{\sigma}}, \tag{8.3.7}$$

where $F(\sigma, \bar{\sigma}, \Delta)$ is the *viscoplastic potential*, $k(\sigma, \bar{\sigma}, \Delta, d) > 0$ is the *viscosity coefficient*, and $\langle A \rangle = \frac{1}{2}(A + |A|)$. The viscosity coefficient may depend also on a parameter d, called the *damage parameter*, which will be defined in Chapter 9.

Since the stress power is non-negative, the function F must satisfy the restriction

$$\frac{\partial F}{\partial \boldsymbol{\sigma}} \cdot \boldsymbol{\sigma} > 0 \tag{8.3.8}$$

everywhere, where $\dot{\boldsymbol{\varepsilon}}^{\mathrm{I}} \neq 0$.

In (8.3.7) it was assumed that $\dot{\boldsymbol{\varepsilon}}^{\mathrm{I}}$ is linear in the argument $1 - W^{\mathrm{I}}/H$; other non-linear dependences on the same argument, as for instance those suggested by Cristescu and Suliciu (1982, Chapter II) may also be considered.

H5. The *constitutive domain* in the stress space $0\sigma_1\,\sigma_2\,\sigma_3$, where the constitutive equations are defined is

$$\begin{aligned} &\sigma_i \geqslant 0 \quad (i = 1, 2, 3) \\ &\bar{\sigma} < R(\sigma, \Delta, d), \end{aligned} \tag{8.3.9}$$

where $\bar{\sigma} = R(\sigma, \Delta, d)$ is the failure condition. Therefore, only compressive stresses will be considered, while $(8.3.9)_2$ can be replaced by another relationship, if some other kind of failure condition is assumed. The failure condition will eventually be included in the constitutive equation.

All the constitutive functions mentioned above must be determined from the experimental data, as well as the elastic moduli $K(\sigma)$ and G, the first one varying very smoothly with σ.

8.4. Constitutive equations

H2 and H4 yield the following quasilinear constitutive equation (see, for instance, Cristescu (1967), and Cristescu and Suliciu (1982) for similar constitutive equations proposed for other materials):

$$\dot{\boldsymbol{\varepsilon}} = \left(\frac{1}{3K} - \frac{1}{2G}\right)\dot{\sigma}\mathbf{1} + \frac{1}{2G}\,\dot{\boldsymbol{\sigma}} + k\left\langle 1 - \frac{W^{\mathrm{I}}}{H}\right\rangle \frac{\partial F}{\partial \boldsymbol{\sigma}}. \tag{8.4.1}$$

For the irreversible part of the volumetric rate of deformation, we obtain

$$\dot{\varepsilon}_{\mathrm{v}}^{\mathrm{I}} = k\left\langle 1 - \frac{W^{\mathrm{I}}}{H}\right\rangle \frac{\partial F}{\partial \boldsymbol{\sigma}} \cdot \mathbf{1}. \tag{8.4.2}$$

Let us assume that at time t_0 and at a certain point of the rock the present stress-state $\boldsymbol{\sigma}^{\mathrm{P}} = \boldsymbol{\sigma}(t_0)$ (called 'primary', since it refers to the undisturbed rock) is a stable stress-state, i.e. it satisfies $H(\sigma^{\mathrm{P}}) = W(t_0)$. Further, if the stress changes for $t > t_0$, due to an excavation, say, to $\boldsymbol{\sigma}(t)$, various cases are possible. If (see Fig. 12.5.1)

$$H(\boldsymbol{\sigma}(t)) < H(\boldsymbol{\sigma}^{\mathrm{P}}) \quad \text{(elastic unloading)}, \tag{8.4.3}$$

the stress variation is an *unloading*, and the constitutive equation reduces to (8.3.2) with no irreversible variation of the volume. If, however,

$$H(\boldsymbol{\sigma}(t)) = H(\boldsymbol{\sigma}^{\mathrm{P}}), \tag{8.4.4}$$

the new stress-state is on the same *stabilization boundary* or on the same *loading/unloading boundary*.

The case $H(\boldsymbol{\sigma}(t)) > H(\boldsymbol{\sigma}^{\mathrm{P}})$ will be called *loading*, since in this case a variation of $\boldsymbol{\varepsilon}^{\mathrm{I}}$ is possible. We can distinguish three subcases according to the sign of $\partial F/\partial \boldsymbol{\sigma} \cdot \mathbf{1}$ or of $\partial F/\partial \sigma$ (if function $F(\sigma, \bar{\sigma})$ only depends on the two invariants shown). Thus

$$\left.\begin{array}{l} H(\boldsymbol{\sigma}(t)) > H(\boldsymbol{\sigma}^{\mathrm{P}}) \\[1ex] \dfrac{\partial F}{\partial \sigma} > 0 \quad \text{(compressibility)}, \\[2ex] \dfrac{\partial F}{\partial \sigma} < 0 \quad \text{(dilatancy)}, \\[2ex] \dfrac{\partial F}{\partial \sigma} = 0 \quad \begin{array}{l}\text{(compressibility/} \\ \text{dilatancy boundary).}\end{array} \end{array}\right\} \begin{array}{l}\text{viscoplastic} \\ \text{deformation} \\ \text{(loading)}\end{array} \tag{8.4.5}$$

The inequalities $(8.4.5)_{1,2}$ are the mathematical definition of irreversible compressibility, $(8.4.5)_{1,3}$ that of irreversible dilatancy, and finally, relations $(8.4.5)_{1,4}$ are the definition of the dilatancy threshold or of the compressibility/dilatancy boundary.

For compressive stresses, i.e. $\sigma > 0$, these inequalities are equivalent to

$$P_{\mathrm{v}}^{\mathrm{I}}(t) = \sigma \dot{\varepsilon}_{\mathrm{v}}^{\mathrm{I}} \begin{cases} > 0 & \text{(compressibility)} \\ < 0 & \text{(dilatancy)} \\ = 0 & \text{(compressibility/dilatancy boundary).} \end{cases} \tag{8.4.6}$$

In the neighbourhood of the stress-states which satisfy $(8.4.5)_{1,4}$, the behaviour of the volume of the rock is very close to linearly elastic.

As already mentioned, the functions $H(\sigma, \bar{\sigma}, \Delta)$ and $F(\sigma, \bar{\sigma}, \Delta)$ are to be determined from experimental data. Most of the experimental data are obtained in 'triaxial' tests in which only two principal stress components can be varied independently and recorded, i.e. σ_1 and $\sigma_2 = \sigma_3$. That is why in the numerical examples given below it is assumed that all constitutive functions depend on the two stress invariants σ and $\bar{\sigma}$ only. In 'triaxial' tests the expressions of these invariants are

$$\bar{\sigma} = |\sigma_1 - \sigma_2|, \qquad \sigma = \tfrac{1}{3}(\sigma_1 + 2\sigma_2). \tag{8.4.7}$$

It will be assumed on the basis of experimental evidence that the invariant Δ influences to a lesser degree the failure surface (in the classical sense), and that is why it is to be expected that this invariant is also of lesser importance in the expressions of the functions H and F, at least when compared to the influence of the invariants σ and $\bar{\sigma}$ (see Fig. 2.4.5).

In order to determine the function $H(\sigma, \bar{\sigma})$, the integration in (8.3.4) is performed following various loading paths chosen to cover systematically the entire constitutive domain. In triaxial tests, first a hydrostatic load is usually applied up to a certain value $\sigma_1^{\mathrm{H}} = \sigma_2^{\mathrm{H}} = \sigma_3^{\mathrm{H}} = \sigma^{\mathrm{H}}$; afterwards, $\sigma_2 = \sigma_3$ are maintained constant and σ_1^{R} only is further increased. Using the decomposition (8.2.6), the irreversible part of the stress power becomes

$$
\begin{aligned}
P^{\mathrm{I}} = {} & \sigma_1^{\mathrm{H}} \left[\dot{\varepsilon}_1^{\mathrm{R}} - \left(\frac{1}{3G} + \frac{1}{9K} \right) \dot{\sigma}_1^{\mathrm{R}} \right] + \\
& + 2\sigma_2^{\mathrm{H}} \left[\dot{\varepsilon}_2^{\mathrm{R}} - \left(-\frac{1}{6G} + \frac{1}{9K} \right) \dot{\sigma}_1^{\mathrm{R}} \right] + \\
& + \sigma_1^{\mathrm{R}} \left[\dot{\varepsilon}_1^{\mathrm{R}} - \left(\frac{1}{3G} + \frac{1}{9K} \right) \dot{\sigma}_1^{\mathrm{R}} \right] + \\
& + (\sigma_1^{\mathrm{H}} + \sigma_1^{\mathrm{R}} + 2\sigma_2^{\mathrm{H}}) \frac{1}{3} \dot{\varepsilon}_{\mathrm{v}}^{\mathrm{IH}} .
\end{aligned}
\tag{8.4.8}
$$

Writing the stress power in this form was suggested by the type of data obtained in the 'triaxial' tests. Let us assume that at $t = T_{\mathrm{H}}$ the stabilization of the volumetric creep occurs, and that at $t = T$ the stabilization of the creep in the second stage of the experiment takes place. Integrating (8.4.8) and taking into account that in the first stage of the test all the components with superscript R are zero, while in the second stage the components with superscript H are constant, we obtain

$$
\begin{aligned}
W^{\mathrm{I}}(T) = {} & \sigma_1^{\mathrm{H}} \varepsilon_1^{\mathrm{R}}(T) + 2\sigma_2^{\mathrm{H}} \varepsilon_2^{\mathrm{R}}(T) - \sigma_1^{\mathrm{H}} \left(\frac{1}{3G} + \frac{1}{9K} \right) \sigma_1^{\mathrm{R}}(T) - \\
& - 2\sigma_2^{\mathrm{H}} \left(-\frac{1}{6G} + \frac{1}{9K} \right) \sigma_1^{\mathrm{R}}(T) + \int_{T_{\mathrm{H}}}^{T} \sigma_1^{\mathrm{R}}(t)\, \dot{\varepsilon}_1^{\mathrm{R}}(t)\, \mathrm{d}t - \\
& - \frac{1}{2} \left(\frac{1}{3G} + \frac{1}{9K} \right) [\sigma_1^{\mathrm{R}}(T)]^2 + \int_0^{T_{\mathrm{H}}} \sigma_1(t)\, \dot{\varepsilon}_{\mathrm{v}}^{\mathrm{I}}(t)\, \mathrm{d}t .
\end{aligned}
\tag{8.4.9}
$$

The integrals involved in (8.4.9) are calculated numerically, using the kind of experimental data mentioned. Thus the value of W^{I} is obtained for various equilibrium states on the stabilization boundaries, i.e. for various values of the stress components on the creep stabilization boundary. Further, using a procedure similar to the one described in Chapter 7, we can determine the function

$H(\sigma, \bar{\sigma})$. If the irreversible compressibility of the volume is negligible, then the last term in (8.4.9) is also negligible.

Direct methods to determine the function $F(\sigma, \bar{\sigma})$ are difficult to imagine. However, in a first approximation one can make the assumption

$$F(\sigma, \bar{\sigma}) \equiv H(\sigma, \bar{\sigma}); \tag{8.4.10}$$

the validity of this assumption can be tested by checking if the relation

$$\frac{\dot{\varepsilon}^{\mathrm{I}}_{ij}}{\dot{\varepsilon}^{\mathrm{I}}_{kl}} = \frac{\dfrac{\partial H}{\partial \sigma_{ij}}}{\dfrac{\partial H}{\partial \sigma_{kl}}} \tag{8.4.11}$$

is in agreement with the experimental data. More generally, the validity of the assumption (8.4.10) can be tested if any other prediction of the constitutive equation (8.4.1) in which (8.4.10) is accepted is in agreement with experimental data.

Finally, the viscosity coefficient k can be determined in creep tests. By scalar multiplication of (8.4.1) by $\boldsymbol{\sigma}$ and integrating for various constant stresses, in the time interval $t \in [t_{\mathrm{i}}, t_{\mathrm{f}}]$, we get

$$\int_{W^{\mathrm{I}}_{\mathrm{i}}}^{W^{\mathrm{I}}_{\mathrm{f}}} \frac{\mathrm{d}W^{\mathrm{I}}}{1 - \dfrac{W^{\mathrm{I}}}{H(\sigma, \bar{\sigma})}} = k \frac{\partial F}{\partial \boldsymbol{\sigma}} \cdot \boldsymbol{\sigma}(t_{\mathrm{f}} - t_{\mathrm{i}}). \tag{8.4.12}$$

(Here the subscripts i and f stand for 'initial' and 'final' moments of the creep process.) Thus, for the determination of k we use the formula

$$k(\sigma, \bar{\sigma}) = -\ln \left| 1 - \frac{W^{\mathrm{I}}(t)}{H(\sigma, \bar{\sigma})} \right| \Bigg|_{t_{\mathrm{i}}}^{t_{\mathrm{f}}} \frac{H(\sigma, \bar{\sigma})}{\dfrac{\partial F(\sigma, \bar{\sigma})}{\partial \boldsymbol{\sigma}} \cdot \boldsymbol{\sigma}} \frac{1}{t_{\mathrm{f}} - t_{\mathrm{i}}}. \tag{8.4.13}$$

Hence the values of k for various stress-states follow. The quantity E/k, where E is the Young's modulus, is strictly speaking the *viscosity coefficient* and is expressed in poise. In solving engineering problems only the order of magnitude of k is of relevance. We have assumed above that k does not depend on the time interval $t_{\mathrm{f}} - t_{\mathrm{i}}$, or at least that this dependence is not significant. In the examples so far considered, this seems to be the case at least within certain limits of variation of the stresses and strains. However, if for certain rocks this does not happen and if the experimental data are quite accurate, then these data can be used to improve the formulation of the constitutive equation; for instance, we can try to determine a non-linear form of it

$$\dot{\boldsymbol{\varepsilon}}^{\mathrm{I}} = k f\left(\left\langle 1 - \frac{W^{\mathrm{I}}}{H} \right\rangle\right) \frac{\partial F}{\partial \boldsymbol{\sigma}}, \tag{8.4.14}$$

the values of k being determined afterwards.

In most examples given below we have disregarded the oscillatory volumetric creep. If we wish to include it in the mathematical model, then we could introduce, for instance, one pair of constitutive functions H_{d}, F_{d}, k_{d} and H_{c}, F_{c}, k_{c}. Here H_{d} is involved in the definition of the stabilization boundary for dilatant volumetric creep. In other words, H_{d} is to be determined by the points marked by crosses in Fig. 8.2.2. In a similar way, H_{c} governs the stabilization of creep, producing compressibility, and is to be determined by the points marked by circles in Fig. 8.2.2. For any stress-state producing loading, we must have

$$\frac{\partial F_{\mathrm{d}}}{\partial \sigma} < 0, \qquad \frac{\partial F_{\mathrm{c}}}{\partial \sigma} > 0. \tag{8.4.15}$$

Therefore, in this case the irreversible rate of deformation becomes

$$\dot{\boldsymbol{\varepsilon}}^{\mathrm{I}} = k_{\mathrm{d}} \left\langle 1 - \frac{W^{\mathrm{I}}}{H_{\mathrm{d}}} \right\rangle \frac{\partial F_{\mathrm{d}}}{\partial \boldsymbol{\sigma}} + k_{\mathrm{c}} \left\langle 1 - \frac{W^{\mathrm{I}}}{H_{\mathrm{c}}} \right\rangle \frac{\partial F_{\mathrm{c}}}{\partial \boldsymbol{\sigma}}. \tag{8.4.16}$$

According to this model the creep describing dilatancy is rather a ‘fast’ creep, when compared with the creep producing compressibility which can be termed ‘slow’ creep. This observation bears on some practical aspect of the model formulation: k_{d} and k_{c} are determined for various time intervals, where either dilatancy or compressibility is the dominant phenomenon, other phenomena being negligible. Other variants of the model which can describe the oscillatory volumetric creep have been given by Cristescu (1982).

Other variants of the previously presented models can be obtained, for instance, by writing the equations of the stabilization boundary in various other forms, such as

$$H(\sigma, \bar{\sigma}, \Delta) = \sqrt{\tfrac{1}{2}\, \boldsymbol{\varepsilon}^{\mathrm{I}}(t) \cdot \boldsymbol{\varepsilon}^{\mathrm{I}}(t)}\,, \tag{8.4.17}$$

or

$$H(\sigma, \bar{\sigma}, \Delta) = \frac{2}{3} \int_0^T \sqrt{\dot{\boldsymbol{\varepsilon}}^{\mathrm{I}}(t) \cdot \dot{\boldsymbol{\varepsilon}}^{\mathrm{I}}(t)}\, \mathrm{d}t. \tag{8.4.18}$$

If we denote the irreversible equivalent strain by

$$\bar{\varepsilon}^{\mathrm{I}}(T) = \frac{2}{3} \int_0^T \sqrt{\dot{\boldsymbol{\varepsilon}}^{\mathrm{I}}(t) \cdot \dot{\boldsymbol{\varepsilon}}^{\mathrm{I}}(t)}\, \mathrm{d}t, \tag{8.4.19}$$

we can write the irreversible rate of strain as

$$\dot{\boldsymbol{\varepsilon}}^{\mathrm{I}} = k \left\langle 1 - \frac{\bar{\varepsilon}^{\mathrm{I}}(t)}{H} \right\rangle \frac{\partial F}{\partial \boldsymbol{\sigma}}, \tag{8.4.20}$$

while for the determination of k, the relationship similar to (8.4.12) is

$$\int_{\bar{\varepsilon}^{\mathrm{I}}(t_{\mathrm{i}})}^{\bar{\varepsilon}^{\mathrm{I}}(t_{\mathrm{f}})} \frac{\mathrm{d}\bar{\varepsilon}^{\mathrm{I}}}{1 - \dfrac{\bar{\varepsilon}^{\mathrm{I}}}{H}} = \frac{2}{3} k \sqrt{\frac{\partial F}{\partial \boldsymbol{\sigma}} \cdot \frac{\partial F}{\partial \boldsymbol{\sigma}}} (t_{\mathrm{f}} - t_{\mathrm{i}}), \tag{8.4.21}$$

from where we get

$$k = -\ln \left| 1 - \frac{\bar{\varepsilon}^{\mathrm{I}}(t)}{H} \right| \Bigg|_{t_{\mathrm{i}}}^{t_{\mathrm{f}}} \frac{H}{\dfrac{2}{3}\left(\dfrac{\partial F}{\partial \boldsymbol{\sigma}} \cdot \dfrac{\partial F}{\partial \boldsymbol{\sigma}}\right)^{1/2}} \frac{1}{t_{\mathrm{f}} - t_{\mathrm{i}}}. \tag{8.4.22}$$

A similar procedure can be used if the equation of the stabilization boundary is written in the form (8.4.17). In some particular cases it may be easier to obtain the expressions of the stabilization boundaries written either in the form (8.4.17) or (8.4.18) than in the form (8.3.3) (see below). If function H coincides with the plastic potential, for dimensional reasons we write $F(\boldsymbol{\sigma}) = E\,H(\boldsymbol{\sigma})$.

Although the work-hardening parameters involved in (8.4.17) and (8.4.18) are easier to handle from a mathematical point of view, it is only the work-hardening parameter $W^{\mathrm{I}}(t)$ which can distinguish accurately irreversibility deformation occurring during compressibility from irreversibility during dilatancy (see Exercise 12).

Formulae such as (8.4.12) are quite useful in solving various problems. Let us consider a more general version. Assume that a loading process of stress variation $\boldsymbol{\sigma}(t)$ with $t \in [t_0, T]$ is prescribed. By scalar multiplication by $\boldsymbol{\sigma}$ on both sides of (8.3.7) we get

$$\dot{W}^{\mathrm{I}} = k \left\langle 1 - \frac{W^{\mathrm{I}}}{H} \right\rangle \frac{\partial F}{\partial \boldsymbol{\sigma}} \cdot \boldsymbol{\sigma}. \tag{8.4.23}$$

Assuming that the bracket in (8.4.23) is non-zero for the process considered, we get

$$W^{\mathrm{I}}(T) = \left\{ \int_{t_0}^{T} k \frac{\partial F}{\partial \boldsymbol{\sigma}} \cdot \boldsymbol{\sigma} \exp\left[\int_{t_0}^{t} \frac{k}{H} \frac{\partial F}{\partial \boldsymbol{\sigma}} \cdot \boldsymbol{\sigma}\, \mathrm{d}\tau \right] \mathrm{d}t + \right.$$

$$\left. + W^{\mathrm{IP}} \right\} \exp\left(- \int_{t_0}^{T} \frac{k}{H} \frac{\partial F}{\partial \boldsymbol{\sigma}} \cdot \boldsymbol{\sigma}\, \mathrm{d}t \right) \tag{8.4.24}$$

where $W^{\mathrm{IP}} = H(\boldsymbol{\sigma}^{\mathrm{P}}) = H(\boldsymbol{\sigma}(t_0))$ is the initial (primary) value of W^{I} at $t = t_0$.

In the particular case when in a certain time interval the stress is constant, as for instance in the creep tests, we obtain the useful formula

$$\frac{W^{\mathrm{I}}(t)}{H} - 1 = \left(\frac{W^{\mathrm{IP}}}{H} - 1\right) \exp\left[-\frac{k}{H}\frac{\partial F}{\partial \boldsymbol{\sigma}} \cdot \boldsymbol{\sigma}(t - t_0)\right] \tag{8.4.25}$$

while stabilization is obtained for $t \to \infty$.

A formal *generalization for finite strains* of the constitutive equation can be obtained quite easily, by expressing it with respect to the actual configuration. The estimate of $W^{\mathrm{I}}(t)$ is certainly more involved, while its 'initial' value is determined, as shown in the following chapters.

8.5. An example for a compressible/dilatant hard rock

Let us give an example for a hard rock, disregarding oscillatory volumetric creep. The example given is only illustrative, since the experimental data used for this purpose have been collected from various sources in order to get all the necessary information for the formulation of a complete model. Even this information is not really complete.

The experimental data for *granite* obtained by Brace (1964, 1965), Brace *et al.* (1966), and Volarovich *et al.* (1974) have been used for the determination of the stabilization boundaries in short-term creep tests. Using the procedure from Chapter 7, we get

$$H(\sigma, \bar{\sigma}) := \left[\frac{a_0}{\left(\sigma - \frac{\bar{\sigma}}{3}\right)\frac{1}{\sigma_*} + a_1} + a_2\right]\left(\frac{\bar{\sigma}}{\sigma_*}\right)^4 + \left[\frac{\left(\left(\sigma - \frac{\bar{\sigma}}{3}\right)\frac{1}{\sigma_*} - b_0\right)^2}{b_1} + b_2\right]\frac{\bar{\sigma}}{\sigma_*} + \begin{cases} c_0 \sin\left(\omega \dfrac{\sigma}{\sigma_*} + \varphi\right) + c_1 & \text{if } 0 \leqslant \sigma \leqslant \sigma_0 \\ c_0 + c_1 & \text{if } \sigma_0 \leqslant \sigma, \end{cases} \tag{8.5.1}$$

with $a_0 = 3.833 \times 10^{-4}$ GPa, $a_1 = 9.225 \times 10^{-3}$, $a_2 = 2.622 \times 10^{-4}$ GPa, $b_0 = 6.005 \times 10^{-3}$, $b_1 = -969.4893$ $(\mathrm{GPa})^{-1}$, $b_2 = 4.508 \times 10^{-4}$ GPa, $c_0 = 1.16454 \times 10^{-4}$ GPa, $c_1 = 1.14645 \times 10^{-4}$ GPa, $\omega = 493.14748°$, $\varphi = -79.889307°$, $\sigma_* = 1$ GPa, $\sigma_0 = 0.3445$ GPa. The curves $H(\sigma, \bar{\sigma}) = \text{const.}$ for various values of this constant, i.e. of W^{I}, are shown as dotted lines in Fig. 8.5.1 in a $\sigma_1 0 \sigma_2$-plane. The following values of this constant (in MPa) have been used: 0.05; 0.1; 0.2; 0.3; 0.4; 0.6; 0.8.

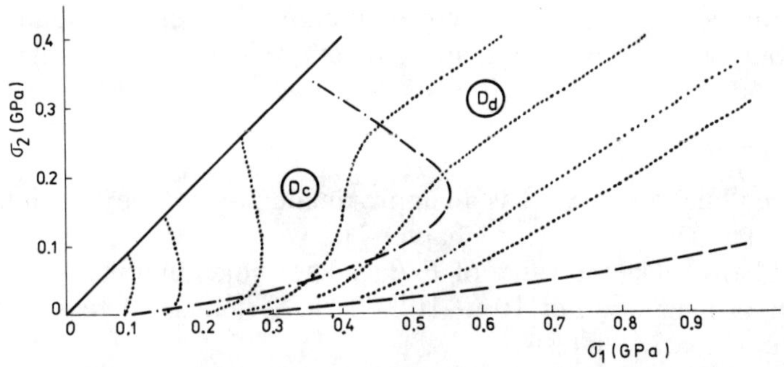

Figure 8.5.1. Constitutive domain for granite, showing: stabilization surfaces H = const. (.......), compressibility/dilatancy boundary $\partial H/\partial\sigma = 0$ (— · — ·), failure surface (— — —), domain of compressibility D_c, and domain of dilatancy D_d.

The irreversible part of the volumetric rate of deformation is obtained from

$$\dot{\varepsilon}_v^{\mathrm{I}} = k\left\langle 1 - \frac{W^{\mathrm{I}}(t)}{F(\sigma,\bar{\sigma})}\right\rangle \left[-\frac{a_0}{\sigma_*}\frac{\left(\dfrac{\bar{\sigma}}{\sigma_*}\right)^4}{\left(\left(\sigma - \dfrac{\bar{\sigma}}{3}\right)\dfrac{1}{\sigma_*} + a_1\right)^2} + \right.$$

$$+ \frac{2\left(\left(\sigma - \dfrac{\bar{\sigma}}{3}\right)\dfrac{1}{\sigma_*} - b_0\right)}{b_1\,\sigma_*}\,\frac{\bar{\sigma}}{\sigma_*} +$$

$$\left. + \begin{cases} \dfrac{\omega c_0}{\sigma_*}\cos\left(\omega\dfrac{\sigma}{\sigma_*} + \varphi\right) & \text{if } 0 \leqslant \sigma \leqslant \sigma_0 \\ 0 \quad \text{if } \sigma_0 \leqslant \sigma, & \end{cases} \right] \tag{8.5.2}$$

while the equation of the compressibility/dilatancy boundary is obtained by equating to zero the term in the large square brackets from (8.5.2). The shape of this boundary is shown in Fig. 8.5.1 by the border line (— · — ·). For the case $\boldsymbol{\sigma}^{\mathrm{P}} = \mathbf{0}$ in the domain D_c, the condition $(8.4.5)_2$ is satisfied, i.e. a creep deformation process will produce compressibility. Along the line $\sigma_1 = \sigma_2 = \sigma_3$, the domain D_c does not extend beyond the point σ_0, since at high hydrostatic pressures $\sigma \geqslant \sigma_0$ volumetric creep is no longer possible, the volumetric deformation taking place according to a linearly elastic (reversible) relationship. D_d is the domain where the deformation by creep will produce a dilatancy of the volume. It will be shown below that volumetric dilatancy is significant for relatively high values of $\bar{\sigma}$ only. For $\sigma > \sigma_0$ and relatively small values of $\bar{\sigma}$, the shape of the surface $H(\sigma, \bar{\sigma})$ = const. approaches that of a circular cylinder and therefore the behaviour of the volume is close to linearly elastic. The behaviour of the volume is rigorously linearly elastic along the compressibility/dilatancy boundary only (and with some approximation in a neighbourhood of this boundary), and also along the hydrostatic line

$$\sigma_1 = \sigma_2 = \sigma_3 \quad \text{for} \quad \sigma \geqslant \sigma_0. \tag{8.5.3}$$

The broken line in Fig. 8.5.1 is an approximate position of the failure surface, as determined in triaxial tests (see Chapter 7).

For most hard rocks the value of E/k as determined in uniaxial creep tests is of the order of magnitude of 10^{15}—10^{16} P. No systematic determinations have been done in triaxial creep tests.

The mean values $K = 51.28$ GPa and $G = 32.26$ GPa give a reasonably good approximation of the slopes of the stress—strain curves during unloading, for $\sigma \leqslant 0.5$ GPa, which are therefore

$$\frac{1}{3G} + \frac{1}{9K} = 0.0125\ (\text{GPa})^{-1}, \qquad -\frac{1}{6G} + \frac{1}{9K} = -0.003\ (\text{GPa})^{-1}.$$

The constitutive equation given above can describe various *time effects* exhibited by rocks. Consider, for instance, the triaxial creep tests for an incremental increase of the axial stress. In the second stage of the experiment when $\sigma_1 = \sigma_1^{\text{H}} + \sigma_1^{\text{R}}$ and $\sigma_2 = \sigma_3 = \sigma^{\text{H}}$, from the model we get at each loading step

$$\varepsilon_1^{\text{R}}(t) = \left(\frac{1}{3G} + \frac{1}{9K}\right)\sigma_1^{\text{R}} + \\ + \frac{\left(1 - \dfrac{W^{\text{I}}(t_0)}{H}\right)\dfrac{\partial H}{\partial \sigma_1}}{\dfrac{1}{H}\left(\dfrac{\partial H}{\partial \sigma_1}\sigma_1 + 2\dfrac{\partial H}{\partial \sigma_2}\sigma_2\right)} \times \\ \times \left\{1 - \exp\left[\frac{k}{H}\left(\frac{\partial H}{\partial \sigma_1}\sigma_1 + 2\frac{\partial H}{\partial \sigma_2}\sigma_2\right)(t_0 - t)\right]\right\}, \tag{8.5.4}$$

and another similar formula for ε_2^{R}. Here, t_0 is the moment of time from which the component σ_1 is kept constant, just after being suddenly increased by a quantity $\Delta\sigma_1^{\text{R}}$. Typical creep curves ε_1—t, ε_2—t, and ε_v—t obtained with such formulae are shown in Fig. 8.5.2 for the case $\sigma_1^{\text{H}} = 0$ and successive axial stress increments $\Delta\sigma_1^{\text{R}} = 0.05$ GPa. Stabilization of the creep after each reloading occurs after 5—10 days. Thus the successive creep curves shown in the figure are obtained for the following values of the relative axial stress σ_1^{R} (in GPa): 0.05; 0.1; 0.15; 0.20; 0.25 GPa. In the first two loading steps, volumetric creep produces compressibility, while in the last three loading steps the instantaneous variation of the volume (the one which occurs just after each stress increment) produces instantaneous elastic compressibility, but this compressibility is subse-

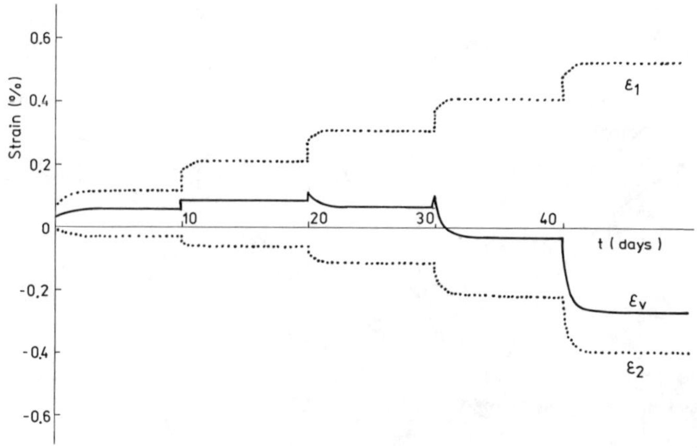

Figure 8.5.2. Creep curves obtained in uniaxial compression test according to the model for granite, showing compressibility at smaller stresses and then dilatancy of the volume at higher stresses.

quently followed by a volumetric creep producing dilatancy. At high stresses this dilatancy is very significant.

Let us describe now how the model can describe the *loading rate influence*, both on the stress—strain curves and on dilatancy and compressibility. In the first example considered a hydrostatic loading is first applied up to $\sigma_1^H = \sigma_2^H = \sigma_3^H = 0.411$ GPa which satisfies (8.5.3), after which only the component σ_1 is subsequently increased with various loading rates (Fig. 8.2.2). In order to follow the same kind of loading programme as that generally used in experimental testing programmes, the relative stress is increased in steps, $\Delta\sigma_1^R = 0.1$ GPa, and after each increase the stress is kept constant for a certain prescribed time interval. By changing the length of this time interval, which is the same after each stress increase (the beginning of such an interval is denoted by t_0), various global loading rates are obtained. The example shown in Fig. 8.5.3 corresponds to the following values of the dimensionless factor $k(t - t_0)$: zero (———), 0.036 (—·—·), and 0.36 (————). The conclusion is the following: an increase of the loading rate raises both curves σ_1^R—ε_1^R and σ_1^R—ε_2^R, starting from the origin. When the loading rate is greatly increased, these curves approach the corresponding linearly elastic response straight line. The same effect is obtained for the σ_1^R—ε_v^R curves describing dilatancy. Though according to the model the dilatancy threshold (onset) here is at $\sigma_1^R = 0$ for all loading rates, the 'apparent' dilatancy threshold (possibly defined as that small volumetric dilatancy which can be established experimentally) increases significantly with the loading rate.

A similar conclusion follows if in the first stage of the experiment the final hydrostatic stress state is $\sigma_1^H = \sigma_2^H = \sigma_3^H = 0.1$ GPa (i.e. a point belonging to the domain D_c with $\sigma < \sigma_0$), and afterwards only σ_1^R is increased. The mean loading rates considered are (Fig. 8.5.4): $\dot{\sigma}_1 \simeq 0$ (—.—.); 2.8 ×10^{-6} GPa s^{-1} (———); 2.8 × 10^{-5} GPa s^{-1} (......), and $\dot{\sigma}_1 \rightarrow \infty$ (———). The dilatancy threshold is $\sigma_1^R = \sigma_D = 0.391$ GPa for all cases. The increase of the loading rate diminishes the compressibility but increases the apparent dilatancy threshold.

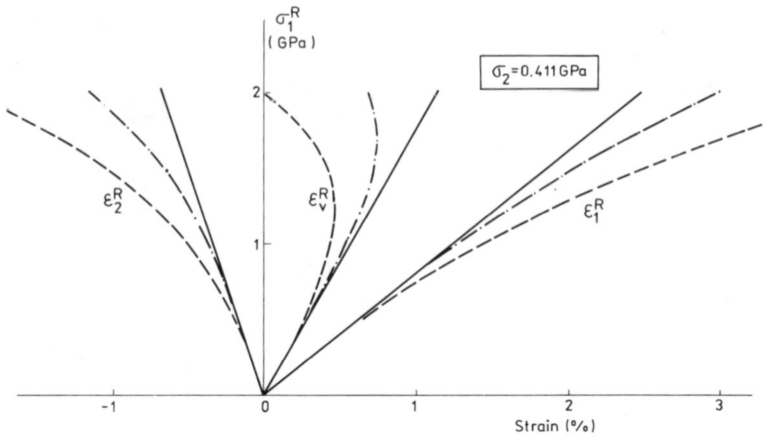

Figure 8.5.3. Loading rate influence on dilatancy in triaxial test.

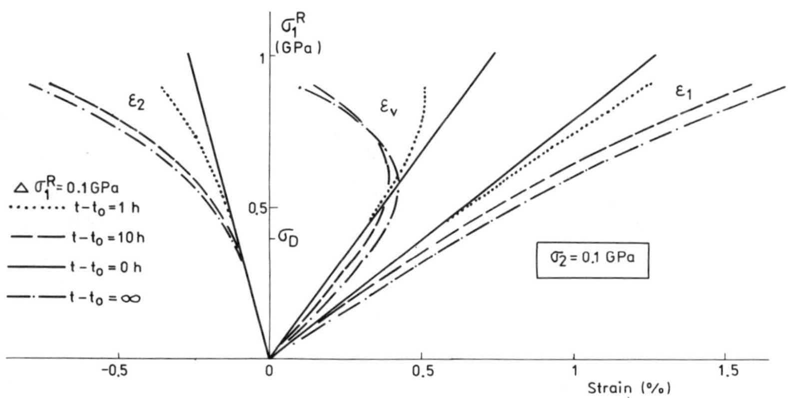

Figure 8.5.4. Incremental triaxial creep tests: stress—strain curves obtained at the end of time intervals shown, during which the stress is held constant. (Influence of loading rate on compressibility and dilatancy is obvious.)

For uniaxial loadings (Fig. 8.5.5) the conclusions are similar. The axial stress is increased by successive steps $\Delta\sigma_1 = 0.025$ GPa, and after each increase the stress is kept constant for various time intervals, distinct (but constant) in various tests. More specifically, the following values for the dimensionless expression $k(t - t_0)$ have been used: zero (curve 1); 0.0036 (curve 2); 0.036 (curve 3); 0.36 (curve 4); infinity (curve 5). The dilatancy threshold is always at $\sigma_1 = 0.0999$ GPa, but the apparent dilatancy threshold increases significantly with an increase of the loading rate while the compressibility diminishes. Maximal compressibility is obtained for $\dot{\sigma}_1 \to 0$, and generally, an influence of the loading rate on compressibility is obvious only for small loading rates. With relatively high loading rates (such as those generally used with universal testing machines), no significant influence on compressibility is observed when this loading rate is slightly changed, and generally, the compressibility may become negligible at high loading rates.

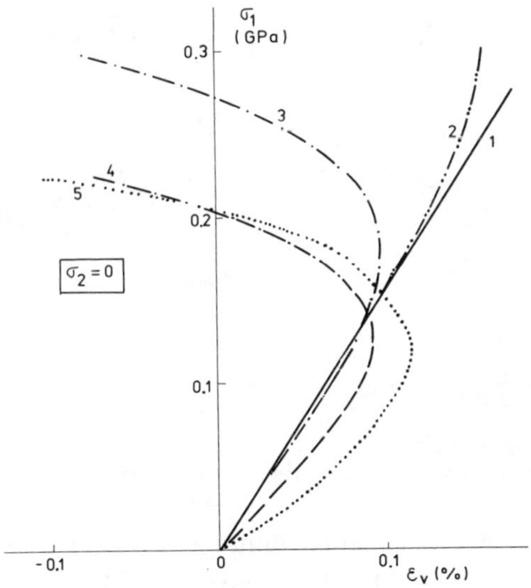

Figure 8.5.5. Influence of the loading rate on compressibility and dilatancy in uniaxial compression tests.

Let us return now to Fig. 8.5.3. We observe that all stress—strain curves depart smoothly from the corresponding elastic straight lines and coincide with these lines for a while. This always happens if the origin of the coordinate axis corresponds to a stress-state represented by a point on the hydrostatic line, with $\sigma \geqslant \sigma_0$. In other words, for $\sigma \geqslant \sigma_0$, *in a quite large domain around the* 0σ*-axis the behaviour of the rock is essentially elastic* and the departure from 'elasticity' is smooth.

Finally, let us see how the model (8.4.1) describes experiments of the type shown in Fig. 8.2.1, i.e. incremental hydrostatic loadings with the increments $\Delta\sigma = 0.1$ GPa, after which, during the period of time when the mean stress is held constant, volumetric creep takes place. The value $k = 10^{-5}\ \mathrm{s}^{-1}$ is used, resulting from $E/k = 4 \times 10^{14}$ P and $E = 41$ GPa. Volumetric creep at four stress levels is shown in Fig. 8.5.6. The instantaneous response is calculated taking into account that the modulus K is variable and its value is obtained from the empirical formula

$$K(\sigma) := \begin{cases} K_0 - K_1\left(1 - \dfrac{\sigma}{\sigma_0}\right)^4, & \text{if } \sigma \leqslant \sigma_0 \\ K_0, \quad \text{if } \sigma_0 \leqslant \sigma, & \end{cases} \tag{8.5.5}$$

with $K_0 = 59$ GPa and $K_1 = 48$ GPa, while $\sigma_0 = 0.3445$ GPa is a limit pressure, since for $\sigma \geqslant \sigma_0$ volumetric creep is no longer possible.

Comparison with experimental evidence. The phenomena described by the above model are typical for many rocks. At least qualitatively these are in pretty good agreement with experimental evidence. Besides our own experimental data, with which the predictions of the model have been compared, there are

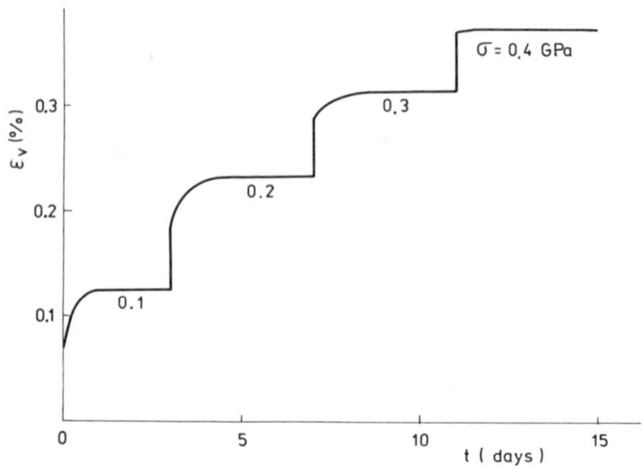

Figure 8.5.6. Hydrostatic volumetric creep curve; for higher pressures the rock is elastic only.

many other interesting experimental findings. Thus Schock *et al.* (1973), and Schock and Heard (1974), for instance, have found that the onset of dilatancy is loading-path-independent, that the inelastic processes are strain-rate-dependent, that dilatancy is diminished at high strain rates, and that relative ductility (of *sandstone*) is much lower at high strain rates than at the normal rate of testing, while, for *granite*, ductile behaviour cannot be expected at high strain rates ($\dot{\varepsilon} > 1\ s^{-1}$, for example). Further, the onset of dilatant behaviour was found for *granite* to be rate of deformation-dependent by Brace *et al.* (1966). Furthermore, when the mean pressure is increased, the degree of dilatancy observed in *sandstone* is sharply curtailed, but at still higher mean stress levels strong dilatancy is again observed and is associated with strong work-hardening (Schock *et al.* (1973)). Generally, for brittle rocks the onset of dilatant behaviour is suppressed (occurs at higher stresses) and the dilatant strain is reduced as the strain rate increases (Schock and Heard (1974), Schock (1977)). One of the interesting aspects of creep as observed in uniaxial test by Kranz (1979) for Barre *granite* 'is that, though both the strain parallel and perpendicular to the maximum stress direction go through the same three creep phases before the rock fails, the radial strain always shows a higher strain rate than the axial strain'.

Note. If the creep experiments which are necessary for the finding of the stabilization boundary have not been carried out for a long enough period of time, one can extrapolate this interval to a certain degree using the model.

The formulation of the mathematical model given above needs a great deal of experimental data, which has to be a coherent and complete set of data. Generally, such a complete set of data is hard to find in the literature. Anyhow, the model cannot be formulated without creep data obtained in triaxial tests.

Another variant of the model can be obtained if the equation of the stabilization boundary is written in the form (8.4.17), i.e.

$$H(\sigma, \bar{\sigma}, \Delta) = (\tfrac{1}{2}\,\boldsymbol{\varepsilon}^{\mathrm{I}} \cdot \boldsymbol{\varepsilon}^{\mathrm{I}})^{1/2}. \tag{8.5.6}$$

In order to estimate the right-hand-side term, the experimental data obtained during the two loading stages involved in the triaxial tests are used. The numerical example which follows is given only to illustrate the procedure to be used, and does not claim to be a very accurate model for a specific kind of rock. Formulae (7.6.1) are used for hydrostatic tests, in which, however, it is assumed for simplicity that during unloading, $K(\sigma) = K_0 = \text{const.}$ (instead of formula (8.5.5)). Thus we get

$$\begin{aligned} \varepsilon_{\mathrm{v}}^{\mathrm{IH}} &= -p + q\left(\frac{\sigma^{\mathrm{H}}}{\sigma_*} + r\right)^{1/2} - \frac{\sigma^{\mathrm{H}}}{K_0} \quad \text{if } 0 \leqslant \sigma \leqslant \sigma_0 \\ \varepsilon_{\mathrm{v}}^{\mathrm{IH}} &= \frac{K_2}{K_0} \quad \text{if } \sigma_0 \leqslant \sigma, \end{aligned} \tag{8.5.7}$$

with $p = 1.7469 \times 10^{-3}$, $q = 0.020112$, $r = 7.544 \times 10^{-3}$, $K_0 = 59$ GPa, $K_2 = 0.2565$ GPa, $\sigma_0 = 0.3445$ GPa.

Using the decomposition $\boldsymbol{\varepsilon}^{\mathrm{I}} = \boldsymbol{\varepsilon}^{\mathrm{IH}} + \boldsymbol{\varepsilon}^{\mathrm{IR}}$, we obtain for the triaxial tests

$$\begin{aligned} \frac{1}{2}\,\boldsymbol{\varepsilon}^{\mathrm{I}} \cdot \boldsymbol{\varepsilon}^{\mathrm{I}} = \frac{1}{2}\Bigg\{ &\frac{1}{3}(\varepsilon_{\mathrm{v}}^{\mathrm{IH}})^2 + \frac{2}{3}\varepsilon_{\mathrm{v}}^{\mathrm{IH}}\left[\varepsilon_1^{\mathrm{R}} - \left(\frac{1}{3G} + \frac{1}{9K}\right)\sigma_1^{\mathrm{R}} + \right. \\ &\left. + 2\varepsilon_2^{\mathrm{R}} - 2\left(-\frac{1}{6G} + \frac{1}{9K}\right)\sigma_1^{\mathrm{R}}\right] + \\ &+ \left[\varepsilon_1^{\mathrm{R}} - \left(\frac{1}{3G} + \frac{1}{9K}\right)\sigma_1^{\mathrm{R}}\right]^2 + \\ &+ 2\left[\varepsilon_2^{\mathrm{R}} - \left(-\frac{1}{6G} + \frac{1}{9K}\right)\sigma_1^{\mathrm{R}}\right]^2\Bigg\}. \end{aligned} \tag{8.5.8}$$

A numerical calculation of this expression is done using experimental data and $G = 31.5$ GPa, $K = K_0$. The results obtained are shown in Fig. 8.5.7. By approximating these curves, the function $H(\sigma, \bar{\sigma})$ is determined as

$$\begin{aligned} H^{\mathrm{R}}(\sigma, \bar{\sigma}) &:= a\left(\frac{\bar{\sigma}}{\sigma_*}\right)^4 + b\,\frac{\bar{\sigma}}{\sigma_*} \\ H^{\mathrm{H}}(\sigma) &:= \begin{cases} \dfrac{1}{\sqrt{6}}\left(-p + q\left(\dfrac{\sigma}{\sigma_*} + r\right)^{1/2} - \dfrac{\sigma}{K_0}\right) & \text{for } 0 \leqslant \sigma \leqslant \sigma_0 \\ \dfrac{1}{\sqrt{6}}\,\dfrac{K_2}{K_0} & \text{for } \sigma_0 \leqslant \sigma \end{cases} \\ H(\sigma, \bar{\sigma}) &= H^{\mathrm{H}}(\sigma) + H^{\mathrm{R}}(\sigma, \bar{\sigma}). \end{aligned} \tag{8.5.9}$$

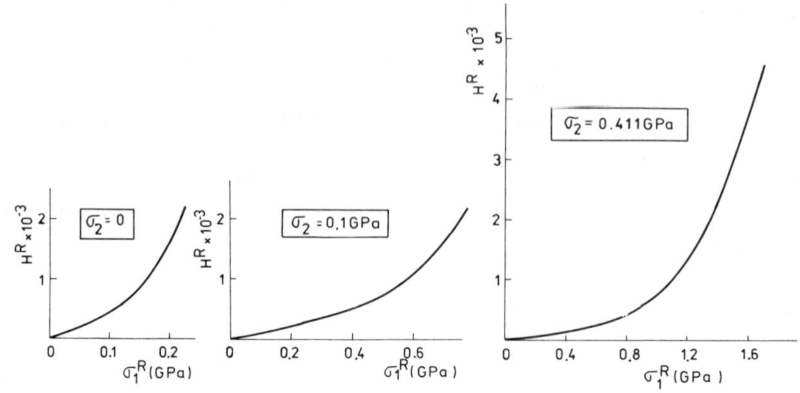

Figure 8.5.7. Variation with stress of the relative part H^R of the function H involved in the stabilization boundary (8.5.6).

The expressions of the coefficients a and b as functions of $\sigma - \bar{\sigma}/3$ are

$$a = \frac{a_0}{\left(\sigma - \frac{\bar{\sigma}}{3}\right)\frac{1}{\sigma_*} + a_1} + a_2, \qquad b = \frac{b_0}{\left(\sigma - \frac{\bar{\sigma}}{3}\right)\frac{1}{\sigma_*} + b_1} + b_2, \qquad (8.5.10)$$

with $a_0 = 4.40813 \times 10^{-4}$, $a_1 = 8.80782 \times 10^{-4}$, $a_2 = -5.70283 \times 10^{-4}$, $b_0 = 1.46501 \times 10^{-4}$, $b_1 = 3.59266 \times 10^{-2}$, $b_2 = -7.7794 \times 10^{-5}$. Thus

$$H(\sigma, \bar{\sigma}) := \left[\frac{a_0}{\left(\sigma - \frac{\bar{\sigma}}{3}\right)\frac{1}{\sigma_*} + a_1} + a_2\right]\left(\frac{\bar{\sigma}}{\sigma_*}\right)^4 +$$

$$+ \left[\frac{b_0}{\left(\sigma - \frac{\bar{\sigma}}{3}\right)\frac{1}{\sigma_*} + b_1} + b_2\right]\frac{\bar{\sigma}}{\sigma_*} + \qquad (8.5.11)$$

$$+ \begin{cases} \frac{1}{\sqrt{6}}\left(-p + q\left(\frac{\sigma}{\sigma_*} + r\right)^{1/2} - \frac{\sigma}{K_0}\right) & \text{if } 0 \leqslant \sigma \leqslant \sigma_0 \\ \frac{1}{\sqrt{6}}\frac{K_2}{K_0} \quad \text{if } \sigma_0 \leqslant \sigma. \end{cases}$$

The compressibility/dilatancy boundary is obtained from $\partial H/\partial\sigma = 0$.

Several surfaces H = const. have been represented in Fig. 8.5.8 for the following values of the dimensionless constant: 0.001; 0.0012; 0.0014; 0.0016;

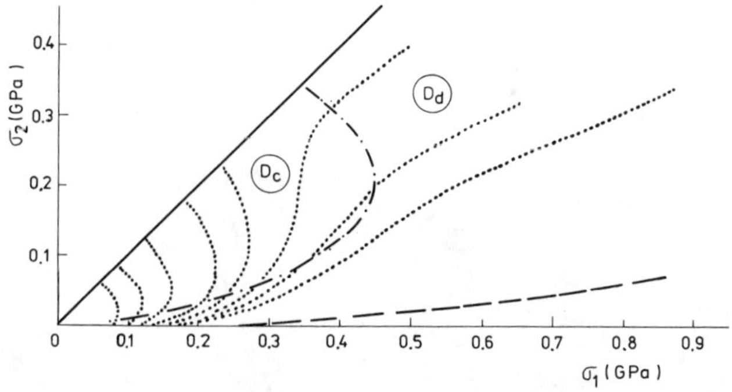

Figure 8.5.8. Constitutive domain showing stabilization boundaries (......), compressibility/dilatancy boundary (— • — •), failure surface (———), domain of compressibility D_c, and domain of dilatancy D_d.

0.0017; 0.0018; 0.0019; 0.002. The compressibility/dilatancy boundary is represented by a border line (— • — •), and an approximate position of the failure surface as obtained in triaxial tests, by a broken line. Let us observe that for rocks which are compressible and/or dilatant, stabilization boundaries of the form (8.5.6) are not very accurate. Such expressions for a stabilization boundary can be used for rocks which are dilatant only, say.

8.6. Examples for softer rocks

Another example will be given for *andesite*, a somewhat softer rock than granite. The density of the rock is between 2.38 g cm^{-3} and 2.70 g cm^{-3}, while the porosity is between 1.1% and 7.84%.

The results of uniaxial creep tests of short duration are shown in Fig. 8.6.1, while for slow triaxial tests done with a lateral pressure of $p = \sigma_2 = 13.7$ MPa, the stabilization boundaries shown in Fig. 8.6.2 are obtained. The results shown in the first figure have been obtained after an initial preloading $\sigma_1 = 5$ MPa. For $p = 0$, failure occurs at the axial stress $\sigma_1 = 58$ MPa, while for $p = 13.7$ MPa, failure occurs at a relative stress between 0.131 and 0.147 GPa.

Starting from these experimental data, the irreversible stress-work W^{I} can be computed, using for the elastic parameters the values $G = 12.5$ GPa, $E = 28.77$ GPa, $K = 15$ GPa, and $\nu = 0.174$ which have been obtained in unloading tests. Since the scatter of experimental data used to determine these values was quite large, the final values have been made precise by comparing them with the slopes indicated by small straight lines in Figs. 8.6.1 and 8.6.2. The irreversible stress-work has been computed using formula (8.4.9). The numerical results are represented in Figs. 8.6.3 and 8.6.4. For the function $H(\sigma, \bar{\sigma})$, the same expression as the one already used for the other rocks considered was used. Therefore, for the function involved in the equation of the stabilization boundary as determined from triaxial tests, the expression

$$H^{\mathrm{R}}(\sigma_1^{\mathrm{R}}, \sigma_2) := a(\sigma_1^{\mathrm{R}})^4 + b\,\sigma_1^{\mathrm{R}} \qquad (8.6.1)$$

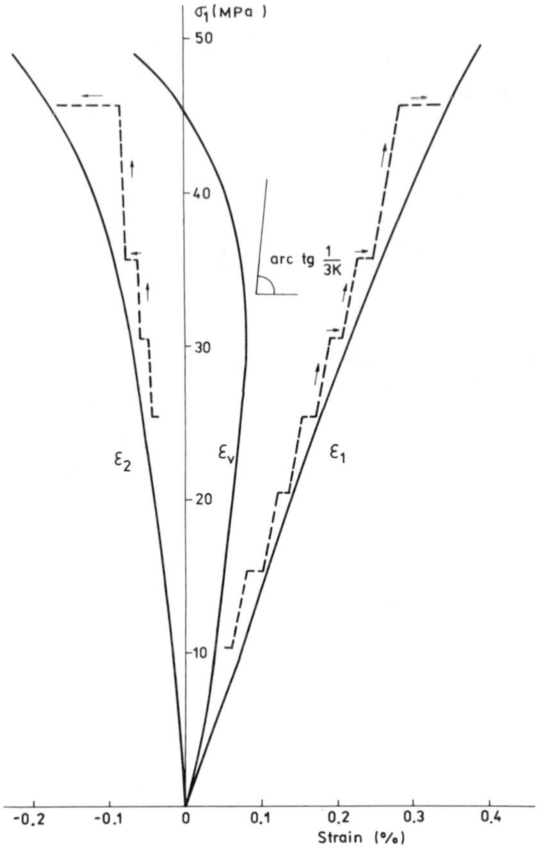

Figure 8.6.1. Incremental uniaxial creep tests of short duration for andesite.

was chosen, where a and b are functions of σ_2^R. By the least squares method we get

$$\sigma_2 = 0 \qquad \begin{cases} a = 7.29025 \pm 0.3105 \text{ GPa} \\ b = 5.0798 \times 10^{-4} \pm 2.56 \times 10^{-5} \text{ GPa} \end{cases}$$
$$\sigma_2 = 0.0137 \text{ GPa} \qquad \begin{cases} a = 0.22736 \pm 0.0114 \text{ GPa} \\ b = 2.09336 \times 10^{-4} \pm 1.28 \times 10^{-5} \text{ GPa.} \end{cases} \tag{8.6.2}$$

For the coefficient functions a and b we obtain

$$a(\sigma_2) = \frac{a_0}{\dfrac{\sigma_2}{\sigma_*} + a_1}, \qquad b(\sigma_2) = \frac{\left(\dfrac{\sigma_2}{\sigma_*} - b_0\right)^2}{b_1}, \tag{8.6.3}$$

with $a_0 = 3.4182 \times 10^{-3}$ GPa, $a_1 = 4.4184 \times 10^{-4}$, $b_0 = 0.0383339$, and $b_1 = 2.892\ (\text{GPa})^{-1}$.

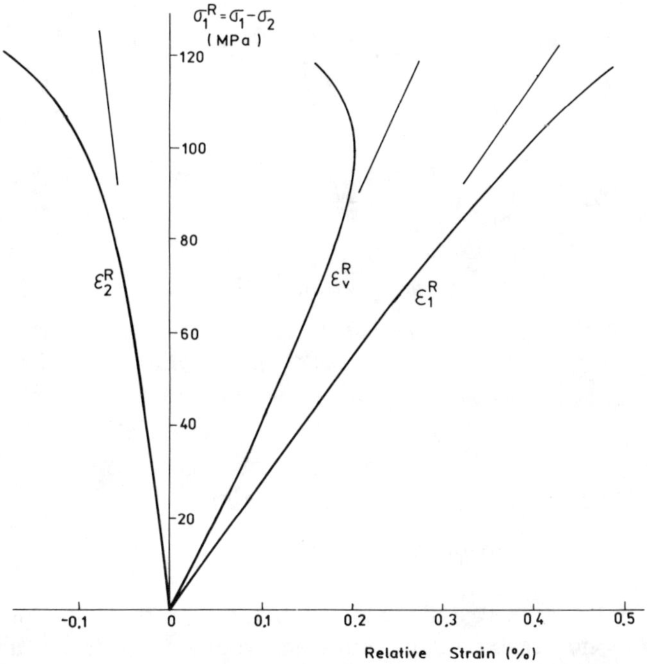

Figure 8.6.2. Stabilization boundaries obtained in triaxial creep tests for andesite.

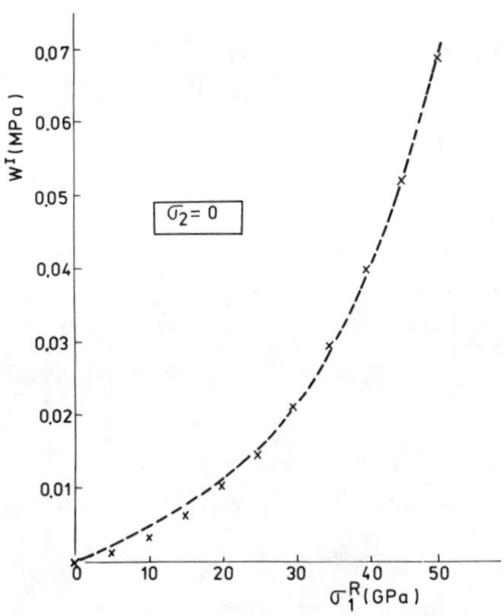

Figure 8.6.3. Irreversible stress-work for andesite in uniaxial tests (crosses) and function $H(\sigma, \bar{\sigma})$, given by (8.6.5), approximating these data (broken lines).

Further, from the hydrostatic compressibility tests we get

$$H^{\mathrm{H}} = \frac{1}{2}\left(\frac{1}{K_0} - \frac{1}{K}\right)\sigma^2 + c\,\sigma, \tag{8.6.4}$$

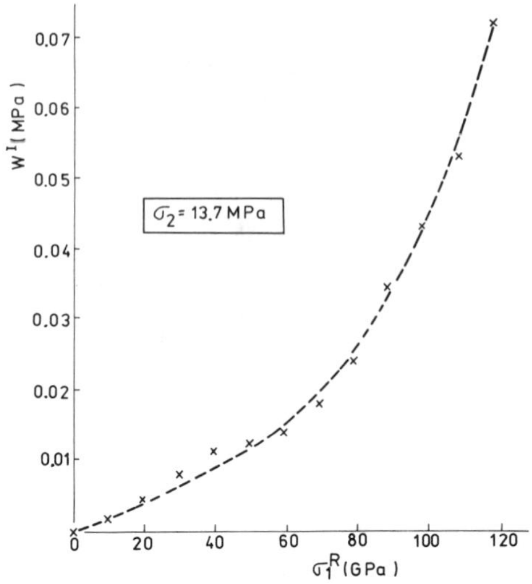

Figure 8.6.4. As in Fig. 8.6.3, but for triaxial test.

with $K_0 = 12$ GPa, $K = 15$ GPa, and $c = 0.6363 \times 10^{-4}$. This formula can be used for depths no greater than 1 km (the experimental data available to the author were incomplete).

Finally, the function H expressed in invariants is

$$H(\sigma, \bar{\sigma}) := \frac{a_0}{\left(\sigma - \frac{\bar{\sigma}}{3}\right)\frac{1}{\sigma_*} + a_1}\left(\frac{\bar{\sigma}}{\sigma_*}\right)^4 + \frac{\left[\frac{1}{\sigma_*}\left(\sigma - \frac{\bar{\sigma}}{3}\right) - b_0\right]^2}{b_1}\frac{\bar{\sigma}}{\sigma_*} + \frac{1}{2}\left(\frac{1}{K_0} - \frac{1}{K}\right)\sigma^2 + c\,\sigma. \tag{8.6.5}$$

The shapes of several surfaces H = const. are shown as dotted lines in Fig. 8.6.5 for the following values of the constant expressed in kPa: 2.94; 4.9; 6.35; 7.84. The border line (—·—·) is the compressibility/dilatancy boundary obtained from $\partial H/\partial\sigma = 0$.

In order to determine completely the constitutive equation, we still have to determine the value of the constant of viscosity k, using formulae (8.4.13) and the creep data (corresponding to the horizontal segments in Fig. 8.6.1). Thus, we get a mean value of $k = 5.69 \times 10^{-4}\,\mathrm{s}^{-1}$, and with $E = 28.77$ GPa, the

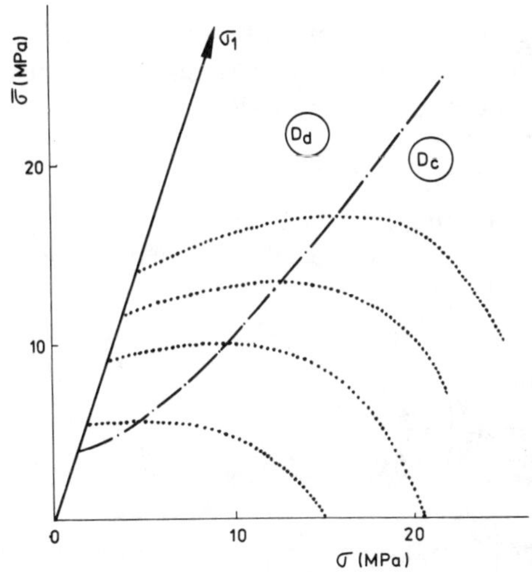

Figure 8.6.5. Constitutive domain for andesite: stabilization boundaries (......), the compressibility/dilatancy boundary (— · — ·), dilatancy domain D_d, and compressibility domain D_c.

value of the viscosity coefficient is $E/k = 5.05 \times 10^{12}$ P. This is a relatively small value, typical for soft rocks.

Finally, the mean compressive strength obtained in standard uniaxial tests for this rock is $\sigma_c = 71.6$ MPa; the mean tensile strength (Brazilian test) is $\sigma_t = 5.37$ MPa.

Failure threshold (or incipient failure) due to dilatancy will, however, be characterized by an energetic criterion: the irreversible volumetric stress-work reaches a certain limit (see the next chapter) which is assumed to be typical for the rock under consideration and may depend independently on the mean stress as well. For the andesite considered, we get $-W_v^I(\text{cr}) \simeq 3.37 \times 10^4$ J/m^3 (J m^{-3} = N m^{-2} = kg m^{-1} s^{-2}). This criterion will be used in the engineering examples given in Chapters 11—13.

Other examples have been given for various kinds of coal. Thus for the *coal from Racoş*, based on uniaxial tests as well as on triaxial tests done for $\sigma_2 = 1.96$ MPa and $\sigma_2 = 2.94$ MPa, we obtain

$$H(\sigma, \bar{\sigma}) := \frac{1}{a_1 + a_2 \dfrac{\sigma}{\sigma_*}} \left(\frac{\bar{\sigma}}{\sigma_*}\right)^2 + b\,\frac{\bar{\sigma}}{\sigma_*} + \\ + \begin{cases} c\left(1 - \cos \omega \dfrac{\sigma}{\sigma_*}\right) & \text{if } 0 \leqslant \sigma \leqslant \sigma_0 \\ 2c \quad \text{if } \sigma_0 \leqslant \sigma \end{cases} \tag{8.6.6}$$

where $a_1 = 586.8$ (MPa)$^{-1}$, $a_2 = 376.8$ (MPa)$^{-1}$, $b = 0.125 \times 10^{-2}$ MPa, $c = 0.46 \times 10^{-3}$ MPa, $\omega = 153.06°$, $\sigma_0 = 1.176$ MPa, $\sigma_* = 1$ MPa, $k = 8.1 \times 10^{-6}$ s^{-1}, $E/k = 3.9 \times 10^{12}$ P, $d_{cr} = 9.4 \times 10^4$ J m^{-3}, $E = 316$ MPa, $G = 107$ MPa, $K = 1922$ MPa.

The values of the constitutive constants were obtained from the experimental data using the procedures described above. The constitutive domain, the compressibility/dilatancy boundary as well as several relaxation boundaries are shown in Fig. 8.6.6, for the following values of $H(\boldsymbol{\sigma}) = 0.5 \times 10^{-3}$ MPa; 1.5×10^{-3} MPa; 2.5×10^{-3} MPa; 3.5×10^{-3} MPa; 4.5×10^{-3} MPa and 5.5×10^{-3} MPa (for increasing values of $\bar{\sigma}$).

For another *coal from Baraolt*, based on uniaxial tests as well on the triaxial tests done for $\sigma_2 = 1.47$ MPa and $\sigma_2 = 2.55$ MPa, the following viscoplastic potential is obtained:

$$H(\sigma, \bar{\sigma}) := \left[a_0 - a_1 \left(\frac{\sigma}{\sigma_*} - \frac{\bar{\sigma}}{3\sigma_*} \right) \right] \left(\frac{\bar{\sigma}}{\sigma_*} \right)^2 + b \frac{\bar{\sigma}}{\sigma_*} + \begin{cases} c \left(1 - \cos \omega \dfrac{\sigma}{\sigma_*} \right) & \text{if } 0 \leqslant \sigma \leqslant \sigma_0 \\ 2c \quad \text{if } \sigma_0 \leqslant \sigma \end{cases} \tag{8.6.7}$$

where $a_0 = 1.245 \times 10^{-3}$ MPa, $a_1 = 4.09 \times 10^{-4}$ MPa, $b = 1.4 \times 10^{-3}$ MPa, $c = 4.9 \times 10^{-3}$ MPa, $\omega = 61.22°$, $\sigma_0 = 2.94$ MPa, $\sigma_* = 1$ MPa, $k = 3.6 \times 10^{-7}$ s^{-1}, $E/k = 1.7 \times 10^{14}$ P, $E = 607.9$ MPa, $G = 209.9$ MPa, $K = 1954.3$ MPa.

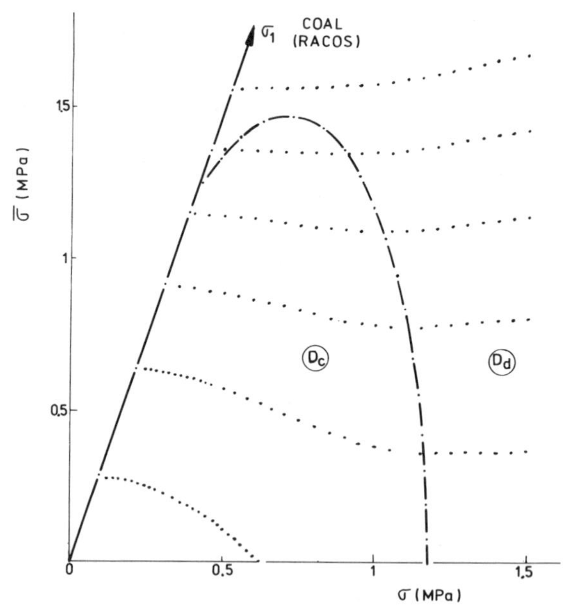

Figure 8.6.6. Constitutive domain for coal (from Racoş) showing stabilization boundaries (......), the compressibility/dilatancy boundary (— • — •), dilatancy domain D_d, and domain of compressibility D_c.

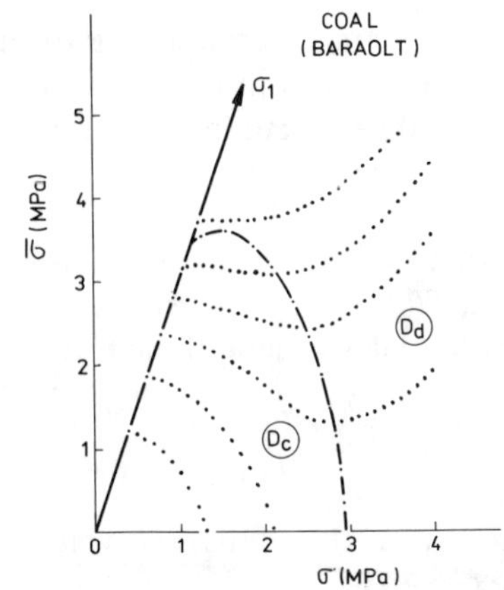

Figure 8.6.7. Constitutive domain for coal (from Baraolt) showing stabilization boundaries (......), the compressibility/dilatancy boundary (— • — •), dilatancy domain D_d, and domain of compressibility D_c.

The constitutive domain is given in Fig. 8.6.7. The relaxation boundaries shown by dotted lines correspond to the values $H(\boldsymbol{\sigma}) = 4 \times 10^{-3}$ MPa; 8×10^{-3} MPa; 12×10^{-3} MPa; 16×10^{-3} MPa; 20×10^{-3} MPa; 26×10^{-3} MPa (for increasing values of $\bar{\sigma}$).

8.7. A uniaxial example

If the hydrostatic pressures involved are low, a simplified (although less accurate) version of the constitutive equation can be obtained from uniaxial creep tests. For this purpose it is useful to write the irreversible stress-work in the form (8.3.6). In the case of uniaxial compressive tests we have for the volumetric and deviatoric parts

$$W_{\mathrm{v}}^{\mathrm{I}}(T) = \frac{1}{3}\int_0^T \sigma_1 \, (\dot{\varepsilon}_1^{\mathrm{I}} + 2\dot{\varepsilon}_2^{\mathrm{I}}) \, \mathrm{d}t$$

$$W_{\mathrm{D}}^{\mathrm{I}}(T) = \frac{2}{3}\int_0^T \sigma_1(\dot{\varepsilon}_1^{\mathrm{I}} - \dot{\varepsilon}_2^{\mathrm{I}}) \, \mathrm{d}t. \tag{8.7.1}$$

Assuming that the elastic constants are already determined, one can obtain the two terms (8.7.1) from experimental data, since

$$\dot{\varepsilon}_1^{\mathrm{I}} = \dot{\varepsilon}_1 - \left(\frac{1}{3G} + \frac{1}{9K}\right)\dot{\sigma}_1, \qquad \dot{\varepsilon}_2^{\mathrm{I}} = \dot{\varepsilon}_2 - \left(-\frac{1}{6G} + \frac{1}{9K}\right)\dot{\sigma}_1. \tag{8.7.2}$$

First, we determine successively the values of $W_{\mathrm{v}}^{\mathrm{I}}$ starting from $\sigma_1 = 0$, where

$W_v^I = 0$ up to the value of $\sigma_1 = 3\sigma_0$ corresponding to the maximal value of W_v^I. An approximate value of σ_0 follows from the value of σ_1 corresponding to this maximum. The shape of the W_v^I—σ_1 curve is similar to the one in Fig. 7.6.1. Further, W_D^I is obtained from the experimental data with the help of formula $(8.7.1)_2$, if $\sigma_1 < 3\sigma_0$, and from W^I—$W_v^I(\max)$ for $\sigma_1 \geqslant 3\sigma_0$. The reason for using this procedure is that W_v^I must increase with σ, and for pressures surpassing a certain limit, σ_0, W_v^I must stay constant.

Further, a viscoplastic potential is sought of the form

$$H(\sigma, \bar{\sigma}) := H^H(\sigma) + H^R(\sigma, \bar{\sigma}), \tag{8.7.3}$$

where $H^H(\sigma)$ is determined from the numerical values of W_v^I, while $H^R(\sigma, \bar{\sigma})$ from those of W_D^I. Thus we obtain for *coal*,

$$H(\sigma, \bar{\sigma}) := \frac{a_0}{a_2 \dfrac{\sigma}{\sigma_*} + a_1} \left(\frac{\bar{\sigma}}{\sigma_*} \right)^4 + b \frac{\bar{\sigma}}{\sigma_*} + \\ + \begin{cases} c_0 \sin \left(\omega \dfrac{\sigma}{\sigma_*} + \varphi \right) + c_1 & \text{if } 0 \leqslant \sigma \leqslant \sigma_0 \\ c_0 + c_1 & \text{if } \sigma \geqslant \sigma_0, \end{cases} \tag{8.7.4}$$

where $a_0 = 4.245 \times 10^{-3}$ MPa, $a_1 = 1$, $a_2 = 50.99952$, $b = 0.00022$ MPa, $c_0 = 4.957 \times 10^{-4}$ MPa, $c_1 = 4.8955 \times 10^{-4}$ MPa, $\omega = 171.927°$, $\varphi = -99.068°$, $\sigma_* = 1$ MPa, and $\sigma_0 = 1.0996$ MPa. The values of the elastic parameters at shallow depths are $G = 300.79$ MPa and $K = 769.614$ MPa. We observe that $a_0/a_2 > 0$ since $\partial H/\partial\sigma < 0$ for $\sigma \geqslant \sigma_0$. The domain of compressibility as well as the compressibility/dilatancy boundary are quite similar in shape to those shown in Fig. 7.2.1 for granite. The viscosity coefficient k is obtained from creep tests, using the formula

$$k = -\ln \left| 1 - \frac{W^I(t)}{H(\sigma, \bar{\sigma})} \right| \Bigg|_{t_i}^{t_f} \frac{H(\sigma, \bar{\sigma})}{\dfrac{\partial H}{\partial \boldsymbol{\sigma}} \cdot \boldsymbol{\sigma}} \frac{1}{t_f - t_i}, \tag{8.7.5}$$

with t_i the initial and t_f the final moments of the creep test, during which $\sigma_1 =$ const. Thus $k = 6 \times 10^{-6}$ s^{-1} and E/k is of the order of 10^{13} P.

Failure threshold due to dilatancy (see the next chapter) takes place at $-W_v^I(\text{cr}) = 4.4 \times 10^3$ J m^{-3}.

Another simplified variant of the model for the same *coal* has been obtained as

$$H(\sigma, \bar{\sigma}) := \frac{a_0}{a_2 \dfrac{\sigma}{\sigma_*} + a_1} \left(\frac{\bar{\sigma}}{\sigma_*} \right)^2 + b_0 \frac{\bar{\sigma}}{\sigma_*} +$$

$$+ \begin{cases} c_0 \sin \left(\omega \dfrac{\sigma}{\sigma_*} + \varphi \right) + c_1 & \text{if } 0 \leqslant \sigma \leqslant \sigma_0 \\ c_0 + c_1 & \text{if } \sigma \geqslant \sigma_0, \end{cases} \tag{8.7.6}$$

where $a_0 = 7.65 \times 10^{-4}$ MPa, $a_1 = 0.55$, $a_2 = 8.1599 \times 10^{-3}$, $b_0 = 0.001$ MPa, $c_0 = 4.957 \times 10^{-4}$ MPa, $c_1 = 4.8955 \times 10^{-4}$ MPa, $\omega = 171.297°$, $\varphi = -99.068°$, $\sigma_* = 1$ MPa, $\sigma_0 = 1.0996$ MPa. The values of the elastic parameters are the same as above. Though the latter model is in poorer agreement with experimental data than the former one, as will be shown in Chapter 13, it is easier to handle in solving mining engineering problems.

Another variant of the model can be obtained if, besides the uniaxial data, we also possess experimental data concerning the hydrostatic (or quasihydrostatic) compressibility of the volume. Let us give an example for *cement concrete*; the volume compressibility was been presented in Chapter 4, for a similar kind of cement concrete. This example is also distinguished from the previous one by the high porosity of cement concrete of about 25%, generally due to pores and not to cracks.

From the experimental data obtained by Constantinescu (1981) concerning the stabilization boundary for volumetric creep, the irreversible volumetric stress-work has been computed, using the formula

$$W_{\mathrm{v}}^{\mathrm{I}}(T) = \sum_{t=0}^{T} \sigma(t)\, \Delta\varepsilon_{\mathrm{v}}(t) - \frac{\sigma^2(T)}{2K}, \tag{8.7.7}$$

where $K = 11.81$ GPa was determined from unloading experiments. Experimental data for three specimens have been used. The results of numerical computation of the expression (8.7.7) are shown in Fig. 8.7.1. The behaviour shown in this figure can be approximated by a simple empirical relationship (broken line)

$$H^{\mathrm{H}}(\sigma) := c \left(1 - \cos \omega \frac{\sigma}{\sigma_*} \right), \tag{8.7.8}$$

where $c = 0.04073$ GPa, $\omega = 156.06°$, and $\sigma_* = 1$ GPa. Thus, the viscoplastic potential for the hydrostatic case is determined. Since only the beginning of this curve is of significance for engineering applications, more care was devoted to the fitting of the data at small stresses.

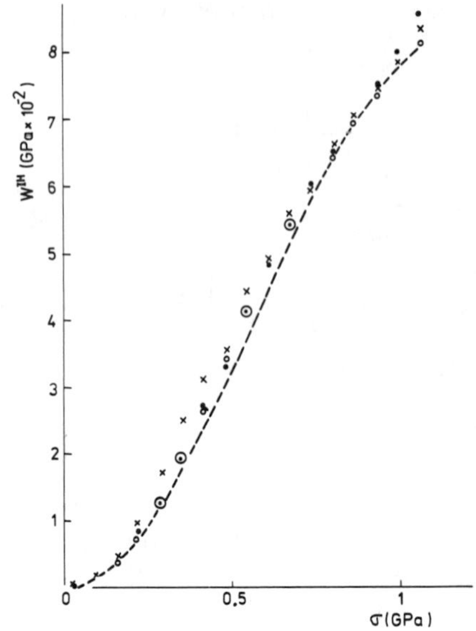

Figure 8.7.1. Irreversible hydrostatic stress-work for cement concrete.

Let us consider now the stabilization boundaries obtained in uniaxial compression tests. These boundaries, as the averages of several tests, are given in Fig. 8.7.2, where, by segments of straight lines, the elastic slopes

$$\frac{1}{3G} + \frac{1}{9K}, \qquad -\frac{1}{6G} + \frac{1}{9K},$$

are shown with $G = 6.95$ GPa. Using the formulae (8.7.1), W^{I} is determined.

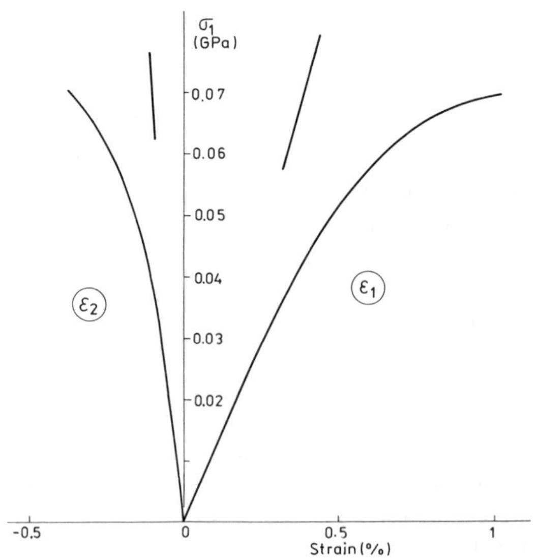

Figure 8.7.2. Stabilization boundaries for cement concrete obtained in uniaxial compression tests.

Then, using the expression already determined for H^H, the difference $W^I - H^H$ for various values of σ is computed. Further, if we approximate the numerical results thus obtained by a simple empirical formula of the form already used (the first two terms from the right-hand side of (8.7.4)), we obtain the second term $H^R(\sigma, \bar{\sigma})$ of the viscoplastic potential, where $a_0 = 1$ GPa, $a_1 = 0.12$, $a_2 = 0.3$, $b = 1 \times 10^{-5}$ GPa, and $\sigma_* = 1$ GPa.

It should be noted that even for the highest pressures involved in the hydrostatic compression tests (over 1.1 GPa), it was not possible to crush all the pores. Thus, it was not possible to determine in the laboratory for this rock-like material the limit pressure σ_0 after which, at even higher pressures, the material would behave elastically in a hydrostatic test. From the extrapolation of (8.7.8) (i.e. by finding the value of σ corresponding to the maximum of $H^H(\sigma)$), one can, however, determine $\sigma_0 \simeq 1.153$ GPa. Furthermore, for all conceivable engineering problems, the viscoplastic potential is of the form

$$H(\sigma, \bar{\sigma}) := \frac{a_0}{a_2 \dfrac{\sigma}{\sigma_*} + a_1} \left(\frac{\bar{\sigma}}{\sigma_*} \right)^4 + b \frac{\bar{\sigma}}{\sigma_*} + c \left(1 - \cos \omega \frac{\sigma}{\sigma_*} \right). \quad (8.7.9)$$

The compressibility/dilatancy boundary, obtained as always from $\partial H / \partial \sigma = 0$, is thus determined for the range of the mean stresses involved in the hydrostatic tests only. This boundary is located in a domain of the stress-space where failure is, generally, already to be expected. It follows that for all stress-states possible in engineering problems, cement concrete is always compressible (never dilatant) to failure. We have also $H(\sigma_0, 0) = 2c$. This conclusion is certainly in full agreement with the experimental data. This last variant of the model was also used for coal, the results being quite similar to the models (8.7.4) and (8.7.6).

Finally, let us observe that *accurate constitutive equations for rock-like materials can be determined in triaxial tests only*; the above simplified models may be used only if complete triaxial data are not available, if the problem considered allows it, and if the model for a very similar rock was obtained from a full set of triaxial data.

8.8. Acoustic emission

During their deformation by creep, rocks emit transient elastic waves. This phenomenon is generally termed *acoustic emission* but some other terms, such as *microseismic activity, seismo-acoustic activity, micro-earthquake activity*, etc. are also used. These sounds are monitored by special devices and are used as diagnostic tests to recognize the stress and strain state existing in the rock. Generally, *acoustic emission* (AE) is defined as the class of phenomena whereby transient elastic waves are generated by rapid release of energy by a localized source or sources within a material, or the transient elastic waves so generated. The importance of AE methods for mining, petroleum, and civil engineering is obvious. It is expected that by in-field source location (Rindorf (1984)) we should be able to predict the possible failure location. The importance of these methods for rock and soil mechanics (geomechanics) is also attested by the organization within a few years of four international conferences devoted to 'acoustic emission/microseismic activity in geologic structures and materials' (see Hardy and Leighton (1977, 1980, 1984, 1988)). These conferences have resulted in the publication of a many papers dealing with either laboratory or field studies. Comprehensive review papers in which a variety of aspects of AE in rocks is discussed are due to Hardy (1972, 1984).

It has long been recognized that one can distinguish between loading and unloading by means of AE. More exactly, the phenomenon termed the *Kaiser effect* can be defined as the absence of detectable AE until previously applied stress levels are exceeded. This effect is now widely studied in various laboratories. For instance, it was found by McElroy *et al.* (1985) that in a cyclic stress test, previous stress levels can be discerned by a sudden AE rate increase, as the stress on a particular cycle exceeds the previously applied stress. Some other authors (Sondergeld and Estey (1981)), however, question the significance and existence of the Kaiser effect in uniaxial stressing of rock. Chen (1979) has shown that, generally, the AE behaviour of rock is strongly related to the stress path. An interesting experimental method of determining the yield surfaces of sandy soils in three-dimensional stress space by AE measurements is due to Tanimoto and Tanaka (1985).

In the following we present some experimental data which are revealing for the *correlation between the mechanical behaviour of the rock and its AE*. First, let us mention that when creep of rock is studied in the laboratory, as well as stresses and strains the AE of the rock can be monitored. Thus we can register the total number of AE events N, their rate $\dot{N}$ (number of events per unit time), possibly also the location inside the specimen from where each event comes, the amplitude of the AE, the energy of the AE (the square of the amplitude), or the total energy of the AE (the sum of the squares of all amplitudes).

In our Geomechanical Laboratory (D. Fotă, unpublished results from 1983) we have measured during short-term creep tests (uniaxial compressive creep tests with incremental loading in which the stress is successively increased in steps; after each such increase the stress level is held constant for 1 to 15 minutes), besides the stress and strain, the noise rate, and the total amount of events. Sometimes the creep tests lasted longer. Only the events between 500 and 2000 Hz have been monitored.

It was found that the total number of events increases during loading throughout the whole experiment. However, in the last period, just before failure, this increase is quite sharp. A *mechanical response — AE correlation* (MR—AE correlation) can be formulated in mathematical form, if the *rate of events is compared to the volumetric strain rate*. Generally, such a correlation must relate the 'mechanical' parameters involved in the constitutive equation to the AE parameters such as N, $\dot{N}$, etc.

It was found that, just after each reloading, the rate $\dot{N}$ sharply increases while in the interval of time in which the stress is constant, $\dot{N}$ steadily decreases and levels off at a small value which does not remain really constant (random small oscillations). Similar behaviour is obtained for $\dot{\varepsilon}_v$ with the following distinction. As already mentioned in previous chapters, when the axial stress increases, generally the volume first decreases (compressibility), then for higher stresses there is only a slight variation of ε_v, and finally at even higher stresses preceding failure, the volume increases. In a similar way, in the first interval of stress variation, $\dot{\varepsilon}_v$ sharply increases at each additional loading and afterwards steadily decreases in time towards zero. Thus the $\dot{\varepsilon}_v$ variation characteristics are somewhat similar to those of $\dot{N}$. In Fig. 8.8.1 the variations in time of $\dot{N}$, $\dot{\varepsilon}_v$, and σ_1 are shown for an *andesitic rock* (for which $K = 13.4$ MPa, $G = 18.2$ MPa, $E = 37.5$ MPa, $\nu = 0.031$). During the second interval of stress variation there is only a small variation of ε_v and also of N (more precisely, for N a disorderly small variation is obtained). During the third interval of stress variation, $\dot{\varepsilon}_v$ decreases sharply at each additional loading, followed by a steady increase towards zero. In other words, in the first interval the variation of $\dot{\varepsilon}_v$ and that of $\dot{N}$ are alike, while in the third one that of $-\dot{\varepsilon}_v$ and $\dot{N}$ are alike. Therefore, a relationship of the form (Cristescu (1983*b*, 1988*a*))

$$|\dot{\varepsilon}_v| = A(\boldsymbol{\sigma})\,\dot{N} + B(\boldsymbol{\sigma}) \tag{8.8.1}$$

could be formulated for an MR—AE correlation. It was found that the coefficients A and B are stress-dependent, but they might depend on some other parameters as well.

Let us consider now the sequence of final values of ε_v and N, as obtained at the end of various periods during which the stress level is held constant in an incremental form of loading. In Fig. 8.8.2 the successive final values of ε_v and those of N are shown for *andesite*, for incremental loading stresses of $\Delta\sigma_1 =$

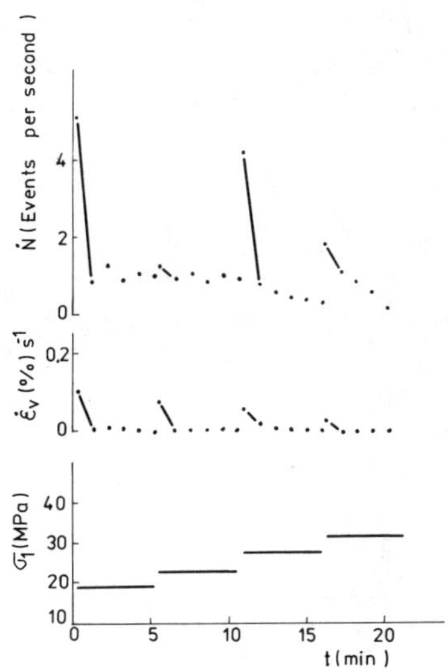

Figure 8.8.1. Variation in time of volumetric strain rate and the rate of the number of events for creep test obtained in incremental loading of andesite.

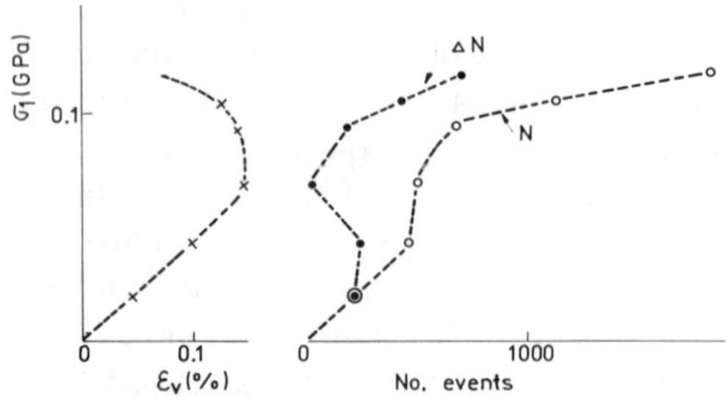

Figure 8.8.2. Variation with applied axial stress of volumetric strain, of the number of the total events N, and the total number of events at various loading levels, for andesite.

24.5 MPa, and time interval of 15 minutes at each constant stress level, during which creep takes place. The rock has an initial porosity $n = 1.84\%$. Thus, in the stress interval during which compressibility takes place, the total number of events ΔN, as recorded at the end of each of the successive loading steps, (so that ΔN is the total number of events for a single loading increment), is higher than that recorded during the loading steps, corresponding to the stress interval when the rock passes from compressibility to dilatancy. Finally, during dilatancy the number of events per loading step increases quite fast from one step to the next. In the same figure the total number of events N is also given: the slowest increase of N takes place during the passage from compressibility to dilatancy. Similar results with those shown in this figure were obtained for dried sandy soils by Tanimoto and Nakamura (1984).

A similar conclusion is obtained for another soft *andesitic rock* of initial porosity $n = 4.7\%$ (Fig. 8.8.3). This time, however, an incremental loading of $\Delta\sigma_1 = 2.6$ MPa was successively applied every minute. The recordings of ΔN were also done after each minute. The variation of both N and ΔN are shown in Fig. 8.8.3*a* and the conclusions which follow are similar. Fig. 8.8.3*b* shows the variations with stress of $\dot{\varepsilon}_v$ and $\dot{N}$; it is obvious that the variation of $\dot{N}$ and that of $|\dot{\varepsilon}_v|$ are quite similar.

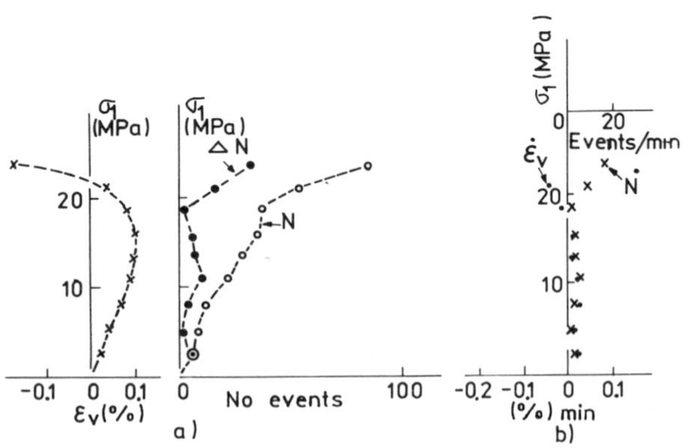

Figure 8.8.3. (*a*) As in Fig. 8.8.2, for another andesite. (*b*) Variation with applied axial stress of the volumetric strain rate and of the rate of the total number of events for andesite.

If the specimen has some imperfections, the variation of N is rather irregular, but the smallest values of $\mathrm{d}N/\mathrm{d}\sigma_1$ always correspond to the passage from compressibility to dilatancy and the largest ones to the advanced stage of dilatancy just preceding failure. A relative maximum of $\mathrm{d}N/\mathrm{d}\sigma_1$ is obtained during non-linear compressibility. It was not possible to make any qualitative distinction between the events produced by the closing of pores and cracks (when $\dot{\varepsilon}_v > 0$) and those due to crack multiplication (when $\dot{\varepsilon}_v < 0$).

The coefficient functions A and B from (8.8.1) are determined from experimental data. From the scarce experimental data so far available it has been found that both coefficients are second-degree polynomials in σ_1. How these coefficients may depend on stress invariants in an invariant form of (8.8.1) is not yet known; such relationships could be obtained by triaxial creep tests. If these coefficients are determined, using the constitutive equation (8.4.1) the full MR—AE correlation is established. Since damage and failure of rocks are also related to dilatancy (see Chapter 9), using this correlation and the AE recorded in the field, one can expect to predict damage and failure around underground openings.

On the other hand, since the constitutive equation (8.4.1) can be used to predict quite accurately where and when around an underground opening, during stress and strain variation, the rock passes from compressibility to dilatancy (the compressibility/dilatancy boundary), by using AE events recorded around underground openings one can predict where and when around these openings the rock will begin to be in a dilatant state.

Another correlation between the variation in time of N and that of ε_1 during creep tests has been obtained by Hardy *et al.* (see also Hardy (1972)): it was observed that the overall characteristics of ε_1—t creep curves are quite close to those of the corresponding N—t curves. Assuming linear viscoelastic behaviour for the rock, a linear relationship between N and ε_1 is established.

A revealing result has been obtained by Lockner and Byerlee (1977, 1980). It was found that during creep a relationship between $\dot{N}$ and N of the form

$$\ln \dot{N} = \beta - \alpha N \tag{8.8.2}$$

holds. To a first approximation, β was found to be nearly constant, but α depends on the stress-state. During primary creep, α seems nearly constant, while as the stress-state approaches failure, α tends towards zero and further, becomes even negative just before failure. From the way in which this coefficient varies with stress, another correlation may possibly be formulated.

A correlation between the inelastic volumetric strain rate during dilatancy and the AE rate was pointed out by Sano (1981), Sano *et al.* (1981), and Sano *et al.* (1982). It was also shown for Westerly *granite* samples (Granryd *et al.* (1983)) that acoustic velocities become highly anisotropic as the loading proceeds.

For many other experimental results, see for instance the proceedings of the four international conferences mentioned above (Hardy and Leighton (1977, 1980, 1984, 1988)).

From the discussion above it follows that the correlation between the irreversible volumetric deformation of the rock and the AE can be obtained as a correlation between ε_v^I and N. Since the correlation (8.8.1) has been obtained in creep tests, a similar relationship could be established between $\dot{W}_v^I$ and $\dot{N}$. As will be shown in the next chapter, $\dot{W}_v^I$ is a parameter directly involved in a criterion determining the damage and possible failure of rocks. The accuracy and amount of the existing experimental data do not allow us to decide at present whether the AE—MR correlation is to be expressed as a relationship between $\dot{N}$ and $\dot{\varepsilon}_v$ or between $\dot{N}$ and $\dot{W}_v^I$. Since $\dot{W}_v^I$ is a dissipating energy which gives a global characterization of the irreversible deformation stage of the rock, it may be that a relation between W_v^I and the AE energy rate (the sum of the energy emitted by all events observed per unit time) would be the appropriate one.

Exercises

1. Make hydrostatic volumetric compressibility tests of your choice. Determine the pressure range where the rock exhibits viscoplastic properties and where only elastic ones (see Fig. 8.2.1).

2. Make uniaxial compression incremental creep tests (see Fig. 8.2.2) of your choice. Determine a relative stabilization boundary (circles in Fig. 8.2.2) by keeping the stress constant at each stress level for a certain fixed amount of time (a few minutes, one hour, one day, etc.).

3. Show for a rock of your choice how important is the influence of the loading rate during loading and unloading on the uniaxial unconfined stress—strain curves (see Fig. 8.2.3). Record both strain components, axial and diametral, and find out whether the rock is dilatant and/or compressible. How significant are the irreversible strains?

4. Show that if the viscoplastic potential $F(\sigma, \bar{\sigma})$ depends on the mean stress and the equivalent stress only, then

$$\frac{\partial F}{\partial \boldsymbol{\sigma}} \cdot \mathbf{1} = \frac{\partial F}{\partial \sigma} \quad \text{and} \quad \frac{\partial F}{\partial \boldsymbol{\sigma}} \cdot \boldsymbol{\sigma} = \frac{\partial F}{\partial \sigma}\sigma + \frac{\partial F}{\partial \bar{\sigma}}\bar{\sigma}.$$

5. For 'triaxial' tests ($\sigma_2 = \sigma_3$, $\sigma_1 > \sigma_2$) derive the formulae (8.4.7)—(8.4.9).

6. Using 'triaxial' creep test data available to you, determine the function $H(\sigma, \bar{\sigma})$ involved in (8.3.3), using the numerical procedure described in Chapters 7 and 8.

7. The same as Exercise 6, but using the function $H(\sigma, \bar{\sigma})$ involved in (8.4.17) or (8.4.18) (see formulae (8.5.6)—(8.5.11)).

8. Derive formula (8.4.13). Find numerical values of k using creep test data of your choice. Check if k depends on the stress state.

9. Derive formula (8.4.22) and use it for the numerical determination of k from experimental data.

10. For the functions $H(\sigma, \bar{\sigma})$ determined in Exercise 6 or 7, derive the compressibility/dilatancy boundary from $\partial H/\partial \sigma = 0$, and find its shape (see Fig. 8.5.1). Also plot several of the curves $H =$ const. as in Fig. 8.5.1. (Hint: as a preliminary exercise use the function H defined by (8.5.1) and try to reproduce Fig. 8.5.1.)

11. For the same function H as determined above, write a complete constitutive equation by finding the elastic parameters and the viscosity coefficient. Then test the model just found in 'numerical' tests such as those described by Figures 8.5.2—8.5.6.

12. Consider the three work-hardening parameters involved in the work-hardening laws (8.3.3), (8.4.17), and (8.4.18). Which of them is the most accurate when we must distinguish between irreversible deformation during compressibility and irreversible deformation during dilatancy?

9

Damage and failure of rocks

9.1. Classical short-time failure-strength criteria

In the last two hundred years several short-time failure-strength criteria have been formulated for rock-like materials. Different empirical formulae have been proposed, starting from experimental evidence. Generally, these criteria depend on the stress-state alone, though it is well known that other factors such as time, humidity, loading history, temperature, etc., may well influence the damage and failure of rocks. Different variants of the 'classical' failure criteria are described in the monographs by Obert and Duvall (1967), Jumikis (1979), Jaeger and Cook (1979), Coates (1981), Vutukuri *et al.* (1974, vol. 1), and Bieniawski (1984), as well as in various papers (Lee and Ingles (1968), Paul (1968), Obert (1972), Hoek (1983), Hunsche (1984*a*), Kim and Lade (1984), Chen and Han (1985), Mier and Borst (1986), etc.).

Let us discuss briefly the main criteria which have been widely used. The oldest is the Coulomb–Navier criterion

$$|\tau| = \tau_f + \mu\sigma, \tag{9.1.1}$$

where τ and σ are the shear and normal stresses acting on the failure surface, τ_f is the shear strength of the rock, and μ the coefficient of internal friction related to the angle of internal friction θ by

$$\tan 2\theta = \frac{1}{\mu}. \tag{9.1.2}$$

A non-linear generalization is the Mohr criterion

$$\tau = f(\sigma) \tag{9.1.3}$$

for compressive stress states and limit tension cut-off planes

$$\sigma_i = -\sigma_{it} \quad (i = 1, 2, 3) \tag{9.1.4}$$

for tensile stress states. Thus for compressive stresses the shear stress at failure depends on the normal stress acting on the failure surface, while for tensile stresses failure takes place when the smallest principal stress equals a limit value.

Further, instead of (9.1.3), a relationship involving the stress magnitude (or equivalent stress $\bar{\sigma}$) and the mean stress can be used:

$$\bar{\sigma} = g(\sigma). \tag{9.1.5}$$

This is known as the Mises (1913) and Schleicher (1926) criterion, which has to be coupled with (9.1.4). A particular case is the Nadai (1950) criterion

$$(\sigma_1 - \sigma_2)^2 + (\sigma_2 - \sigma_3)^2 + (\sigma_3 - \sigma_1)^2 = 2\,[a_0(\sigma_1 + \sigma_2 + \sigma_3) + a_1]^2, \tag{9.1.6}$$

where the physical meaning of the constants a_0 and a_1 are obtained from uniaxial compression tests, when $\sigma_3 = \sigma_c$, $\sigma_1 = \sigma_2 = 0$, and uniaxial tensile tests when $\sigma_1 = -\sigma_t$, $\sigma_2 = \sigma_3 = 0$. Thus we get

$$a_0 = \frac{\sigma_c - \sigma_t}{\sigma_t + \sigma_c}, \qquad a_1 = \frac{2\,\sigma_c\,\sigma_t}{\sigma_t + \sigma_c}, \qquad \sigma_c > 0, \quad \sigma_t > 0. \tag{9.1.7}$$

Another generalization of this criterion is due to Birger (1977) and involves the equivalent stress and the principal stresses in the form

$$\lambda_0\,\bar{\sigma} + \lambda_1\,\sigma_1 + \lambda_2\,\sigma_2 + \lambda_3\,\sigma_3 = -\sigma_t, \tag{9.1.8}$$

where all coefficients are constant (some may be zero). Giving various particular values to these constants, all the classical failure criteria can be obtained, including the tension cut-off planes.

The above-mentioned criteria describe reasonably well the short-time failure of rocks in many particular cases. Generally, all these strength criteria are formulated in terms of stresses. In an invariant form all the criteria can be expressed as

$$F(\sigma, \bar{\sigma}, \Delta) = 0, \tag{9.1.9}$$

where σ is the mean stress, $\bar{\sigma}$ the equivalent stress, and $\Delta = \det \boldsymbol{\sigma}$ the third stress invariant. Below, the validity of these criteria will be discussed in conjunction with the experimental evidence, and some other possible criteria which may be appropriate for rock-like materials will be presented. The presentation follows the paper by Cristescu (1986).

9.2. Some experimental evidence

Several objections can arise, however, with respect to the formulation of the failure condition in the form (9.1.9), i.e. solely in stress invariants. First, this relationship is independent of the loading rate. It is well known, however, that when a higher loading rate is used in standard uniaxial compression tests, then the rock specimen fails at a much higher stress than when the loading rate is smaller. For instance, in Fig. 9.2.1 three stress—strain curves obtained in uniaxial tests with three loading rates shown, are given for *schist*; the cross at the end of the curves denotes failure. It is obvious that the magnitude of the loading rate has a significant influence on failure: with increasing loading rate the ultimate stress at failure increases, while the strain at failure generally

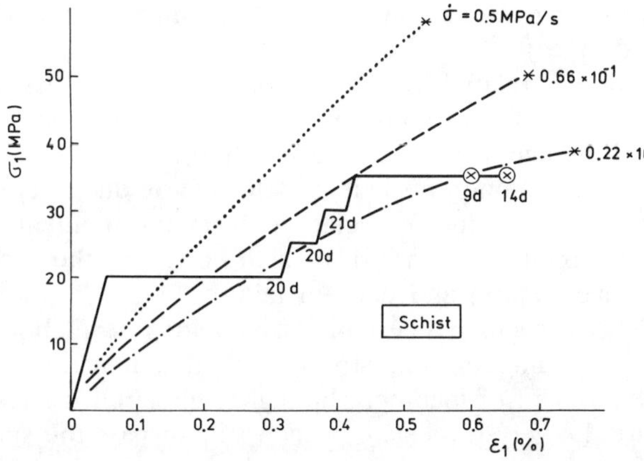

Figure 9.2.1. Three stress–strain curves for schist showing the influence of the loading rate on failure (crosses). The results of two creep tests are also given, showing that strain at failure is much smaller in such tests.

decreases. This property has been observed in many rocks. In order to give a second example, in Fig. 9.2.2 five stress–strain curves for *limestone* in uniaxial compression tests are given, for five different loading rates. Similar conclusions have been obtained by Rusch (1960) for *concrete*, by Dreyer (1973) for *halitic salt*, by Wawersik and Hannum (1980) for *rock salt*, by Schock and Heard (1974), and Genevois and Prestininzi (1979) for *granite*, and others. As early as 1965, Rinehart (1965) pointed out that the dynamic strength of rocks can exceed the static strength by one order of magnitude. Using a servo-controlled testing machine in order to determine the failure behaviour of brittle rock, Rummel and Fairhurst (1970) have found that deformation in the pastpeak load can be developed slowly and progressively if the test is conducted at a sufficiently slow strain rate and that the complete stress–strain curve is strain-rate-dependent. Thus, a more accurate failure criterion has to take into account the *influence of the loading rate on failure* as well, since in engineering problems

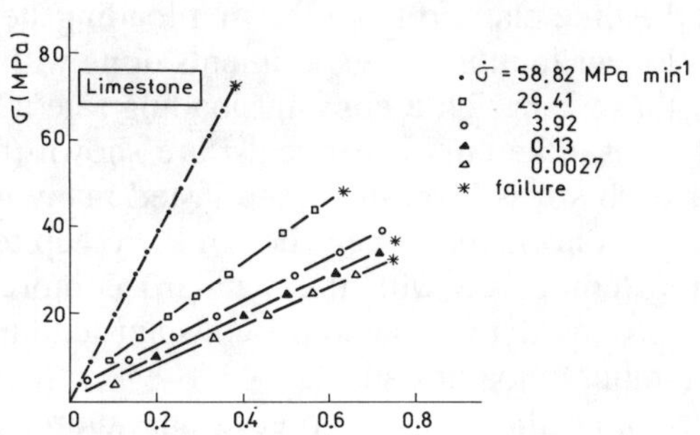

Figure 9.2.2. Stress–strain curves in uniaxial compression tests obtained with five loading rates, showing the influence of the loading rates on failure: with increasing loading rates stress at failure increases while strain decreases.

a wide range of loading rates is involved (creep failure or failure due to rock blasting are the extremes).

Another objection is related to *rock failure in creep test.* As is well known, rocks deform by creep, under constant applied stresses. The phenomenon is more pronounced at higher applied stresses. If the stress does not surpass a certain limit, the (generally non-negligible) deformation due to creep comes to a stable state after a certain time has elapsed from the moment of loading. If, however, this stress limit is surpassed, then the creep of the rock lasts a long time (months or even years) and ends in failure. Failure is therefore not the consequence solely of reaching a certain constant stress level, but the result of a combination of a certain constant stress level in conjunction with the time elapsed from the moment of loading. The higher the applied stress, the shorter the time to failure. Lower stress-states which still surpass the stated limit may lead to failure after a very long time interval, during which the rock deforms by creep. Similar conclusions are reported by Kranz (1980) for *granite* tested in triaxial tests: the time to fracture increases with decreasing stress difference at every confining pressure, and 'increases in pressure at constant stress difference increase the time required for creep rupture'; also 'higher pressures require more volumetric strain to accumulate prior to the onset of instability' (see also Kranz (1979), Carter *et al.* (1981), Kranz *et al.* (1982), Schmidtke and Lajtai (1985), Lajtai and Schmidtke (1986)). By testing small circular openings in samples of jointed *coal*, Kaiser *et al.* (1985) have found that rupture process is time-dependent and more readily detectable by observing creep deformation than from the instantaneous response to loading. Thus, failure of rocks occurring during creep tests cannot be described by a criterion such as that given by equation (9.1.9).

In connection with the last remark, one can point out some other peculiarities as well. It has been observed experimentally that failure is strongly *loading-history-dependent*, in the following sense. If the final loading level is reached in a great number of small successive steps of constant stress, failure occurs generally for higher strains than in the case where the same stress level is reached in fewer loading steps. However, in the case of fewer loading steps, the time of failure (i.e. the time elapsed from the first loading step to the moment of failure) is generally longer than in the experiments done with a larger number of loading steps or in those in which a constant loading rate is used. For instance, in Fig. 9.2.1 the staircase-like curves for schist are shown (the results obtained in creep tests). For each stress level the creep lasted many days, as shown by a number marked on each horizontal line. Though the creep test lasted altogether for a very long time (compared with the standard compression tests), failure occurred at smaller strains than in standard constant loading rate experiments, due to the smaller number of loading steps.

Failure of rocks, and the damage preceding failure, have some typical peculiarities, distinguishing rocks of relatively small porosity (due mainly to microcracks, and not to pores) from most other solids. When a cylindrical specimen is tested in uniaxial compressive tests, for instance, the typical behaviour of most rocks is shown in Fig. 9.2.3 which has been obtained for

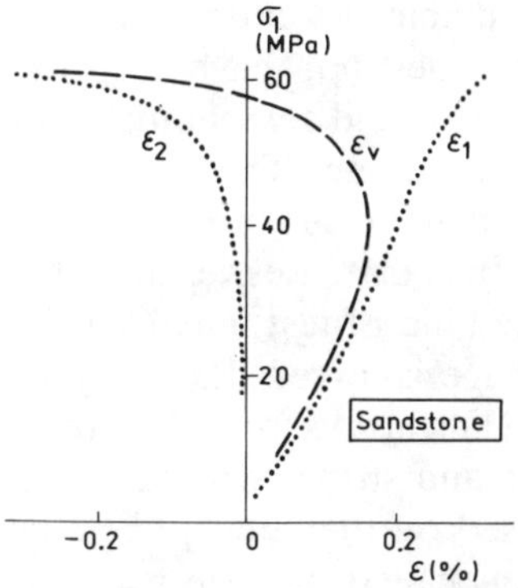

Figure 9.2.3. Stress—strain curves for sandstone obtained in uniaxial unconfined compression test showing significant dilatancy before failure.

sandstone. Thus, (as already discussed at Chapter 4), for most rocks the significant volume dilatancy preceding failure is typical, and it is due to the *formation of a large number of new microcracks over the whole volume of the specimen*. One can conclude that damage and failure of rocks are progressive and *closely related to the same mechanisms which produce dilatancy*. Damage of the rock specimen begins when dilatancy of the rock starts to increase. Thus it is very important to determine quite accurately, even for general triaxial stress states, the time at which during deformation, the rock passes from the compressibility state into the dilatancy one, i.e. the location in the stress space of the compressibility/dilatancy boundary.

Schock (1976) has found for Climax Stock *granodiorite* that 'failure is always preceded by dilatant behavior' and that dilatant strain would be a better indication of failure than a $\bar{\sigma} - \sigma$ envelope, because accumulated shear strain is assumed to be directly related to failure. Surface deformation studies of Westerly granite have shown that the detection of the location of the future rupture cone is as late as the tertiary creep stade (Kurita *et al.* (1983)).

Similar peculiarities are also observed in triaxial tests, with the only difference being the initial non-linear portion of the σ_1—ε_v curve steadily less significant, as the initial hydrostatic pressure is increased (see Chapter 4).

The observation that loading produces damage of the rock specimen over its entire volume is also confirmed by AE studies. Thus the distribution of microseismic hypocentres in the specimen has been found to be random over the entire volume of the specimen during most of the loading period (see Scholz (1968*b*), Mogi (1968), Lockner and Byerlee (1977), Yanagidani *et al.* (1985), etc.). This distribution becomes less random, as faulting is approached, and a clustering of microseismic events in several zones occurs only in the moments just preceding faulting (Scholz (1968*a*), Sondergeld and Estey (1981), Yanagidani *et al.* (1985)).

The mathematical models describing damage and failure of metals or other materials in creep tests, such as those initiated by Kachanov (1960) and further developed by many authors, cannot be applied to rock-like materials, since in these models it is assumed that a fictitious undamaged state exists (an initial damage-free 'virgin state' does not have any meaning for rocks) and that damage is dependent on stress and temperature but independent of hardening, deformation history, dilatancy, etc. There are a great many creep-to-failure cumulative-damage laws which are all empirical and established independently of the constitutive equation. In most cases, each of these cumulative-damage laws is formulated for a specific kind of test only (see Zhang and Brinson (1986)). A theory of cumulative damage incurred during cyclic loading was developed by Costin and Holcomb (1981) starting from the growth of the length of a discrete crack due to cyclic fatigue and stress corrosion. Another theory of degradation of rock-like materials, based on the concept of damage surface and of damage tensor obtained *via* an 'associated damage rule', is due to Mróz and Angelillo (1982). An experimental method for detecting damage surfaces in brittle materials by AE is due to Holcomb and Costin (1986). In an advanced stage of deterioration the process has the tendency to become anisotropic (see Appendix 1) so that the concept of *damage tensor* may be a useful one (Murakami (1983, 1987), Lemaitre (1986), Lemaitre and Chaboche (1985, Chapter 7)).

In order to describe the process of damage leading to failure, we shall use the constitutive equation developed in the previous chapter. We recall that work-hardening by dilatancy and/or compressibility is described by

$$H(\sigma, \bar{\sigma}, \Delta) = W^{\mathrm{I}}(t), \tag{9.2.1}$$

where

$$W^{\mathrm{I}}(T) = \int_0^T \boldsymbol{\sigma}(t) \cdot \dot{\boldsymbol{\varepsilon}}^{\mathrm{I}}(t)\, \mathrm{d}t \tag{9.2.2}$$

is the irreversible stress work. The compressibility/dilatancy boundary is defined by

$$\frac{\partial H}{\partial \boldsymbol{\sigma}} \cdot \mathbf{1} = 0. \tag{9.2.3}$$

When during the stress evolution the boundary (9.2.3) is crossed, the same relationship marks the *onset of dilatancy and damage of the rock*. Thus we would like to *describe damage and further failure, using elements already existing in the constitutive equation, but only those related to dilatancy.*

An alternative procedure would be to start from some properties of the function H alone. For instance, assuming H to be a function of two invariants only, $H(\sigma, \bar{\sigma})$, we observe that this function has the property that when $\sigma =$ const. while $\bar{\sigma}$ increases greatly, then H increases asymptotically with $\bar{\sigma}$. This increase signifies the imminence of failure. Thus we could determine the function $H(\sigma, \bar{\sigma})$ in the form (see (7.6.7))

$$H(\sigma, \bar{\sigma}) := \left[\frac{a(\sigma, \bar{\sigma})}{F(\sigma, \bar{\sigma})} + b(\sigma, \bar{\sigma}) \right] \bar{\sigma} + H^{\mathrm{H}}(\sigma), \tag{9.2.4}$$

so that the stress-states satisfying $F(\sigma, \bar{\sigma}) = 0$ produce failure. Thus failure can be included in the constitutive equation, but this way of describing failure has all the disadvantages discussed in the previous section. In a similar way a failure criterion in terms of strains (irreversible volumetric strain, for instance) could be included in the constitutive equation, again with essentially the same disadvantages.

9.3. The energetic damage parameter

Besides the failure criteria formulated in stress invariants which have already been mentioned, some other criteria have been proposed taking into account the specific characteristics possessed by rocks when damage and failure are produced by the applied stresses.

For instance Kranz and Scholz (1977) have proposed a criterion for the transition from the secondary to tertiary creep, i.e. for the instability in the final stage of the deformation by creep, a certain limit value of the irreversible volumetric strain ε_v^I. Since tertiary creep appears in creep tests during the very last period of deformation (just before failure), the irreversible volumetric strain might possibly be used as a criterion to predict failure. It should be pointed out, however, that in performing creep experiments with rocks, it is sometimes difficult to record the tertiary region at all. In other cases, however, the tertiary creep period can be observed for several days. For instance, in the two creep tests shown in Fig. 9.2.1, at the last loading stage, one of the specimens was clearly in the tertiary stage for about 4 days (out of a total of 9 days), and the other one for about 6 days (out of a total of 14 days). Thus a fairly long time elapsed from the beginning of the tertiary creep to failure, and therefore a criterion involving ε_v^I alone cannot be used. Another remark can be made in conjunction with the data discussed in §9.2: irreversible volumetric deformation alone cannot characterize failure since with increasing loading rate the rock fails at smaller values of $-\varepsilon_v^I$.

By testing *granite* at room temperature, Kranz (1980) has shown that the dilatant volumetric strain at the onset of instability leading to failure increases with pressure. This increase is more significant in creep tests than in constant rate (of 10 MPa s^{-1}) fracture tests.

The influence of the loading rate on the failure mechanism has been studied also by Sano *et al.* (1981), (1982), who have also used the AE technique. It was found that for *granite* the *failure is loading-rate-dependent* in the sense that the ratio $\Delta\varepsilon_v/\Delta\sigma_1$ is dependent on the loading rate (or rate of deformation). Another conclusion pointed out by these authors is that there is a critical magnitude of the irreversible volumetric strain which is independent of the rate of strain and which characterizes the beginning of the acceleration of the deformation by creep producing failure. This magnitude of the irreversible volumetric strain seems to be a characteristic constant of a particular rock.

Several authors (see Schock *et al.* (1973), Schock and Heard (1974), Schock (1977)) have shown that an increase of the strain rate raises the failure surface, i.e. the strength increases. Also for ductile rocks exhibiting work-hardening, the failure surface is not unique (Schock (1977)). Another theory of time-dependent damage and failure is due to Costin (1983, 1985).

In the paper by Lindholm *et al.* (1974) it was shown that Dresser *basalt* shows a strong dependence of the ultimate or fracture strength on both temperature and rate of deformation. The combined temperature-rate dependence of the strength was such as to indicate that the fracture was controlled by a thermal activation process. This led the authors to the formulation of a fracture criterion of the form

$$\frac{\sigma_1}{S_C(0)} + \frac{S_C(0) - S_{BC}(0)}{S_C(0)\,S_{BC}(0)}\,\sigma_2 - \frac{\sigma_3}{S_T(0)} = 1 - \beta T(A - \ln \dot{\varepsilon}),$$

where T is the absolute temperature, $S_C(0)$, $S_T(0)$, and $S_{BC}(0)$ are suitably determined constant strength values at zero absolute temperature, β and A are constants involving the activation energy, volume, and frequency, while $\dot{\varepsilon}$ is some averaged strain rate.

Several authors have discussed the decrease of the ratio v_S/v_P (see (6.2.1)) prior to an earthquake and have found in it strong evidence for the *occurrence of dilatancy* in situ (Nur (1972), Nur *et al.* (1973), Nur and Schultz (1973)).

In the following an *energetic damage parameter*, which describes the history of damage, will be proposed for rocks. Let us recall that the irreversible stress work W^I can be decomposed into two parts:

$$W^I(T) = \int_0^T \boldsymbol{\sigma}(t)\cdot\dot{\boldsymbol{\varepsilon}}^I(t)\,dt = \int_0^T \sigma(t)\,\dot{\varepsilon}_v^I(t)\,dt + \int_0^T \boldsymbol{\sigma}'(t)\cdot\dot{\boldsymbol{\varepsilon}}'^I(t)\,dt$$

$$= W_v^I(T) + W_D^I(T), \tag{9.3.1}$$

corresponding to the volumetric deformation and to the change in shape (deviatoric). Here the prime denotes 'deviator'. Let us consider the first part, $W_v^I(t)$. For most rocks during standard uniaxial tests or triaxial tests, $W_v^I(t)$ first increases (during the period of compressibility) and then decreases (during dilatancy). The total decrease of the stress work W_v^I starting from its maximal value $W_{v(max)}^I$ will be used as a measure of the damage of the rock due to loading. Let us introduce the notation

$$d(t) = W_{v(max)}^I - W_v^I(t) \tag{9.3.2}$$

for the *damage parameter of the rock* at moment t. Here t is a moment of time during the period when dilatancy takes place and therefore $t > t_{(max)}$, the time when W_v^I reaches its maximum. Thus *the damage parameter is a measure of the energy release due to microcracking when dilatancy takes place.* If for some rocks, $W_{v(max)}^I$ is very small, then in an advanced stage of dilatancy, $W_{v(max)}^I$ can be neglected with respect to $W_v^I(t)$, so that

$$d(t) = -W_v^I(t) \quad \text{with } t \gg t_{(max)}. \tag{9.3.3}$$

The *damage rate* is defined by the evolution law

$$\dot{d}(t) = -\dot{W}^{\mathrm{I}}_{\mathrm{v}}(t) = -\sigma(t)\,\dot{\varepsilon}^{\mathrm{I}}_{\mathrm{v}}(t)$$

$$= -k\left\langle 1 - \frac{W^{\mathrm{I}}(t)}{H}\right\rangle \frac{\partial H}{\partial \sigma}\,\sigma \quad \text{if} \quad \frac{\partial H}{\partial \sigma} < 0, \tag{9.3.4}$$

and therefore the damage of the rock due to a loading history is described by the constitutive equation itself, more exactly by that part of the constitutive equation describing the irreversible deformation of the volume by dilatancy. Thus d depends on the stress-state and on the loading history.

The *energetic parameter which characterizes the failure threshold* will be chosen as the *total past energy release due to microcracking during the whole dilatancy process*,

$$d_{\mathrm{f}} = -W^{\mathrm{I}}_{\mathrm{v}}(\text{failure}) \quad \text{or} \quad d_{\mathrm{f}} = W^{\mathrm{I}}_{\mathrm{v(max)}} - W^{\mathrm{I}}_{\mathrm{v}}(\text{failure}), \tag{9.3.5}$$

depending on whether $W^{\mathrm{I}}_{\mathrm{v(max)}}$ is or not negligible with respect to $W^{\mathrm{I}}_{\mathrm{v}}$(failure). The magnitude of d_{f} is thought to be typical for each particular rock. A criterion relating failure to the damage rate $\dot{d}$ can also be considered, but the experimental data available at present do not seem to support this criterion.

In order to make clear these concepts, let us analyse the experimental data of Sano *et al.* (1981) for *granite*. In their paper the results of uniaxial tests carried out with mean loading rates of $\dot{\sigma}_1 = 1.98$ MPa s^{-1}, 1.92×10^{-1} MPa s^{-1}, 1.68×10^{-2} MPa s^{-1}, and 1.76×10^{-3} MPa s^{-1} are reported. The last specimen was unloaded, but the first three specimens failed at the following stresses and times: failure at 199 MPa after 98 s, failure at 186 MPa after 928 s, and failure at 172 MPa after 9936 s. Thus, if the loading rate is decreased, the stress at failure also decreases and the loading duration increases, as already mentioned in the first section. Using these data, the term $W^{\mathrm{I}}_{\mathrm{v}}$ was computed using the formula

$$W^{\mathrm{I}}_{\mathrm{v}}(T) = \frac{1}{3}\int_0^T \sigma_1(t)\,\dot{\varepsilon}_{\mathrm{v}}(t)\,\mathrm{d}t - \frac{\sigma_1^2(T)}{18K}, \tag{9.3.6}$$

where σ_1 is the only non-zero stress component and the last term is the reversible part of the stress work. The value of $W^{\mathrm{I}}_{\mathrm{v}}$ at various stress levels is shown in Fig. 9.3.1 for the four loading rates. The failure of the three specimens is marked by vertical arrows. $W^{\mathrm{I}}_{\mathrm{v}}$ first increases during the period of compressibility and then decreases. Toward the end of the experiment, close to failure, the decrease is abrupt. d_{f} is the total decrease of $W^{\mathrm{I}}_{\mathrm{v}}$ starting from $W^{\mathrm{I}}_{\mathrm{v(max)}}$ to failure. The estimate of d_{f} from the three tests furnishes values ranging from 1.4×10^{-4} GPa to 1.97×10^{-4} GPa, with a mean value of 1.67×10^{-4} GPa corresponding to 1.67×10^{5} J m^{-3}. Though the exact estimate of d_{f} is quite difficult and depends essentially on the last data registered before failure, as well as on the fidelity of the measuring device, it seems that d_{f} is invariant to a change of the loading rate.

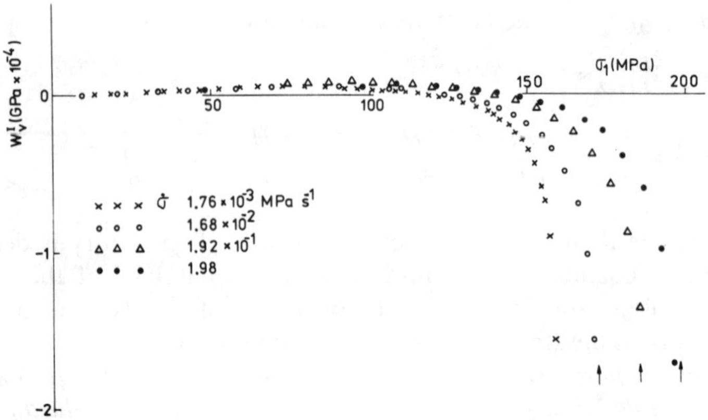

Figure 9.3.1. The computed values of W_v^I for granite tested in uniaxial compression tests with four loading rates.

In the experiments done in the triaxial tests the numerical estimation of W_v^I is more involved. We recall that in triaxial tests the specimen is usually first loaded hydrostatically up to a time t_h, and thereafter a single principal stress component σ_1 is increased, the other two principal stress components being kept constant. Thus, we have

$$W_v^I(T) = \int_0^{t_h} \sigma(t)\,\dot{\varepsilon}_v(t)\,dt - \frac{\sigma^2(t_h)}{2K} + \\ + \frac{1}{3}\int_{t_h}^{T} \sigma_1^R(t)\,[\dot{\varepsilon}_1^R(t) + 2\dot{\varepsilon}_2^R(t)]\,dt - \frac{1}{9K}\frac{(\sigma_1^R(T))^2}{2}. \tag{9.3.7}$$

As an example, for *andesite* we get $d_f = 2.84 \times 10^4$ J m^{-3} in uniaxial tests and $d_f = 3.89 \times 10^4$ J m^{-3} in triaxial tests, with $\sigma_2 = 0.014$ GPa. The two values are not too far apart, but the paucity of experimental data does not allow a conclusion concerning a possible dependency of d_f on the mean stress. *d_f seems to depend on the confining pressure.*

Let us make some comments concerning the *reference configuration* for the concept of damage. During the second stage of a triaxial test, $W_v^I(t)$ first increases (during the period of compressibility) and afterwards decreases (during dilatancy). In (9.3.5) $W_{v(max)}^I$ is positive and corresponds to the maximum value of $W_v^I(t)$ during compression (microcrack closure). The absolute maximum $W_{v(Max)}^I$ of W_v^I is reached in hydrostatic compression when $\sigma = \sigma_0$. For any hydrostatic stress-state for which $\sigma \geqslant \sigma_0$, we have $W_v^I = W_{v(Max)}^I = \text{const.}$ Therefore as far as the definition of damage is concerned, the most *natural reference configuration* for a rock is that corresponding to the hydrostatic stress-state $\sigma = \sigma_0$, at which W_v^I reaches its absolute maximum. Relative maxima for W_v^I are reached during triaxial tests at the crossing of the compressibility/dilatancy boundary.

Now let us return to the constitutive equation. In order to include failure produced by loading in the constitutive equation, it will be assumed that the constitutive parameters involved in the constitutive equation are dependent on the damage parameter d as well, i.e. on the past microcracking. Of all constitutive parameters, the viscosity parameter k is assumed to be the one most sensitive to rock damage. Furthermore, k will be assumed to be possibly dependent on the stress invariants, but also on the damage parameter. If a dependence of k on d is prescribed, then the evolution law (9.3.4) becomes quite involved. The simplest idea however, is to define the dependence of k on d by

$$k(d) = \begin{cases} k_1 & \text{if } 0 \leqslant d < d_t \\ k_2 & \text{if } d_t \leqslant d < d_f \\ +\infty & \text{if } \quad d = d_f. \end{cases} \tag{9.3.8}$$

Here d_t is the value of d at the onset of tertiary creep, and $k_2 > k_1$. A continuous dependence of k on d can certainly also be formulated. Our experimental data have shown that during primary and even secondary creep the value of k does not vary significantly.

When a problem concerning creep of the rock around an underground structure is solved, the irreversible work is continuously estimated at all times and at all locations, and so is W_v^I. Thus, deterioration is continuously recorded according to the evolution described by the constitutive equation for the specific problem under consideration and the failure threshold can be predicted (timing and location). Several examples will be given in the following chapters.

During rock deterioration by stress loading, it is to be expected that some other constitutive parameters may also change, for instance the elastic parameters may also depend on the past microcracking (as shown, for example by ultrasonic tests). That is why some authors (Grady and Kipp (1980), Taylor *et al.* (1986)) have defined the damage in dynamic tests by the variation of elastic moduli. In slow static tests, however, it is thought that the influence of damage on the constitutive equation bears mainly on the flow properties and thus on the viscosity parameter. The dependence of k on d can be determined in creep tests, and the dependency of elastic parameters on d in dynamic tests.

Finally, we mention that the energy produced by microcracking of granite has been estimated by Wong (1982), using a simplified formula and the experimental data of Taponnier and Brace (1976) obtained with a confined pressure of 50 MPa. This energy is estimated to range between 6.67×10^3 J m^{-3} and 6.67×10^4 J m^{-3} and thus is rather smaller than the one given above. No data are available for confining pressures approaching σ_0. We know, however, that according to the constitutive equation $W^I_{v(\text{Max})} = c_0 + c_1$ and thus $W^I_{v(\text{Max})}$ is of the order 3.31×10^5 J m^{-3}.

9.4. Numerical examples

As a first example let us consider the failure of the rock specimen subjected to uniaxial creep tests as described by the above model. For this purpose the model for *andesitic rock* given in Chapter 8 will be used. Thus, we use the constitutive equation (8.4.1) with the viscoplastic potential (8.6.5) and the values of the constitutive constants listed in §8.6. The various possible values for the viscosity coefficient will be given below. We recall that the experimental data available allowed the determination of this model for moderate mean stresses only, so that this model can be used for depths not greater than 1 km, say.

In a uniaxial creep test the stress σ_1 is increased in steps; after each increase, it is kept constant for a longer or shorter time interval. In the so-called 'short-term' creep tests, the length of these intervals is conventionally chosen so that at the end of the test, staircase-like stress—strain curves are obtained with a desired mean loading rate.

At each increment of the stress, $\Delta\sigma_1$, applied at time t_0, the corresponding strain increments are obtained from

$$\Delta\varepsilon_1 = \left(\frac{1}{3G} + \frac{1}{9K}\right)\Delta\sigma_1 + \frac{H\left(1 - \dfrac{W^{\mathrm{I}}(t_0)}{H}\right)\dfrac{\partial H}{\partial \sigma_1}}{\dfrac{\partial H}{\partial \boldsymbol{\sigma}}\cdot\boldsymbol{\sigma}} \times \left\{1 - \exp\left[\frac{k}{H}\frac{\partial H}{\partial \boldsymbol{\sigma}}\cdot\boldsymbol{\sigma}(t_0 - t)\right]\right\}$$

$$\Delta\varepsilon_2 = \left(\frac{1}{9K} - \frac{1}{6G}\right)\Delta\sigma_1 + \frac{H\left(1 - \dfrac{W^{\mathrm{I}}(t_0)}{H}\right)\dfrac{\partial H}{\partial \sigma_2}}{\dfrac{\partial H}{\partial \boldsymbol{\sigma}}\cdot\boldsymbol{\sigma}} \times \left\{1 - \exp\left[\frac{k}{H}\frac{\partial H}{\partial \boldsymbol{\sigma}}\cdot\boldsymbol{\sigma}(t_0 - t)\right]\right\}, \tag{9.4.1}$$

where the first terms correspond to the instantaneous elastic response occurring at moment t_0 and the last terms correspond to the creep which occurs in the time interval $t - t_0$. The last terms are also used for the calculation of $\varepsilon_v^{\mathrm{I}}$ involved in the damage parameter. When the damage parameter reaches the limit value (9.3.5) during the test, the rock specimen fails.

As an example, Fig. 9.4.1 shows the curves $\sigma_1 - \varepsilon_1$ and $\sigma_1 - \varepsilon_v$ for this andesite for three mean loading rates given in Table 9.1 (the last three cases). The instantaneous increment of the stress for all steps and in all cases is $\Delta\sigma_1 = 0.005$ GPa, and the mean loading rates for the three cases are also given in Table 9.1. For a particular value of k, the time intervals during which the stress is held constant are given in Table 9.1. With an increase of the loading rate the

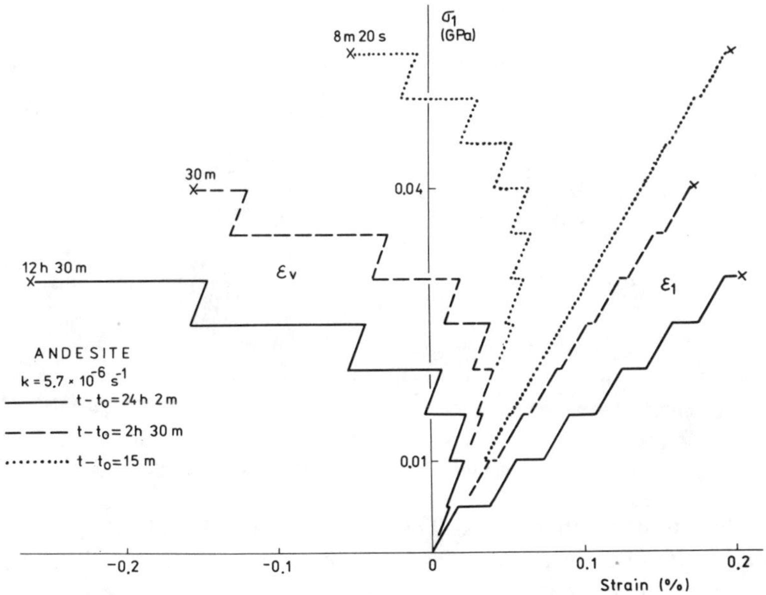

Figure 9.4.1. Incremental creep test for andesite; stress at various levels is kept constant, certain conventional time interval shown. Crosses mark failure according to criteria proposed.

Table 9.1. Data used for plotting of Fig. 9.4.1 and 9.4.2

No.	$\dot{\sigma}_1$ (GPa s^{-1})	$k(t - t_0)$	$t - t_0$ for $k = 5.7 \times 10^{-6}$ s^{-1}	Time elapsed from last loading up to failure
1	6.61×10^{-9}	5.13	10 days 10 hours	50 hours
2	6.06×10^{-8}	0.513	24 hours 2 mins	12 hours 30 mins
3	6.17×10^{-7}	0.0513	2 hours 30 mins	30 mins
4	5.7×10^{-6}	0.00513	15 mins	8 mins 20s

curves $\sigma_1 - \varepsilon_1$, as well as $\sigma_1 - \varepsilon_v$, are rising; the apparent dilatancy threshold (the first deformation by dilatancy which is observable, say) is also rising significantly, but the *total dilatancy at failure decreases.* Let us mention here that Kranz (1979) found that 'inelastic volumetric strain is as large or greater at the onset of tertiary creep than the inelastic volumetric strain at failure in constant strain-rate tests on Barre granite', and that 'more subcritical crack growth will occur the slower the test'.

The *failure* is marked at the end of each curve in Fig. 9.4.1 by a cross and is determined from the condition $k \to \infty$ when $d \to d_f$. Of the three cases under consideration, in the first one the deformation by creep which follows after the last loading lasts 12 hours 30 minutes before failure. This time duration is given in Table 9.1 and marked in Fig. 9.4.1 and Fig. 9.4.2; also given is the duration

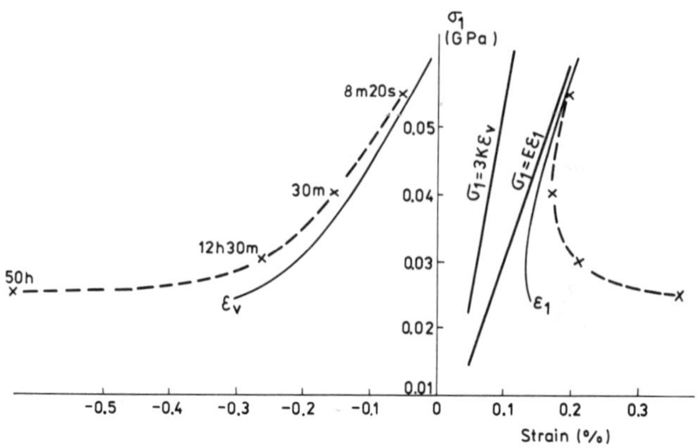

Figure 9.4.2. Dependence of failure on the loading rate.

corresponding to the other cases. Thus the model given above describes the *dependence of failure on the loading rate*, since the stress at failure increases very much with an increase of the loading rate, while the strain ε_1 at failure decreases very much as does dilatancy at failure.

In order to illustrate the *dependence of the failure threshold on the loading history*, Fig. 9.4.2 shows the points where failure occurs for two loading histories. The broken line marks the failure threshold as obtained in the loading path described above, i.e. loading in successive steps of $\Delta\sigma_1 = 0.005$ GPa. The crosses correspond to the failure already shown in Fig. 9.4.1. The numbers adjacent to the crosses again show the time elapsed between the moment of the last loading and failure. Failure at 50 hours after the last loading corresponds to the overall mean loading rate of $\dot{\sigma}_1 = 6.61 \times 10^{-9}$ GPa s^{-1}. If the loading history is changed so that a single loading step is done to the desired ultimate stress level and further, the stress is kept constant to failure, then the loci of the stress and strain state at failure are shown by thin full lines in Fig. 9.4.2. For stresses below $\sigma_1 < 0.024$ GPa, the deformation by creep which follows after a single loading finally stabilizes so that no failure is possible. For higher stresses failure occurs (for the same stress level) at much smaller strains than in the previously described path (loading in several steps). Generally, if a certain stress level is reached in fewer steps, then *failure occurs faster, and the longitudinal deformation at failure and the total dilatancy are smaller.* Finally, if the stress is increased very fast in a single step, then failure occurs in a very short time interval and at very small strains (Fig. 9.4.2).

In *conclusion*, the model formulated above describes qualitatively all the failure peculiarities, as revealed by the experimental data. The influence of other parameters such as humidity, temperature, mean stress, etc., remains to be considered. Some of these parameters, for instance temperature, sometimes have a remarkable influence on dilatancy and on the mechanical properties of rocks generally. Furthermore, their influence on damage and failure is to be expected.

How the above model can be used to predict the failure threshold and further, the evolving process of failure either around a vertical shaft or around a horizontal tunnel, will be discussed in the Chapters 11—13.

There are also rocks or rock-like materials which do not exhibit any perceptible dilatancy up to stress-states which are very close to failure. For instance, *cement concrete*, a hard rock-like material with 25% porosity generally due to pores and not to cracks, is compressible only (see (8.7.9)). A long-term failure criterion for such materials is still to be established.

Exercises

1. Make uniaxial compression tests with various loading rates. Determine the influence of the loading rate on stress and strain at failure. Use several kinds of rocks.

2. Make uniaxial compression creep tests with incremental successive increasing loading steps. Try first to determine, using a single loading step of increasing magnitude, at what loading level the creep of the rock comes ultimately to a stable state, and at what loading level the creep becomes ultimately unstable. Try now to reach the same stress level by two (then three, four, etc.) loading steps, and see if it makes any difference. Then try to determine if the strain at failure produced at a certain loading level changes if this level is reached in one, two, three, etc. loading steps.

3. Repeat the above two exercises but in triaxial tests, using various levels of lateral confining pressures.

4. Make uniaxial compression tests with various loading rates. Determine the influence of the loading rate on the volumetric irreversible strain. See if failure can be characterized by a constant value of ε_v^I for all loading rates.

5. For the same data used in Example 4, compute the parameter d according to (9.3.2). See if it depends on the loading rate. Then determine d_f according to (9.3.5). See if $W^I_{v(max)}$ is sometimes negligible with respect to W^I_v(failure).

6. Using an elastic/viscoplastic constitutive equation of your choice (determined by you or one of those proposed in Chapter 8), make numerical tests of the kind presented in §9.4. First do incremental uniaxial creep tests and plot a figure similar to Fig. 9.4.1. Determine the stress and strain at failure according to various criteria. Compare qualitatively with experimental data available. Further, determine the influence of the loading rate on failure (see Fig. 9.4.2).

10

Stress-states in situ

10.1. Primary stress-state

The undisturbed *in situ* rock before an excavation is already under a certain stress. This stress-state is due to a variety of causes: gravity, stresses due to tectonic motions, the strata orientation, the creep and relaxation produced during geological periods, possible local unloading due to a relatively fast erosion, thermal contractions, etc.

This initial stress-state or natural stress-state will be called the *primary stress-state* or *far-field stresses*. In what follows this stress-state will be imagined as the final result of a very long evolution on a geological time-scale which has resulted in the present quasistable state. Thus, it will be assumed that in any point of the rock the primary stresses satisfy the Cauchy equilibrium equation. It will also be assumed that the primary stress-states depend on the depth alone.

Various experimental methods have been developed in order to measure the primary stress-state, by measuring either the stresses necessary to prevent a rock displacement, or the displacements produced by an excavation, and further using an assumed form of the constitutive equation in order to get the stresses. Though none of the methods is very accurate, since these methods themselves disturb this stress-state in the rock, they can be used to estimate the vertical stress σ_v and the horizontal stress σ_h at various depths. As a rule, the horizontal stress is not independent of the orientation in the horizontal plane (Lindner and Halpern (1978)). Generally, it is assumed that σ_v and σ_h are principal stresses and that σ_h is a mean horizontal stress, though these are merely simplifying assumptions, sometimes in disagreement with the data. A new technique which uses AE monitoring in conjunction with a borehole rockjack is now being tested in order to get an indication of some forms of *in situ* stress (Mc Elroy *et al.* (1985)). Another new technique for measuring the complete stress-state of *in situ* rock in a single borehole was developed by Sugawara and Obara (1986) and Sugawara *et al.* (1986).

As precise an estimate as possible of the primary stresses is essential for the calculation of the stress-state created by an underground excavation: the primary stresses play the role of initial stresses. Let us consider a Cartesian coordinate system with the $0z$-axis directed vertically downward. From the above assumptions we get

$$\sigma_{xx}(h) = \sigma_h(h), \qquad \sigma_{yy}(h) = \sigma_h(h), \qquad \sigma_{xy} = \sigma_{yz} = \sigma_{zx} = 0$$
$$\sigma_{zz} = \int_0^h \rho g \, \mathrm{d}z = \sigma_v(h), \tag{10.1.1}$$

with h = const. the depth of the excavation. Thus σ_v is the weight of a column of unit area of the overburden. If the rock along this column is not homogeneous, then a certain mean value of the integral (10.1.1) is to be estimated. Sometimes a *mean value for the density* is used, and formula $(10.1.1)_7$ is written as

$$\sigma_v = K\rho h = \gamma h, \tag{10.1.2}$$

with $K = 9.86 \times 10^{-6}$, the density ρ expressed in kg m^{-3}, the depth h in metres, and the stress in MPa. For instance (Blackwood (1979)), if $\rho = 2300$ kg m^{-3}, we get $\sigma_v = 0.023$ h.

The primary stress-state is called *lithostatic* if it is hydrostatic, i.e. $\sigma_h = \sigma_v$, and if σ_v is obtained from (10.1.1) (Jaeger and Cook (1979, Chapter 14)). The assumption that the primary stresses are lithostatic is called *Heim's rule*.

Many measurements of underground stresses have been reported and discussed by various authors (see, for instance, Panet (1969), Abou-Sayed *et al.* (1978), Brown and Hoek (1978), Blackwood (1979), Duncan Fama and Pender (1980), Goodman (1980), Jamison and Cook (1980), Hoek and Brown (1980), Wawersik and Stone (1986), and the references mentioned in these papers). All the authors have found a *linear* relationship between σ_v and depth. For instance, Brown and Hoek (1978) give

$$\sigma_v = 0.027\,h, \tag{10.1.3}$$

with h in metres and σ_v in MPa. This relationship is in good agreement with the experimental data for depths up to 2500—3000 m, though at shallow depths there is a great scatter of the data. Sometimes σ_v is *much smaller* and other times *much greater* than the value obtained by the overburden weight (formula (10.1.3)).

Concerning the horizontal stress, it has long been assumed that at shallow depths $\sigma_h < \sigma_v$, and only at great depths is $\sigma_h \approx \sigma_v$. *In situ* measurements obtained in recent decades have shown, however, that generally at shallow depths $\sigma_h > \sigma_v$, as will be shown below.

The experimental determination of the horizontal stress σ_h, sometimes as a mean stress in the horizontal plane and sometimes without this assumption, has been carried out by several authors who report a variety of results. Thus, Haimson (1978) from measurements done in the USA arrives at the relationship (average value for σ_h)

$$\sigma_h = 4.90 + 0.020\,h, \qquad \sigma_v = 0.025\,h, \tag{10.1.4}$$

while Worotnicki and Denham (1976) from measurements done in Australia arrive at the following formula for the average horizontal stress:

$$\sigma_h = 7.26 + h(0.0215 \pm 0.0028). \tag{10.1.5}$$

Similar results have also been obtained by other authors. It should be pointed out, however, that even at shallow depths all authors have found sites where $\sigma_h \simeq \sigma_v$, and others in which $\sigma_h < \sigma_v$.

Brown and Hoek (1978) have analysed the results obtained by various authors and have found that the ratio between the mean horizontal stress and the vertical one ranges between the limits (see Fig. 10.1.1)

$$\frac{100}{h} + 0.30 \leqslant \frac{\sigma_h}{\sigma_v} \leqslant \frac{1500}{h} + 0.50, \tag{10.1.6}$$

or taking into account (10.1.3),

$$2.7 + 0.008\,h \leqslant \sigma_h \leqslant 40.5 + 0.014\,h. \tag{10.1.7}$$

Similar results have been reported by Jamison and Cook (1980), Saxena *et al.* (1979), and others.

Thus, from (10.1.7) and Fig. 10.1.1 it results that for a given depth the interval of variation of σ_h is quite large, mainly at shallow depths. A more exact

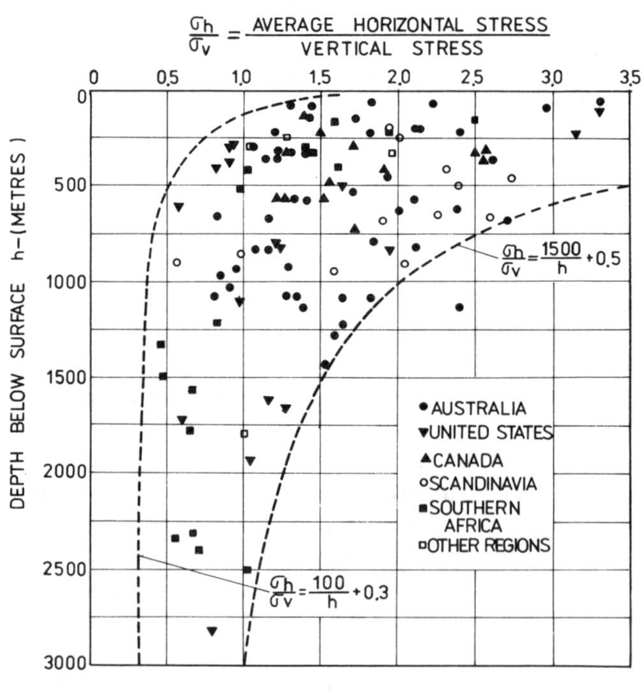

Figure 10.1.1. Variation of the ratio σ_h/σ_v with depth as obtained by measurements *in situ* by several authors.

estimate of the ratio σ_h/σ_v can be done only at specific locations, where empirical formulae for the variation of σ_h/σ_v with depth can be formulated. For instance, from (10.1.3) and (10.1.5) it follows that

$$\frac{\sigma_h}{\sigma_v} = \frac{269}{h} + 0.80, \tag{10.1.8}$$

while from (10.1.4) we get

$$\frac{\sigma_h}{\sigma_v} = \frac{200}{h} + 0.80. \tag{10.1.9}$$

Therefore, in all existing variants the relationship between σ_h and σ_v is of the form

$$\frac{\sigma_h}{\sigma_v} = \frac{a}{h} + b, \tag{10.1.10}$$

or with (10.1.2),

$$\sigma_v = \gamma h, \qquad \sigma_h = b\sigma_v + \gamma a. \tag{10.1.11}$$

This straight line is shown in Fig. 10.1.2. Thus all possible primary stress states are represented by points on this solid line. We have, therefore,

$$\begin{aligned} \sigma_h &> \sigma_v \quad \text{if } h < \frac{a}{1-b} \\ \sigma_h &< \sigma_v \quad \text{if } h > \frac{a}{1-b} \\ \sigma_h &= \sigma_v \quad \text{if } h = \frac{a}{1-b}. \end{aligned} \tag{10.1.12}$$

Using the numerical coefficients from (10.1.9) it follows that the last equality occurs at $h = 1$ km, where $\sigma_v = 25$ MPa ($= 0.25$ kbar $= 255$ kgf cm^{-2}).

At greater depths, in the neighbourhood of $h = 1$ km and below this depth, the ratio σ_h/σ_v is close enough to unity.

The main *conclusions* are the following:

— generally, σ_v can be obtained from the overburden weight (10.1.2);

— generally, $\sigma_h > \sigma_v$, mainly at shallow depths, and can be obtained from (10.1.11);

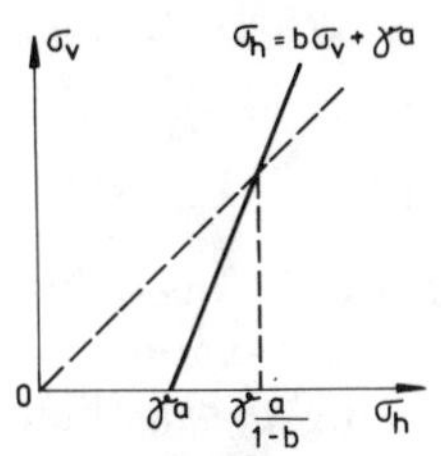

Figure 10.1.2. The linear relationship between σ_h and σ_v.

— in special locations σ_v may not satisfy (10.1.2), being either larger or smaller than this value;

— at shallow depths there is a great scatter of the values of σ_h; depending on location, all three cases $\sigma_h > \sigma_v$, $\sigma_h = \sigma_v$, and $\sigma_h < \sigma_v$ are possible, and sometimes $\sigma_h \gg \sigma_v$.

In the engineering applications presented in the next few chapters it is assumed that before any excavation is carried out, the rock *in situ* is under a certain primary stress. No restrictions concerning their order of magnitude are made. In order to make some mathematical simplifications, however, we assume that

— σ_h is uniform in any direction in the horizontal plane;

— σ_v is a principal component, generally obtained from (10.1.2);

— σ_h and σ_v depend on depth alone and vary gently with h, so that around an excavation of sizes much smaller than h, both σ_h and σ_v can be considered to be constant;

— σ_h and σ_v must be compatible with the balance equations of mechanics (balance of momentum) and the constitutive equation (see the following paragraphs).

10.2. Secondary and relative stress fields

An excavation carried out in a competent rock, such as a tunnel, a shaft, an oil well, galleries of various forms and inclinations, cavities of various forms, etc., will alter the primary stress-state $\boldsymbol{\sigma}^{P}(X)$ initially existing in the virgin rock. This stress field will be altered in the neighbourhood of the opening only. As a result of the excavation around the underground opening, a new stress field is produced. This new stress field $\boldsymbol{\sigma}^{S}(X, t)$ will be called the *secondary* stress field; while $\boldsymbol{\sigma}^{P}$ is assumed to be constant in each particle X, the secondary stress field may vary in time in a fixed particle.

If during or after excavation the deformation of the walls is not prevented by artificial supports (linings, etc.), then due to the variation of the stress (both at the time of the excavation and in the long period which follows), the strain field around the opening will also vary. The strains are produced by the stress difference

$$\boldsymbol{\sigma}^{R}(X, t) = \boldsymbol{\sigma}^{S}(X, t) - \boldsymbol{\sigma}^{P}(X) \tag{10.2.1}$$

which will be called the *relative* stress. Similar formal relations can also be written for the strains:

$$\boldsymbol{\varepsilon}^{R}(X, t) = \boldsymbol{\varepsilon}^{S}(X, t) - \boldsymbol{\varepsilon}^{P}(X),$$

with the significant difference that $\boldsymbol{\varepsilon}^{P}(X)$ is only a conventional quantity without any physical meaning. Due to the excavation the deformation of the rock with respect to the configuration existing *in situ* before excavation is expressed by the *relative strain* $\boldsymbol{\varepsilon}^{R}(X, t)$. Generally, the relative strains vary in time, if the rock deforms by creep after excavation.

The secondary field must satisfy certain conditions. These follow from the condition that the perturbation of the primary state produced by the excavation should be negligible at distances d far enough from the opening, i.e. for any moment of time we must have:

$$\lim_{d \to \infty} \boldsymbol{\sigma}^{\mathrm{S}} = \boldsymbol{\sigma}^{\mathrm{P}}, \qquad \lim_{d \to \infty} \boldsymbol{\varepsilon}^{\mathrm{S}} = \boldsymbol{\varepsilon}^{\mathrm{P}}, \tag{10.2.2}$$

or

$$\lim_{d \to \infty} \boldsymbol{\sigma}^{\mathrm{R}} = 0, \qquad \lim_{d \to \infty} \boldsymbol{\varepsilon}^{\mathrm{R}} = 0, \qquad \lim_{d \to \infty} \mathbf{u}^{\mathrm{R}} = 0, \tag{10.2.3}$$

where $\mathbf{u}^{\mathrm{R}}(X, t)$ is the *relative* displacement, i.e. the one which is obtained from $\boldsymbol{\varepsilon}^{\mathrm{R}}(X, t)$, using the definition of $\mathbf{u}$, while $\boldsymbol{\varepsilon}^{\mathrm{R}}$ results from $\boldsymbol{\sigma}^{\mathrm{R}}(X, t)$ with the help of the constitutive equation. This relative displacement is just the displacement which it is possible to observe with respect to the state *in situ* before excavation. In a similar way we can introduce the *secondary* displacement $\mathbf{u}^{\mathrm{S}}$, which does not have a real physical meaning, and generally $\lim_{d \to \infty} \mathbf{u}^{\mathrm{S}}$ is not finite.

In conclusion, after excavation the stress-state in the rock is the secondary stress-state $\boldsymbol{\sigma}^{\mathrm{S}}(X, t)$, the deformation with respect to the configuration existing before excavation is $\boldsymbol{\varepsilon}^{\mathrm{R}}(X, t)$, and the particle displacement again with respect to the configuration before excavation is $\mathbf{u}^{\mathrm{R}}(X, t)$. The strain $\boldsymbol{\varepsilon}^{\mathrm{R}}$, and therefore $\mathbf{u}^{\mathrm{R}}$ as well, are obtained from $\boldsymbol{\sigma}^{\mathrm{R}}$ with the help of the constitutive equation.

In the next few chapters examples will be given of the calculation of the secondary and relative fields. If after excavation the surface of the opening is reinforced, these reinforcements will produce a variation in the secondary stress and strain-states. The new stress-state thus obtained in the rock will be called *tertiary*. The tertiary stress-state in the rock depends not only on the mechanical properties of the rock but also on the mechanical properties (the constitutive equation) of the materials used in the support and on the kind of support used, i.e. on the rock—support interaction (see Chapter 13).

10.3. Initial stresses and strains for the linear elastic model

The theoretical estimate of the magnitude of the ratio $\sigma_{\mathrm{h}}/\sigma_{\mathrm{v}}$, where σ_{v} is obtained from the weight of the overburden, has been made by several authors. A linear elastic constitutive equation has been used, together with an assumption of kinematic nature: due to the fact that the rock extends to infinity in any horizontal plane, there are no horizontal displacements. Thus the only strain component produced by the primary stresses is the vertical one.

Let us choose a Cartesian orthogonal coordinate system with the $0z$-axis directed vertically downward. Assuming that no horizontal strains are possible, we have $\varepsilon_x = \varepsilon_y = 0$. Then from Hooke's law it follows that

$$\sigma_x = \nu(\sigma_y + \sigma_z), \qquad \sigma_y = \nu(\sigma_z + \sigma_x) \qquad \text{or } \sigma_h = \nu(\sigma_h + \sigma_v),$$

since $\sigma_x = \sigma_y = \sigma_h$ and $\sigma_z = \sigma_v$. Therefore

$$\sigma_h = \frac{\nu}{1-\nu}\,\sigma_v. \tag{10.3.1}$$

It follows that the ratio σ_h/σ_v depends on Poisson's ratio alone. Since generally for most rocks $0.11 \leqslant \nu \leqslant 0.33$, it follows that

$$0.124 \leqslant \frac{\sigma_h}{\sigma_v} \leqslant 0.50. \tag{10.3.2}$$

Smaller values of ν correspond to smaller values of this ratio. For the limit value accepted in isotropic linear elasticity theory $\nu = 0.5$, we certainly obtain

$$\sigma_h = \sigma_v, \tag{10.3.3}$$

i.e. the stress state is hydrostatic.

If we compare the numerical values from (10.3.2) with *in situ* measurements, we come to the conclusion that numerical values of the ratio σ_h/σ_v as given by (10.3.2) are very seldom met in the field at all depths. Condition (10.3.3) is certainly met at greater depths.

The primary strain field can be obtained from Hooke's law for the primary stresses satisfying (10.3.1). These strains do not have any physical meaning, and in fact we do not need them when the strain around an underground opening is estimated.

Generally, when we wish to use Hooke's law to determine the stresses and strains in the rock around an underground opening, we must first determine the primary stresses $\boldsymbol{\sigma}^{\mathrm{P}}$ from *in situ* measurements. Further, we determine the secondary stresses $\boldsymbol{\sigma}^{\mathrm{S}}$ produced by the excavation. Finally, we introduce the relative stresses $\boldsymbol{\sigma}^{\mathrm{R}} = \boldsymbol{\sigma}^{\mathrm{S}} - \boldsymbol{\sigma}^{\mathrm{P}}$ in Hooke's law written in relative components:

$$\varepsilon_{xx}^{\mathrm{R}} = \frac{1}{E}\,[\sigma_{xx}^{\mathrm{R}} - \nu(\sigma_{yy}^{\mathrm{R}} + \sigma_{zz}^{\mathrm{R}})], \ldots, \qquad \varepsilon_{xy}^{\mathrm{R}} = \frac{\sigma_{xy}^{\mathrm{R}}}{2G}, \ldots, \tag{10.3.4}$$

in order to determine the relative strains $\boldsymbol{\varepsilon}^{\mathrm{R}}$ and also the displacements $\mathbf{u}^{\mathrm{R}}$. The elastic moduli here have a 'mean' value for the considered depth.

In the above-mentioned procedure the primary strains are not involved at all, while the primary stresses can in fact even be arbitrary, with the restriction that the momentum balance equation is satisfied (with assumption (10.1.1) these are always satisfied). The primary state must in principle satisfy Hooke's law as well, but we do not actually need to write it down.

More precisely, the primary strains are conventional quantities which, together with primary stresses, must satisfy the constitutive equation, while the *primary stresses* (or *far-field stresses*) *must be obtained from* in situ *measurements.*

10.4. Primary states for the elasto-plastic constitutive equation

As we did for the model described previously, let us assume that the primary stress-state is a quasistable state which must satisfy the constitutive equation, this time an elasto-plastic one of the form (7.3.5). We assume that this constitutive equation,

$$\dot{\boldsymbol{\varepsilon}} = \frac{\dot{\boldsymbol{\sigma}}}{2G} + \left(\frac{1}{3K} - \frac{1}{2G}\right)\dot{\sigma}\mathbf{1} + \frac{\left\langle \frac{\partial H}{\partial \boldsymbol{\sigma}} \cdot \dot{\boldsymbol{\sigma}} \right\rangle}{\frac{\partial F}{\partial \boldsymbol{\sigma}} \cdot \boldsymbol{\sigma}} \frac{\partial F}{\partial \boldsymbol{\sigma}} \qquad \text{if } H = W^{\mathrm{I}}$$

$$\dot{\boldsymbol{\varepsilon}} = \frac{\dot{\boldsymbol{\sigma}}}{2G} + \left(\frac{1}{3K} - \frac{1}{2G}\right)\dot{\sigma}\mathbf{1} \qquad \text{if } H < W^{\mathrm{I}}, \tag{10.4.1}$$

has been determined using laboratory tests.

Let us assume that the primary stress-state is a 'plastic' state, i.e. satisfies at the present time t the work-hardening condition

$$H(\sigma, \bar{\sigma}, \Delta) = W^{\mathrm{I}}(t). \tag{10.4.2}$$

Since the function H is known, for the primary state we obtain the value of this function $H(\sigma^{\mathrm{P}}, \bar{\sigma}^{\mathrm{P}}, \Delta^{\mathrm{P}}) = W^{\mathrm{IP}}$, i.e. the conventional value W^{IP} of W^{I} for the primary state.

A stress increment with respect to the primary state which satisfies the inequality

$$\frac{\partial H}{\partial \boldsymbol{\sigma}} \cdot \dot{\boldsymbol{\sigma}} > 0 \tag{10.4.3}$$

will produce a 'loading'. The variation of the strains results from (10.4.1), taking into account that the work-hardening condition is written in the form

$$H(\sigma, \bar{\sigma}, \Delta) = W^{\mathrm{IP}} + \int_0^S \boldsymbol{\sigma}(s) \cdot \dot{\boldsymbol{\varepsilon}}^{\mathrm{I}}(s)\, \mathrm{d}s, \tag{10.4.4}$$

where the moment $s = 0$ corresponds to the beginning of the excavation process, i.e. of the rock deformation.

Therefore, in the constitutive equation the stresses are $\boldsymbol{\sigma}^{\mathrm{S}}(X, t) = \boldsymbol{\sigma}^{\mathrm{P}}(X) + \boldsymbol{\sigma}^{\mathrm{R}}(X, t)$, and therefore, $\dot{\boldsymbol{\sigma}} = \dot{\boldsymbol{\sigma}}^{\mathrm{R}}$ and $\dot{\boldsymbol{\varepsilon}} = \dot{\boldsymbol{\varepsilon}}^{\mathrm{R}}$. The primary stress-state is involved in the value of the conventionally defined quantity W^{IP} as well.

10.5. Primary states for the linear viscoelastic model

Let us consider now the linear viscoelastic model

$$\dot{\boldsymbol{\varepsilon}}' = -k\left(\boldsymbol{\varepsilon}' - \frac{1}{2G_0}\boldsymbol{\sigma}'\right) + \frac{1}{2G}\dot{\boldsymbol{\sigma}}'$$

$$\dot{\varepsilon} = -k_{\mathrm{v}}\left(\varepsilon - \frac{1}{3K_0}\sigma\right) + \frac{1}{3K}\dot{\sigma}. \tag{10.5.1}$$

If we wish to use this model in order to describe the deformation of the rocks surrounding the underground opening, then the primary state must be imagined as an equilibrium stress—strain state which must satisfy the constitutive equation. This equilibrium state is the result of a long process of loading over geological time, and is quasistable with respect to the time intervals usually involved in mining or petroleum engineering.

After a very long elapsed time since a possible last loading, from (10.5.1) we obtain for the relaxed state

$$\boldsymbol{\varepsilon}'^{\mathrm{P}} = \frac{1}{2G_0}\,\boldsymbol{\sigma}'^{\mathrm{P}}, \qquad \varepsilon^{\mathrm{P}} = \frac{1}{3K_0}\,\sigma^{\mathrm{P}}. \tag{10.5.2}$$

These relations define the conventionally introduced primary strains, if the primary stresses are known. When solving a problem, we do not really need to know the primary strains, since the constitutive equation (10.5.1), being linear, can be written straightforwardly in relative components,

$$\begin{aligned} \dot{\boldsymbol{\varepsilon}}'^{\mathrm{R}} &= -k\left(\boldsymbol{\varepsilon}'^{\mathrm{R}} - \frac{1}{2G_0}\,\boldsymbol{\sigma}'^{\mathrm{R}}\right) + \frac{1}{2G}\,\dot{\boldsymbol{\sigma}}'^{\mathrm{R}} \\ \dot{\varepsilon}^{\mathrm{R}} &= -k_{\mathrm{v}}\left(\varepsilon^{\mathrm{R}} - \frac{1}{3K_0}\,\sigma^{\mathrm{R}}\right) + \frac{1}{3K}\,\dot{\sigma}^{\mathrm{R}}. \end{aligned} \tag{10.5.3}$$

In many engineering problems we can determine the stress $\boldsymbol{\sigma}^{\mathrm{S}}$ and therefore, $\boldsymbol{\sigma}^{\mathrm{R}}$. Then from (10.5.3) we get $\boldsymbol{\varepsilon}^{\mathrm{R}}$, i.e. the deformation of the rock, and also $\mathbf{u}^{\mathrm{R}}$, i.e. the displacements which follow the excavation. We recall once more that only the strain $\boldsymbol{\varepsilon}^{\mathrm{R}}$ and the displacement $\mathbf{u}^{\mathrm{R}}$ have physical meaning.

10.6. Primary states for the elastic/viscoplastic model

Let us now consider the model (8.4.1) which will be used extensively in the following chapters:

$$\begin{aligned} \dot{\boldsymbol{\varepsilon}} &= \frac{\dot{\boldsymbol{\sigma}}}{2G} + \left(\frac{1}{3K} - \frac{1}{2G}\right)\dot{\sigma}\mathbf{1} + \\ &+ k(\sigma, \bar{\sigma}, \Delta, d)\left\langle 1 - \frac{W^{\mathrm{I}}(t)}{H(\sigma, \bar{\sigma}, \Delta)}\right\rangle \frac{\partial F}{\partial \boldsymbol{\sigma}}. \end{aligned} \tag{10.6.1}$$

The primary stress-state is assumed again to be quasistable, i.e. a relaxed state satisfying

$$H(\sigma^{\mathrm{P}}, \bar{\sigma}^{\mathrm{P}}, \Delta^{\mathrm{P}}) = W^{\mathrm{IP}}. \tag{10.6.2}$$

This relationship defines the conventional value W^{IP} of the irreversible stress work W^{I}. For any loading, i.e. for any stress-state satisfying the inequality $H(\sigma, \bar{\sigma}, \Delta) > W^{\mathrm{IP}}$, the constitutive equation (10.6.1) can be used. During viscoplastic deformation the irreversible stress work is obtained from

$$W^{\mathrm{I}}(T) = W^{\mathrm{IP}} + \int_0^T \boldsymbol{\sigma}(t) \cdot \dot{\boldsymbol{\varepsilon}}^{\mathrm{I}}(t)\,\mathrm{d}t, \tag{10.6.3}$$

where the moment $t = 0$ corresponds to the beginning of the viscoplastic deformation. Therefore, the constitutive equation (10.6.1) will involve:

— the conventionally defined irreversible primary stress work W^{IP},

— the relative strains in $\dot{\boldsymbol{\varepsilon}} = \dot{\boldsymbol{\varepsilon}}^{\mathrm{R}}$ (as well as in W^{I}),

— the relative stresses in $\dot{\boldsymbol{\sigma}} = \dot{\boldsymbol{\sigma}}^{\mathrm{R}}$,

— the total (secondary) stresses $\boldsymbol{\sigma} = \boldsymbol{\sigma}^{\mathrm{P}} + \boldsymbol{\sigma}^{\mathrm{R}}$ involved in H, F, k, and W^{I}, and possibly also in K.

10.7. Stresses and strains around underground openings

The computation of the stress-state around underground openings is a problem of major importance in applied rock mechanics. It is also important to know the evolution in time of the shape of the underground openings, the opening closure, and sidewall instability. There are various kinds of artificial underground openings. Vertical shafts and oil wells have a rather regular circular cylindrical shape. Some horizontal or oblique tunnels may have the same circular cylindrical shape, but they can also have various other shapes. Other underground openings or systems of openings have complex and sometimes non-regular shapes. Underground excavations have been made for railway stations, nuclear power plants, storage facilities (for gases or liquids or radioactive wastes), etc. (see Bergman (1978)). Sometimes these underground openings are quite large, and sometimes they are located at great depths: for instance, the caverns used for storage facilities may have a diameter of several tens or even hundreds of metres. Some openings are located at a depth of several kilometres, and some oil wells are more than 10 km deep (see Hoek and Brown (1980) for a bibliography on large underground excavations).

Concerning the behaviour of underground openings used as storage facilities, three aspects have to be mentioned (Berest and Nguyen (1983*a*)): the volume of these caverns can diminish continuously during the years of use (up to a volume loss of 40%), the deformation of the walls may continue for several months or even years, and the convergence by creep of the vertical walls is more important in the case of small internal pressures in the cavern (these pressures can sometimes be variable). The bottom of the caverns may also move upward by creep. Caverns excavated in rock salt mainly close themselves by creep, sometimes many years after excavation (Dawson and Munson (1983), Boucly (1984), Crotogino (1984), Dreyer (1984), Horseman and Passaris (1984), Staupendahl and Schmidt (1984)).

Convergence of the walls of various tunnels has been observed many years after their excavation; some caverns have been excavated just for the study of closure by creep, with or without internal pressure. Some rheological models have also been proposed to describe this creep process (Passaris (1979), Borchert *et al.* (1979), Baoshen (1979), Gioda and Cividini (1981), Panet

(1979*a*), Popović *et al.* (1979), Wawersik *et al.* (1982), Horseman and Passaris (1984), Lux and Rokahr (1984), Preece and Wawersik (1984), and others).

Temperature effects on the creep of the walls of a deep cavity have been studied by Berest *et al.* (1979), Maxwell (1984), and others.

Photoelastic methods have also been used by many authors for the study of the stress-state around an underground opening (see, for instance, Filatov *et al.* (1983)).

Several authors have used the finite-element method, the boundary-element method, or a combination of the two methods, the method of weighted residuals, or other numerical methods for the study of stress and strain distribution around underground openings. In most cases the main constitutive assumption has been derived either from Mohr—Coulomb or from Drucker—Prager failure conditions; sometimes simply Hooke's law has been used. For various numerical approaches, see Kovári (1977), Wittke (1977), Niwa *et al.* (1979), Brady and Wassyng (1981), Brady (1979), Brady and Bray (1978*a*, *b*), Wittke *et al.* (1978), Ottosen and Krenk (1979), Egger and Gindroz (1979), Wunderlich *et al.* (1981), Gioda and Maier (1980), Borm (1983), Gioda (1984), Duffaut and Feuga (1985), Kaneko *et al.* (1985), Kobayashi (1985), Varadarajan *et al.* (1985), Xu *et al.* (1985), Zhu *et al.* (1985), Carini and Gioda (1986). Non-linear constitutive equations have also been considered in computer calculations for the stability of underground structures (Wallner and Wulf (1982), Wallner (1984)), or for a temperature distribution around an underground opening (Inada (1985), Ishizuka *et al.* (1985), and Carter and Booker (1985)).

Exercises

1. Establish formula (10.1.1) from the assumptions mentioned in §10.1.
2. Derive formula (10.3.1) and give various numerical values of Poisson's ratio; compare the results with the experimental data shown in Fig. 10.1.1.
3. Give a detailed derivation of formula (10.4.4).
4. Give a detailed proof of (10.5.3).
5. Condition (10.6.2) is assumed if the primary state is a relaxed one; what are the other possible cases? What constitutive equation is then appropriate? (Hint: there are two other possible cases).

11

Creep and dilatancy/compressibility of rocks around vertical shafts and oil wells

11.1. Formulation of the problem

In the present chapter the distribution of stresses, strains, and displacements around a deep vertical shaft or oil well will be studied, assuming that the depth of the well is much larger than its diameter. For the mathematical formulation of the problem, the well is considered to be a circular cylindrical vertical orifice drilled in a semi-infinite space. Cylindrical coordinates r, θ, z will be used, with the $0z$-axis coinciding with the axis of the well and directed vertically downward. The initial raidus of the well is $r = a$.

Various aspects of the problem have been approached by several authors. In most cases, linear elasticity has been used, and the assumption of 'plane stress-state' (see, for instance, Obert and Duvall (1967), Jumikis (1979), Dreyer (1973), and others). In the following the problem is assumed to be a 'plane strain' problem. Several constitutive equations will be used in order to describe the creep of the rock around the well, the stress variation, and the possible damage and failure of the rock. The papers by Massier and Cristescu (1981), and Cristescu (1985*a, b*) will be followed.

After the drilling, the wall of the well is sometimes stress-free, i.e. for

$$\left.\begin{array}{l} r = a \\ \text{any } z \end{array}\right\} : \quad (\boldsymbol{\sigma}^{\mathrm{S}})^{\mathrm{T}}\mathbf{n} = \mathbf{0}. \tag{11.1.1}$$

Sometimes, however, the well is filled up with a gas or liquid under pressure, i.e. at a certain depth we have

$$r = a: \quad (\boldsymbol{\sigma}^{\mathrm{S}})^{\mathrm{T}}\mathbf{n} = p\mathbf{n}, \tag{11.1.2}$$

where p is the pressure of the gas or of the liquid at that particular depth. This pressure can be either constant or variable in time, but in most cases this variation is very slow. Three particular cases are generally considered: an empty well, a well filled with a gas, or a homogeneous liquid of density γ_1, therefore

$$p = 0 \quad \text{or } p = \text{const.} \neq 0 \quad \text{or } p = \gamma_1 z. \tag{11.1.3}$$

The boundary conditions on the surface $r = a$ of the well are

$$\begin{array}{ll} r = a \\ \text{fixed } z \end{array} \quad \left\{ \begin{array}{l} \sigma_r^{\mathrm{R}} = \sigma_r^{\mathrm{S}} - \sigma_r^{\mathrm{P}} = p - \sigma_{\mathrm{h}} \\ \sigma_{rz}^{\mathrm{R}} = \sigma_{r\theta}^{\mathrm{R}} = 0 \end{array} \right. , \tag{11.1.4}$$

with p derived from (11.1.3). The boundary conditions at infinity are written in the form (10.2.3), i.e. $\boldsymbol{\sigma}^{\mathrm{R}} \to \mathbf{0}$, when the distance from the well opening increases greatly.

The initial conditions will be discussed below for each particular constitutive equation to be considered.

The problem of finding the stress and strain distribution will be solved in the neighbourhood of a certain fixed depth h, where it is assumed that p, σ_{v}, and σ_{h} are all constant: the values of these parameters at that particular depth. Thus the depth (or cylindrical coordinate z) will be involved in the problem *via* the particular constant values of p, σ_{v}, and σ_{h} only.

From the symmetry of the domain and that of the boundary conditions it follows that in the equations of this problem the only cylindrical coordinate involved is r and the only displacement component involved is the radial one, u. The stress components σ_r, σ_θ, σ_z will be assumed to be principal components. From the above assumptions it follows that the only strain components involved are ε_r, ε_θ, and ε_z and these are in fact principal components:

$$\varepsilon_r^{\mathrm{R}} = \frac{\partial u^{\mathrm{R}}}{\partial r}, \qquad \varepsilon_\theta^{\mathrm{R}} = \frac{u^{\mathrm{R}}}{r}, \qquad \varepsilon_z^{\mathrm{R}} = 0. \tag{11.1.5}$$

With the simplifying assumptions mentioned, the equilibrium equations reduce to

$$\frac{\partial \sigma_r^{\mathrm{R}}}{\partial r} + \frac{\sigma_r^{\mathrm{R}} - \sigma_\theta^{\mathrm{R}}}{r} = 0, \tag{11.1.6}$$

where the superscript R reminds us that all components are relative.

The only compatibility condition involved in this case is

$$\frac{\partial \varepsilon_\theta^{\mathrm{R}}}{\partial r} + \frac{\varepsilon_\theta^{\mathrm{R}} - \varepsilon_r^{\mathrm{R}}}{r} = 0. \tag{11.1.7}$$

11.2. The linear elastic solution

Let us assume first that for the rock considered and at the depth considered the constitutive equation can be written in the linear elastic form

$$\boldsymbol{\varepsilon}'^{\mathrm{R}} = \frac{1}{2G} \boldsymbol{\sigma}'^{\mathrm{R}}, \qquad \sigma^{\mathrm{R}} = 3K\, \varepsilon^{\mathrm{R}}, \tag{11.2.1}$$

where the values of the elastic parameters G and K are considered at that particular depth, and all components are relative (see also §5.3 and §10.2). Introducing (11.2.1) into (11.1.7), we get

$$\frac{\partial}{\partial r}(\sigma_r^{\mathrm{R}} + \sigma_\theta^{\mathrm{R}}) = 0 \tag{11.2.2}$$

or

$$\sigma_r^{\mathrm{R}} + \sigma_\theta^{\mathrm{R}} = 2A, \tag{11.2.3}$$

with A an integration constant. Further, from (11.1.6) we get

$$\sigma_r^{\mathrm{R}} = A + \frac{B}{r^2}, \qquad \sigma_\theta^{\mathrm{R}} = A - \frac{B}{r^2}, \tag{11.2.4}$$

with B another integration constant. By using also the boundary conditions (10.2.3), (11.1.4), and (11.2.1), we finally obtain

$$\sigma_r^{\mathrm{R}} = (p - \sigma_{\mathrm{h}})\frac{a^2}{r^2}, \qquad \sigma_\theta^{\mathrm{R}} = -(p - \sigma_{\mathrm{h}})\frac{a^2}{r^2}, \qquad \sigma_z^{\mathrm{R}} = 0. \tag{11.2.5}$$

The corresponding strains follow from (11.2.1), if (11.2.5) is used. First, we get for the mean stress and mean strain

$$\sigma^{\mathrm{R}} = 0, \qquad \varepsilon^{\mathrm{R}} = 0. \tag{11.2.6}$$

Thus, although the stress-state passes from primary to secondary by the drilling of the well, in each plane z = const. the point representing the stress-state remains in the same octahedral plane ($\boldsymbol{\sigma}^{\mathrm{P}}$ and $\boldsymbol{\sigma}^{\mathrm{S}}$ belong to the same octahedral plane). This is shown in Fig. 11.2.1 for the case $\sigma_{\mathrm{h}} > \sigma_{\mathrm{v}}$ (shallow depths). Point P represents the primary stress-state and point S the secondary one. The equivalent stress

$$\bar{\sigma}^2 = \frac{3}{2}(\sigma_r'^2 + \sigma_\theta'^2 + \sigma_z'^2) \tag{11.2.7}$$

is calculated for the primary stress-state, when

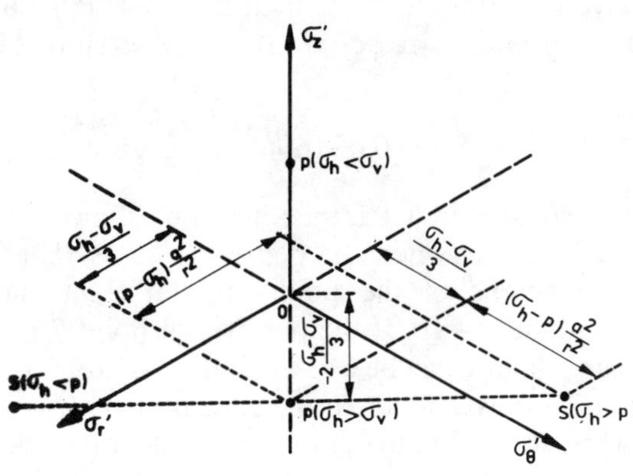

Figure 11.2.1. Deviatoric plane showing primary stress-state *P* and secondary stress-state *S*.

$$\sigma_r'^{\mathrm{P}} = \sigma_\theta'^{\mathrm{P}} = \frac{\sigma_{\mathrm{h}} - \sigma_{\mathrm{v}}}{3}, \qquad \sigma_z'^{\mathrm{P}} = \frac{2}{3}(\sigma_{\mathrm{v}} - \sigma_{\mathrm{h}}), \tag{11.2.8}$$

as

$$(\bar{\sigma}^{\mathrm{P}})^2 = (\sigma_{\mathrm{h}} - \sigma_{\mathrm{v}})^2. \tag{11.2.9}$$

(11.2.8) are just the coordinates of point P in the octahedral plane. In Fig. 11.2.1 all quantities must be multiplied by $\sqrt{2/3}$ (see Chapter 4, Exercise 8).

For the secondary stress-state at S we get

$$\begin{aligned} \sigma_r'^{\mathrm{S}} &= (p - \sigma_{\mathrm{h}})\frac{a^2}{r^2} + \frac{\sigma_{\mathrm{h}} - \sigma_{\mathrm{v}}}{3} \\ \sigma_\theta'^{\mathrm{S}} &= -(p - \sigma_{\mathrm{h}})\frac{a^2}{r^2} + \frac{\sigma_{\mathrm{h}} - \sigma_{\mathrm{v}}}{3} \\ \sigma_z'^{\mathrm{S}} &= \frac{2}{3}(\sigma_{\mathrm{v}} - \sigma_{\mathrm{h}}), \end{aligned} \tag{11.2.10}$$

and also (see (11.2.6))

$$\begin{aligned} \sigma^{\mathrm{S}} &= \sigma^{\mathrm{P}} = \frac{2\sigma_{\mathrm{h}} + \sigma_{\mathrm{v}}}{3} \\ (\bar{\sigma}^{\mathrm{S}})^2 &= 3(p - \sigma_{\mathrm{h}})^2\frac{a^4}{r^4} + (\sigma_{\mathrm{v}} - \sigma_{\mathrm{h}})^2. \end{aligned} \tag{11.2.11}$$

Therefore, the drilling of the well produces a 'loading' of the rock, since $\bar{\sigma}^{\mathrm{S}} \geqslant \bar{\sigma}^{\mathrm{P}}$. Let us observe that $\bar{\sigma}^{\mathrm{S}}$ is maximum for $r = a$, while for $r \to \infty$ we have $\bar{\sigma}^{\mathrm{S}} \to \bar{\sigma}^{\mathrm{P}}$. We recall that besides a multiplicative constant, $\bar{\sigma}^{\mathrm{P}}$ represents the distance from 0 to P, and $\bar{\sigma}^{\mathrm{S}}$ the distance from 0 to S.

Let us also observe that with an increase in depth, when the value of σ_{v} approaches that of σ_{h}, the equivalent stress $\bar{\sigma}$ calculated at P becomes smaller. For $\sigma_{\mathrm{h}} = \sigma_{\mathrm{v}}$, we have $\bar{\sigma} = 0$. If $\sigma_{\mathrm{h}} < \sigma_{\mathrm{v}}$, then the point P is on the positive side of the $0\sigma_z'$-axis. If the primary stresses are obtained from empirical formulae of the form

$$\sigma_{\mathrm{v}} = 0.027\,h, \qquad \sigma_{\mathrm{h}} = 5.4 + 0.0216\,h, \tag{11.2.12}$$

with h in metres and stresses in MPa, then it can be shown that point S departs very much from the $0\sigma_z'$-axis when the depth increases. This is shown in Fig. 11.2.2, where the positions of the points P and S at various depths are represented for $r = a$ and $p = 0$. If the position of point S is very far from the origin 0, then a possible failure can be expected; this is considered below.

Point S in Fig. 11.2.1 corresponds to the case $\sigma_{\mathrm{h}} > p$. If $\sigma_{\mathrm{h}} = p$, then point S coincides with P, while if $\sigma_{\mathrm{h}} < p$, the point S is on the other side of the $0\sigma_z'$-axis on the straight line $\sigma_z' = \text{const}$.

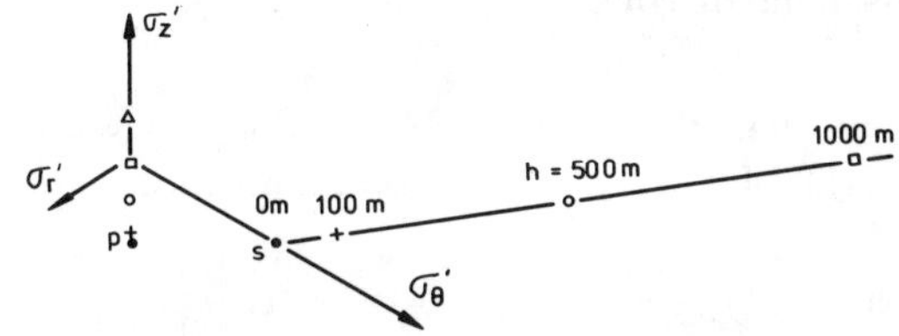

Figure 11.2.2. Primary and secondary stress-states at various depths.

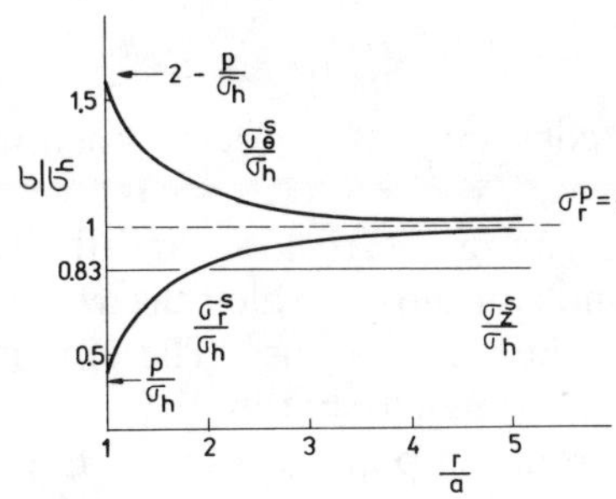

Figure 11.2.3. Stress distribution with distances from the surface of the well.

The relative strains are obtained from (11.2.5) and (11.2.1) as

$$\varepsilon_r^R = \frac{p - \sigma_h}{2G} \frac{a^2}{r^2}, \qquad \varepsilon_\theta^R = -\frac{p - \sigma_h}{2G} \frac{a^2}{r^2}, \qquad \varepsilon_z^R = 0, \tag{11.2.13}$$

while from $(11.1.5)_2$ we get the displacement

$$u^R = -\frac{p - \sigma_h}{2G} \frac{a^2}{r}. \tag{11.2.14}$$

The stress distribution around the well is represented in Fig. 11.2.3 for $h = 500$ m, according to the formulae

$$\begin{aligned} \frac{\sigma_r^S}{\sigma_h} &= \left(\frac{p}{\sigma_h} - 1\right)\frac{a^2}{r^2} + 1 \\ \frac{\sigma_\theta^S}{\sigma_h} &= -\left(\frac{p}{\sigma_h} - 1\right)\frac{a^2}{r^2} + 1 \\ \frac{\sigma_z^S}{\sigma_h} &= \frac{h}{200 + 0.80\,h}. \end{aligned} \tag{11.2.15}$$

The order of magnitude of the principal components of the secondary stresses is as follows: in the immediate vicinity of the well we have $\sigma_r^S < \sigma_z^S < \sigma_\theta^S$, then for $r^2 = a^2((p - \sigma_h)/(\sigma_v - \sigma_h))$, we have $\sigma_r^S = \sigma_z^S < \sigma_\theta^S$, and finally for $r^2 > a^2((p - \sigma_h)/(\sigma_v - \sigma_h))$, we have $\sigma_z^S < \sigma_r^S < \sigma_\theta^S$.

11.3. The linear viscoelastic rock

Let us now assume that the rock obeys a linear viscoelastic constitutive equation of the form (D. Massier, personal communication)

$$\begin{aligned}\dot{\boldsymbol{\varepsilon}}'^{\mathrm{R}} &= -k\left(\boldsymbol{\varepsilon}'^{\mathrm{R}} - \frac{1}{2G_0}\boldsymbol{\sigma}'^{\mathrm{R}}\right) + \frac{1}{2G}\dot{\boldsymbol{\sigma}}'^{\mathrm{R}} \\ \dot{\varepsilon}^{\mathrm{R}} &= -k_{\mathrm{v}}\left(\varepsilon^{\mathrm{R}} - \frac{1}{3K_0}\sigma^{\mathrm{R}}\right) + \frac{1}{3K}\dot{\sigma}^{\mathrm{R}},\end{aligned} \tag{11.3.1}$$

where all coefficients are constant and are determined for the particular depth considered.

In order to study the creep of the rock around the well, two time intervals are considered: the time interval during which the well is drilled, and a further time interval during which the well is in use. The time interval $(0, t_0)$ in which the well is drilled is generally much shorter than the time interval (t_0, t_1) in which the well is in use. From the point of view of the model used we assume that in the first time interval the response of the rock is 'instantaneous', while in the second time interval the deformation continues slowly through time by creep, with or without stress variation.

Using a procedure which is standard in linear viscoelasticity (Boley and Weiner (1960)), we obtain for the instantaneous response

$$\begin{aligned}\boldsymbol{\sigma}'^{\mathrm{R}_0} &\equiv \boldsymbol{\sigma}'^{\mathrm{S}_0} - \boldsymbol{\sigma}'^{\mathrm{P}} = 2G\,\boldsymbol{\varepsilon}'^{\mathrm{R}_0} \\ \sigma^{\mathrm{R}_0} &\equiv \sigma^{\mathrm{S}_0} - \sigma^{\mathrm{P}} = 3K\,\varepsilon^{\mathrm{R}_0},\end{aligned} \tag{11.3.2}$$

where the subscript in R_0 and S_0 shows that this quantity corresponds to the moment of time t_0, i.e. to the initial moment when the deformation by creep begins. Using (11.3.2) and the boundary condition (11.1.4) and (10.2.3), and following the same procedure as in §11.2, we get

$$\begin{aligned}&\sigma_r^{\mathrm{R}_0} = \sigma_r^{\mathrm{S}_0} - \sigma_r^{\mathrm{P}} = (p - \sigma_{\mathrm{h}})\frac{a^2}{r^2} \\ &\sigma_\theta^{\mathrm{R}_0} = -(p - \sigma_{\mathrm{h}})\frac{a^2}{r^2}, \qquad \sigma_z^{\mathrm{R}_0} = 0,\end{aligned} \tag{11.3.3}$$

$$\begin{aligned}&\varepsilon_r^{\mathrm{R}_0} = \varepsilon_r^{\mathrm{S}_0} - \varepsilon_r^{\mathrm{P}} = \frac{p - \sigma_{\mathrm{h}}}{2G}\frac{a^2}{r^2} \\ &\varepsilon_\theta^{\mathrm{R}_0} = -\frac{p - \sigma_{\mathrm{h}}}{2G}\frac{a^2}{r^2}, \qquad \varepsilon_z^{\mathrm{R}_0} = 0,\end{aligned} \tag{11.3.4}$$

$$u^{\mathrm{R}_0} = -\frac{p - \sigma_{\mathrm{h}}}{2G}\frac{a^2}{r}, \qquad w^{\mathrm{R}_0} = 0. \tag{11.3.5}$$

These formulae are similar to the formulae (11.2.5), (11.2.13), and (11.2.14) obtained for the linear elastic case.

If we follow now the procedure described in detail at §12.3, the stresses remain constant during the creep period (t_0, t_1), i.e.:

$$\sigma_r^S = \sigma_r^{S_0}, \qquad \sigma_\theta^S = \sigma_\theta^{S_0}, \qquad \sigma_z^S = \sigma_z^{S_0} = \sigma_v, \tag{11.3.6}$$

and the strain during this period varies according to

$$\begin{aligned} \varepsilon_r^R &= -\varepsilon_\theta^R = \left[\left(\frac{1}{2G} - \frac{1}{2G_0}\right)\exp(-k(t-t_0)) + \frac{1}{2G_0}\right](p - \sigma_h)\frac{a^2}{r^2} \\ \varepsilon_z^R &= 0. \end{aligned} \tag{11.3.7}$$

The corresponding relative displacements are

$$\begin{aligned} u^R &= -\left[\left(\frac{1}{2G} - \frac{1}{2G_0}\right)\exp(-k(t-t_0)) + \frac{1}{2G_0}\right](p - \sigma_h)\frac{a^2}{r} \\ w^R &= 0. \end{aligned} \tag{11.3.8}$$

Note that the stress distribution (11.3.6) is similar to the one for the elastic case, with the difference that now the constants from the instantaneous response are involved. The relative deformations satisfy the conditions mentioned in Chapter 10, and a stabilization in time of the creep deformation also takes place, i.e.

$$\lim_{r\to\infty} \boldsymbol{\varepsilon}^R = \mathbf{0}, \qquad \lim_{t_1\to\infty} \boldsymbol{\varepsilon}^R = \frac{1}{2G_0}\boldsymbol{\sigma}^{R_0}. \tag{11.3.9}$$

Similarly, for the displacement we have

$$\lim_{r\to\infty} u^R = 0, \qquad \lim_{t_1\to\infty} u^R = -\frac{1}{2G_0}(p - \sigma_h)\frac{a^2}{r}. \tag{11.3.10}$$

When $t_1 \to \infty$, the ultimate displacement of the orifice $r = a$ is

$$\lim_{t_1\to\infty} u^R\Big|_{r=a} = \frac{\sigma_h - p}{2G_0}\,a. \tag{11.3.11}$$

Returning now to formula $(11.3.8)_1$ we first recall that the instantaneous modulus is larger than the relaxed modulus $G > G_0$. Therefore, if $p < \sigma_h$, then the radius of the well decreases in time; if $p = \sigma_h$, this radius does not vary in time; finally, if $p > \sigma_h$, then the radius of the well increases in time under the internal applied pressure.

To illustrate the solution given previously, Fig. 11.3.1 shows the variation with distance of the displacement, at various moments of time, using for dolomite the following constants determined in the laboratory: $2G = 26.5$ GPa, $2G_0 = 5.3$ GPa, $1/k = 5 \times 10^5$ s $= 5.78$ days, and using the primary stress distribution (11.2.12). A more accurate description of the variation in time of the displacement can be obtained if these constants (mainly k) are determined by measurements *in situ*.

If the internal constant pressure p is non-zero, then there is a depth, h, where

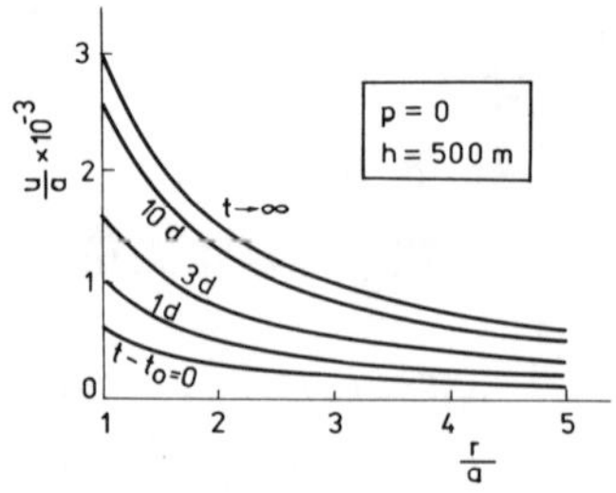

Figure 11.3.1. Variation in time of radial displacement.

$\sigma_h = p$, and therefore $u = 0$. Above this depth the walls of the orifice are enlarging; below it, they are converging.

The variation in time of the strain components can be represented in a similar way.

The assumption at the beginning of this section that the deformation of the rock around the well can be studied in two time intervals of quite distinct length (a very short one followed by a very long one) has been made in order to get a solution which could be expressed by very simple formulae, and in order to use the same procedure with the elastic/viscoplastic constitutive equation as well. With linear viscoelastic constitutive equations other approaches are certainly possible. While the linear viscoelastic constitutive equation can describe the creep of the rock around the well, it cannot describe some other fundamental properties possessed by most rocks, for instance the dilatancy of the volume, etc.

11.4. The elastic/viscoplastic rock

In order to describe more accurately from both a qualitative and a quantitative point of view the creep of the rock around the well, let us consider now an elastic/viscoplastic constitutive equation of the form

$$\dot{\boldsymbol{\varepsilon}} = \left(\frac{1}{3K} - \frac{1}{2G}\right)\dot{\sigma}\mathbf{1} + \frac{1}{2G}\dot{\boldsymbol{\sigma}} + \\ + k(\sigma, \bar{\sigma}, \Delta, d)\left\langle 1 - \frac{W^{\mathrm{I}}(t)}{H(\sigma, \bar{\sigma}, \Delta)}\right\rangle \frac{\partial H}{\partial \boldsymbol{\sigma}} \tag{11.4.1}$$

which describes dilatancy and/or volume compressibility as well (see Chapter 8). Here the stabilization boundary for the creep and relaxation is

$$H(\sigma, \bar{\sigma}, \Delta) = W^{\mathrm{I}}(t), \tag{11.4.2}$$

with the irreversible stress-work

$$W^{\mathrm{I}}(t) = \int_0^t \boldsymbol{\sigma}(\tau) \cdot \dot{\boldsymbol{\varepsilon}}^{\mathrm{I}}(\tau)\, \mathrm{d}\tau. \tag{11.4.3}$$

The deformation of the volume will be described by

$$\dot{\varepsilon}_{\mathrm{v}} = \frac{\dot{\sigma}}{K} + k \left\langle 1 - \frac{W^{\mathrm{I}}(t)}{H(\sigma, \bar{\sigma})} \right\rangle \frac{\partial H}{\partial \sigma}, \tag{11.4.4}$$

assuming that H depends on the two invariants σ and $\bar{\sigma}$ alone.

A procedure similar to the one used for the linear viscoelastic model will again be used: a very short time interval $(0, t_0)$ is first considered, during which the well is drilled. At the end of this short time interval the 'instantaneous' stress $\boldsymbol{\sigma}^{\mathrm{S}_0}$ and strain $\boldsymbol{\varepsilon}^{\mathrm{S}_0}$ satisfy the same relations (11.3.2). Thus, the stresses, strains, and displacement at time t_0 are obtained from (11.3.3)–(11.3.5). In the long time interval (t_0, t_1) which follows, the deformation by creep of the rock around the well takes place. During this last time interval there is no reason to assume that the stresses are constant; rather, in fact, a stress variation is to be expected.

In the following, two solutions of this problem will be considered. First, a simplified solution is obtained with the assumption that in the time interval (t_0, t_1) all stresses are constant; subsequently we present a second solution in which the variation of the stress during the creep period will also be considered. In most examples we have considered $t_0 = 0$.

11.4.1. The simplified solution

Let us assume that during creep, in the time interval (t_0, t_1) all stress components are constant. Then from (11.4.1) we have the differential equation for W^{I}

$$\frac{\mathrm{d}W^{\mathrm{I}}}{1 - \dfrac{W^{\mathrm{I}}}{H}} = k \frac{\partial H}{\partial \boldsymbol{\sigma}} \cdot \boldsymbol{\sigma}\, \mathrm{d}t \tag{11.4.5}$$

in which $\boldsymbol{\sigma} = \boldsymbol{\sigma}^{\mathrm{S}_0}$ are constant. Since at time $t = t_0 = 0$ we have $W(t_0) = W^{\mathrm{IP}}$, from (11.4.5) it follows that

$$1 - \frac{W^{\mathrm{I}}(t)}{H} = \left(1 - \frac{W^{\mathrm{IP}}}{H}\right) \exp\left[\frac{k}{H}\frac{\partial H}{\partial \boldsymbol{\sigma}} \cdot \boldsymbol{\sigma}(t_0 - t)\right], \tag{11.4.6}$$

and further, for the strains we obtain

$$\boldsymbol{\varepsilon} = \boldsymbol{\varepsilon}^{\mathrm{R}_0} + \frac{\left(1 - \dfrac{W^{\mathrm{IP}}}{H}\right)\dfrac{\partial H}{\partial \boldsymbol{\sigma}}}{\dfrac{1}{H}\dfrac{\partial H}{\partial \boldsymbol{\sigma}} \cdot \boldsymbol{\sigma}} \left\{1 - \exp\left[\frac{k}{H}\frac{\partial H}{\partial \boldsymbol{\sigma}} \cdot \boldsymbol{\sigma}(t_0 - t)\right]\right\}, \tag{11.4.7}$$

if the initial conditions

$$t = t_0 : \begin{cases} \boldsymbol{\varepsilon}^{\mathrm{I}} = \mathbf{0} \\ \boldsymbol{\varepsilon} = \boldsymbol{\varepsilon}^{\mathrm{R}_0} = \left(\dfrac{1}{3K} - \dfrac{1}{2G}\right) \sigma^{\mathrm{R}_0}\,\mathbf{1} + \dfrac{1}{2G}\,\boldsymbol{\sigma}^{\mathrm{R}_0} \end{cases} \tag{11.4.8}$$

are also used. The displacement is then obtained from $u = \varepsilon_\theta r$.

Let us observe that (11.4.7) is not a rigorously obtained solution since, for instance, the compatibility condition (11.1.7) is generally not satisfied. Thus the simplified solution has to be considered as only a first estimate of the solution.

As a *numerical example* we consider the constitutive equation for granite given in Chapter 8. From the expression of the viscoplastic potential (8.5.1) it is easy to obtain all the expressions involved in (11.4.7) as $\partial H/\partial \boldsymbol{\sigma}$, $(\partial H/\partial \boldsymbol{\sigma}) \cdot \boldsymbol{\sigma}$, etc. The primary stress distribution (11.2.12) is further used in the expression of the initial secondary stress-state obtained from (11.3.3),

$$\begin{aligned} \sigma_r^{\mathrm{S}_0} &= (p - \sigma_{\mathrm{h}}) \frac{a^2}{r^2} + \sigma_{\mathrm{h}} \\ \sigma_\theta^{\mathrm{S}_0} &= -(p - \sigma_{\mathrm{h}}) \frac{a^2}{r^2} + \sigma_{\mathrm{h}} \\ \sigma_z^{\mathrm{S}_0} &= \sigma_{\mathrm{v}}. \end{aligned} \tag{11.4.9}$$

The variation of function H with distance from the orifice is shown in Fig. 11.4.1 for a depth of 700 m and two values of the pressure p. It follows that at a relatively small distance from the orifice (3—4 radii), the value of H practically coincides with that from the primary state.

The variation of $W^{\mathrm{I}}(t)$ according to formula (11.4.6) depends essentially on the value of k. For this hard rock, taking $E = 41$ GPa and the viscosity coefficient $E/k = 10^{15}$ P, it follows that $k = 2.26 \times 10^{-6}\ \mathrm{s}^{-1}$. This would be the value of k as determined from uniaxial creep tests performed in the laboratory at room temperature. However, at great depths, i.e. at high hydrostatic pressures

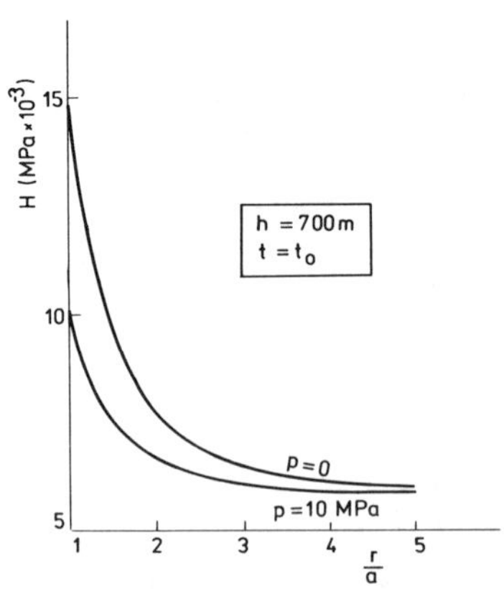

Figure 11.4.1. Variation of function H with distance from the orifice, with and without internal pressure.

and perhaps high temperatures, the value of the viscosity coefficient may differ from this by several orders of magnitude. For this reason several values for k were also considered. For $k = 2.26 \times 10^{-5}\ \mathrm{s}^{-1}$, W^{I} varies according to formula (11.4.6) during a single day only (Fig. 11.4.2), for $k = 2.26 \times 10^{-6}\ \mathrm{s}^{-1}$, the creep of the rock lasts for 10 days, for $k = 2.26 \times 10^{-7}\ \mathrm{s}^{-1}$, the creep lasts for 70 days and finally, for $k = 2.26 \times 10^{-8}\ \mathrm{s}^{-1}$, the creep may last for three years. Generally, if all the other constitutive parameters are determined in the laboratory, an appropriate value for k can be obtained simply by recording the creep of the rock around an underground opening.

Further, the variation of the strains in time is obtained from formulae (11.4.7). The strain components ε_r and ε_θ are generally small (of the order of magnitude 10^{-3} or possibly 10^{-2} at greater depths), while the component ε_z is smaller than the others by one order of magnitude. The fact that this strain component is non-zero certainly contradicts the initial hypothesis of radial displacement only. A variant of the solution was also obtained, assuming $\varepsilon_z = 0$, i.e. $\varepsilon_z^{\mathrm{E}} + \varepsilon_z^{\mathrm{I}} = 0$, and further, using the constitutive equation to determine the variation of σ_z which will ensure that $\varepsilon_z = 0$ at any time. Thus, σ_z must satisfy the differential equation

$$\frac{1}{3}\left(\frac{1}{3K} + \frac{1}{G}\right)\dot{\sigma}_z^{\mathrm{R}} = -k\left(1 - \frac{W^{\mathrm{I}}}{H}\right)\frac{\partial H}{\partial \sigma_z}. \tag{11.4.10}$$

The other stress components are still assumed to be constant during the creep period. From the expression of the derivative $\partial H/\partial \sigma_z$ it follows that there are particular cases when $\dot{\sigma}_z^{\mathrm{R}} \simeq 0$. For instance, in the neighbourhood of the compressibility/dilatancy boundary, when $\dot{\varepsilon}_{\mathrm{v}}^{\mathrm{I}} = 0$, and if $\sigma_{\mathrm{h}} \simeq \sigma_{\mathrm{v}}$, it follows that $\dot{\sigma}_z^{\mathrm{R}} \simeq 0$. Generally, however, $\dot{\sigma}_z^{\mathrm{R}} \neq 0$. In these cases the differential equation

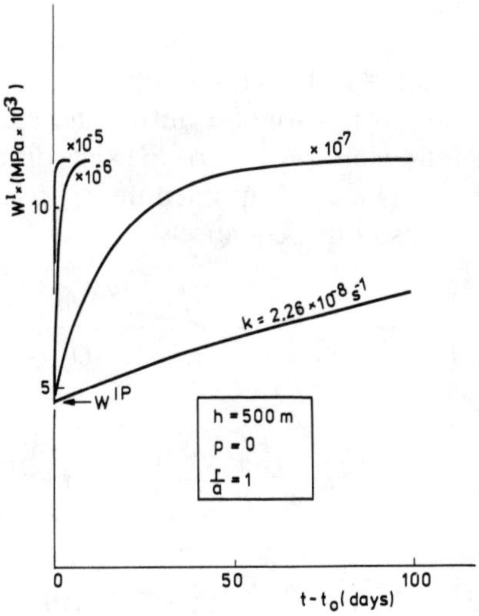

Figure 11.4.2. Variation of irreversible stress-work with time for various viscosity coefficients.

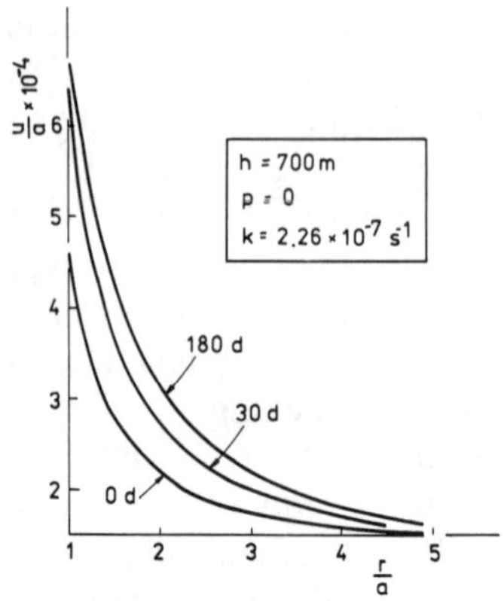

Figure 11.4.3. Distribution of radial displacement at various times after excavation.

(11.4.10) is to be integrated by numerical methods simultaneously with (11.4.1) (in which $\dot{\sigma}_z \neq 0$), with the initial condition at $t = t_0 : \sigma_z^{\mathrm{R}} = 0$. Due to this zero initial condition it is to be expected that σ_z^{R} is generally much smaller than σ_r^{R} and $\sigma_\theta^{\mathrm{R}}$. The numerical example shows that at shallow depths (less than 200 m, say) the variation of σ_z is significant. Below this depth this variation is negligible.

The radial displacement is obtained from $(11.1.5)_2$. To give an example, Fig. 11.4.3 shows the variation with the distance from the surface of the opening, of the displacement at various times for $h = 700$ m. The moment $t_0 = 0$ corresponds to the beginning of the deformation by creep, just after the 'instantaneous' elastic response of the rock. With the value of k considered, the creep lasts for about 180 days. Generally, the radial displacement is small for hard rocks.

11.4.2. The general solution

Let us now consider a numerical solution obtained without any restrictive assumption concerning the stress variation during the creep period. Therefore, we have to solve the following system of partial differential equations by numerical methods: the balance of momentum (11.1.6), the compatibility equation (11.1.7) and the constitutive equations

$$\dot{\varepsilon}_r = \left(\frac{1}{3K} - \frac{1}{2G} \right) \dot{\sigma} + \frac{1}{2G} \dot{\sigma}_r + k \left\langle 1 - \frac{W^{\mathrm{I}}}{H} \right\rangle \frac{\partial H}{\partial \sigma_r}$$

$$\dot{\varepsilon}_\theta = \left(\frac{1}{3K} - \frac{1}{2G} \right) \dot{\sigma} + \frac{1}{2G} \dot{\sigma}_\theta + k \left\langle 1 - \frac{W^{\mathrm{I}}}{H} \right\rangle \frac{\partial H}{\partial \sigma_\theta} \qquad (11.4.11)$$

$$0 = \left(\frac{1}{3K} - \frac{1}{2G} \right) \dot{\sigma} + \frac{1}{2G} \dot{\sigma}_z + k \left\langle 1 - \frac{W^{\mathrm{I}}}{H} \right\rangle \frac{\partial H}{\partial \sigma_z}.$$

This is a hyperbolic—parabolic system with five characteristic lines: $r =$ const.

(three times) and t = const. (twice). The initial conditions are formulated for $t = t_0 = 0$: $\boldsymbol{\sigma} = \boldsymbol{\sigma}^{S_0}$ according to (11.4.9), $\boldsymbol{\varepsilon} = \boldsymbol{\varepsilon}^{R_0}$ according to (11.3.4), and $u = u^{R_0}$ according to (11.3.5). The boundary conditions are formulated at the orifice contour

$$r = a, \qquad t \geqslant t_0 : \sigma_r = p \tag{11.4.12}$$

and at infinity

$$\left.\begin{array}{l} r \to \infty \\ t \geqslant t_0 \end{array}\right\} : \begin{array}{ll} \sigma_r^S = \sigma_\theta^S = \sigma_h, & \sigma_z^S = \sigma_v \\ \varepsilon_r^R = \varepsilon_\theta^R = \varepsilon_z^R = 0. & \end{array} \tag{11.4.13}$$

The finite difference method is used to integrate the system of equations in the quarter of the plane $a \leqslant r < \infty$, $t \geqslant 0$, using a mesh of characteristics, with the time interval Δt first increasing with t and then remaining constant, and space interval Δr increasing with r. Successive up-and-down sweeping along the lines t = const. ensures the matching of the boundary conditions (11.4.12) and (11.4.13). At each mesh point a variant of the 'predictor—corrector' method is used (see Appendix 2).

The *numerical example* has been done for *coal* (see (8.7.6)) of which the mathematical model was given in §8.7 (with $k = 6 \times 10^{-6}\ \text{s}^{-1}$) for the hydrostatic primary stress distribution and $\sigma_h = 1.96$ MPa ($h = 100$ m). The main *conclusions* are the following:

1. In the time interval during which the creep of the rock takes place, the stress component σ_r is undergoing a *relaxation* in the sense of a slow variation, bringing its initial value $\sigma_r^{S_0}$ slightly closer to the corresponding primary value σ_h. Figure 11.4.4 shows the distribution with the distance of the component σ_θ: the initial distribution, i.e. $\sigma_\theta^{S_0}$ as a function of distance, is shown as a dotted line, while the full line shows the distribution of σ_θ at the end of the creep

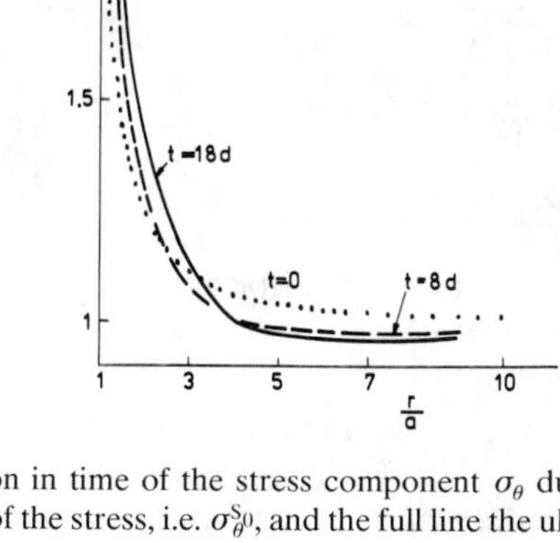

Figure 11.4.4. Slow variation in time of the stress component σ_θ during creep: dotted lines show the initial distribution of the stress, i.e. $\sigma_\theta^{S_0}$, and the full line the ultimate stress distribution.

period. Thus in the neighbourhood of the surface $r = a$, at the end of the creep period, σ_θ is higher than its initial value $\sigma_\theta^{S_0}$, but at farer distances it is lower. The maximum value of σ_θ does not take place on $r = a$, but at a certain small distance from the surface ($r = 1.1\ a$). We recall that the duration of the creep period, and therefore that of the slow stress variation, depends essentially on the value of k; for the particular value considered here, this time interval is about 18 days. In Fig. 11.4.4 the broken line corresponds to the time $t = 8$ days elapsed from excavation. The variation of σ_z during the same period is small, and for $r > 1.2\ a$, this component is close to its primary value, $\sigma_z \simeq \sigma_v$.

2. The absolute values of the *strain components* and that of the *radial displacement* are certainly *increasing by creep* in the time interval $t \in [0, t_1]$, and this increase is more significant than the one obtained with the simplified solution (11.4.7) which assumes the constancy of the stresses during this period. In order to illustrate this conclusion, Fig. 11.4.5 shows the variation with distance of the radial displacement: the dotted line is the initial distribution u^{R_0} according to formula (11.3.5), the broken line is the displacement distribution according to the simplified solution, and the full line shows the present numerical solution. Thus, the simplified solution underestimates the radial displacement near the surface $r = a$ but overestimates the radial displacement at greater distances. Again, the simplified solution must be considered merely a first approximation to a more rigorously obtained solution.

3. The irreversible deformation of the volume is obtained from

$$\varepsilon_v^I = \varepsilon_r^I + \varepsilon_\theta^I + \varepsilon_z^I, \tag{11.4.14}$$

with

$$\begin{aligned} \varepsilon_r^I &= \varepsilon_r - \varepsilon_r^E = \varepsilon_r - \left(\frac{1}{3K} - \frac{1}{2G}\right)\frac{\Delta\sigma_\theta + \Delta\sigma_z}{3} - \\ &\quad - \frac{1}{3}\left(\frac{1}{3K} + \frac{1}{G}\right)\Delta\sigma_r \\ \varepsilon_\theta^I &= \varepsilon_\theta - \varepsilon_\theta^E = \varepsilon_\theta - \left(\frac{1}{3K} - \frac{1}{2G}\right)\frac{\Delta\sigma_r + \Delta\sigma_z}{3} - \\ &\quad - \frac{1}{3}\left(\frac{1}{3K} + \frac{1}{G}\right)\Delta\sigma_\theta \\ \varepsilon_z^I &= -\varepsilon_z^E, \end{aligned} \tag{11.4.15}$$

while $\Delta\sigma_r = \sigma_r(t) - \sigma_r^{S_0}$, etc. are the variations in time of the stress components. Thus, we obtain at moment $t \in [0, t_1]$

$$\varepsilon_v^I(t) = \varepsilon_r(t) + \varepsilon_\theta(t) - \frac{1}{3K}\left[\sigma_r(t) + \sigma_\theta(t) + \sigma_z(t) - (2\sigma_h + \sigma_v)\right]. \tag{11.4.16}$$

This formula allows us to determine the domains around the well, the locations

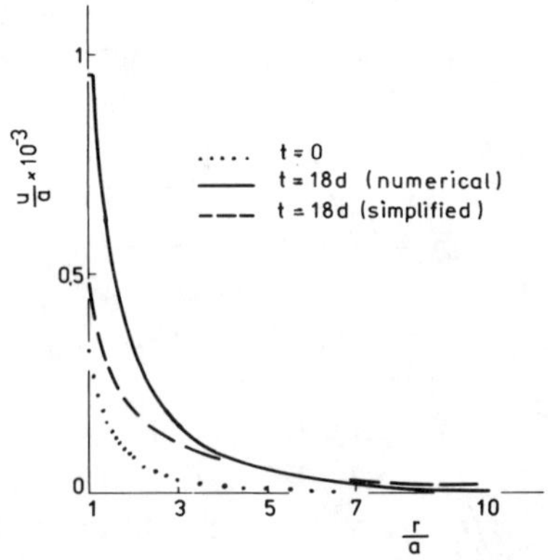

Figure 11.4.5. Distribution with distance of the radial displacement according to the numerical solution (……, initial distribution; ——— ultimate distribution) and the simplified solution (– – – –, ultimate distribution).

where the rock is in a dilatant state and where it is compressible. This subject will be discussed in the next section. Generally, just around the well, the rock is dilatant, and further away it is compressible, while at greater depths it is everywhere dilatant.

11.5. Dilatancy/compressibility and damage around a well (Cristescu (1985*a, b*))

It was shown in §11.2 and illustrated in Fig. 11.2.2 that just after the drilling of the well the value $\bar{\sigma}^{S_0}$ of the equivalent stress is much larger than its primary value $\bar{\sigma}^{P}$. Thus, it is possible to represent the secondary stress $\boldsymbol{\sigma}^{S_0}$ by a point in the dilatancy domain. For instance, for the example discussed in §11.4.1 for *granite*, and assuming the distribution of primary stresses according to (11.2.12) to be valid up to very great depths, we get the primary and secondary stress-states shown in Fig. 11.5.1, for $r = a$ and $p = 0$. The border line (— · — ·) is the compressibility/dilatancy boundary. The primary stresses are represented by circles at the depths $h = 0.5$ km; 1 km; 5 km; 9km; 11 km. The square corresponding to the same value of σ represents the initial secondary stresses at that same depth. Thus at shallow and moderate depths the initial secondary stresses lie in the compressible domain, but at very great depths they are in the dilatancy domain. At greater depths it is only at points situated at greater distances from the orifice $r = a$ that the secondary stress-state in the rock is still in the compressible domain.

Thus, if a deep well is drilled in a hard rock, at shallow depths the surrounding rock deforms elasto/viscoplastically — staying in the compressible domain, however. This is shown in Fig. 11.5.2. At greater depths there is a boundary AB separating domain D_c, where compressibility occurs, from the domain D_d, where the rock is dilatant. Thus, dilatancy appears first at point A, while at greater depths, curve AB has an asymptote $\sigma^P = \sigma_0$. This asymptote

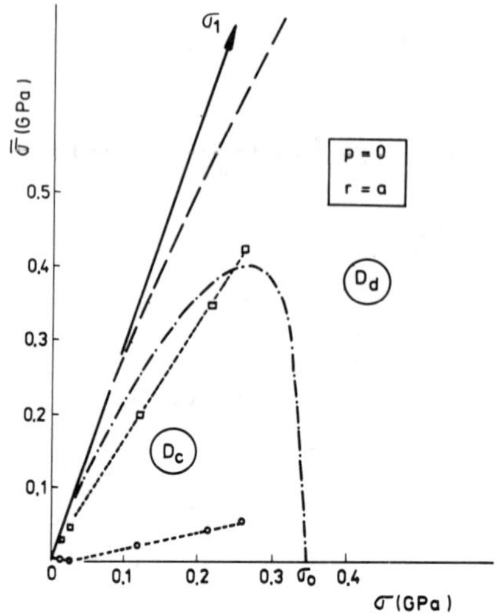

Figure 11.5.1. Primary stresses (circles) and secondary ones (squares) at various depths, showing that with increase in depth there is a tendency for the secondary state to be dilatant.

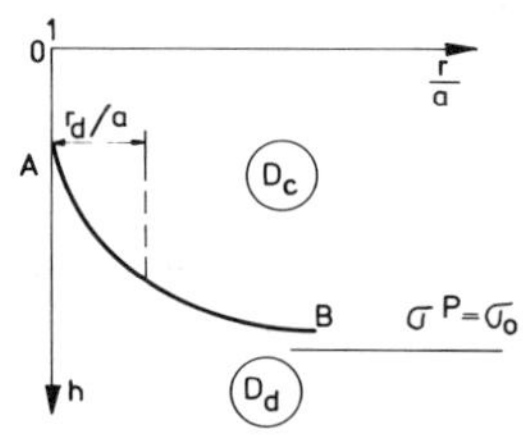

Figure 11.5.2. Boundary AB between domain of compressibility D_c and domain of dilatancy D_d, as depth increases.

corresponds to the primary mean stress equal to the one which closes all the microcracks existing in the rock. The example given above is illustrative only. An exact determination of the compressibility/dilatancy boundary around the well can be made only if the primary stresses are carefully determined for the location considered, if the variations with depth of various constitutive coefficients are also determined, etc.

Let us now give another example for a softer rock, the *andesite* considered in §8.6. The primary stresses according to (11.2.12) and the secondary stresses $\boldsymbol{\sigma}^{S_0}$ around the well are shown in Fig. 11.5.3. The primary stresses at depths 100 m; 300 m; 500 m; 1 km are shown by triangles. If the well is empty, i.e. $p = 0$, then the secondary stresses $\boldsymbol{\sigma}^{S_0}$ for $r = a$ are represented by circles. Therefore, at all depths the secondary stresses $\boldsymbol{\sigma}^{S_0}$ are situated in the dilatancy domain, and therefore the rock around the well is in a dilatant state, even at shallow depths. If, however, the well is filled with water, then the corresponding secondary stresses $\boldsymbol{\sigma}^{S_0}$ are represented by crosses: at shallow depths the rock is again dilatant, but at greater depths the rock is in a compressible state due to water pressure. In the case of a well filled with drilling mud, the inner pressure

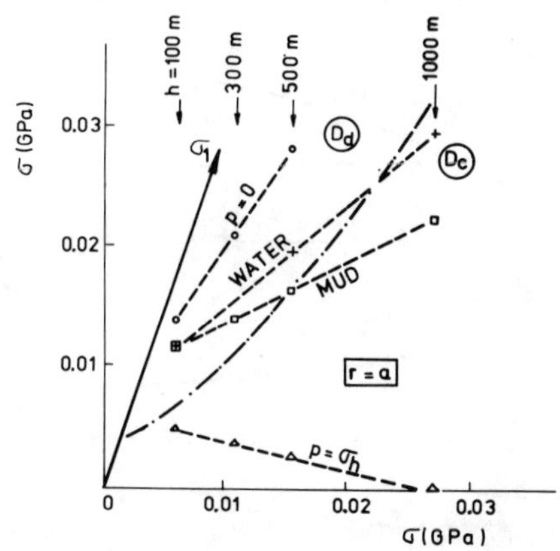

Figure 11.5.3. Primary (triangles) and secondary stresses around a well, for empty well (circles), for well filled with water (crosses), and well filled with drilling mud (squares).

increases by 14 kPa for each metre of depth (Coates (1981, p. 3.6)), and the secondary stresses $\boldsymbol{\sigma}^{S_0}$ are represented by squares. This time, for $h < 0.5$ km, the rock is dilatant around the well, and for $h > 0.5$ km it is compressible.

The secondary stress state shown in Fig. 11.5.3 is only the initial secondary stress state $\boldsymbol{\sigma}^{S_0}$ at $r = a$. If we compute ε_v^I according to the formula (11.4.16) at various distances r for depth $h = 500$ m, we get the irreversible volumetric deformation distribution shown in Fig. 11.5.4, after 6 hours (dotted line), 2.5 days (broken line), and at the end of the creep period (full line). Therefore, in the neighbourhood of the orifice $r = a$, the rock is dilatant during the whole period of deformation by creep, and the irreversible volumetric deformation ε_v^I

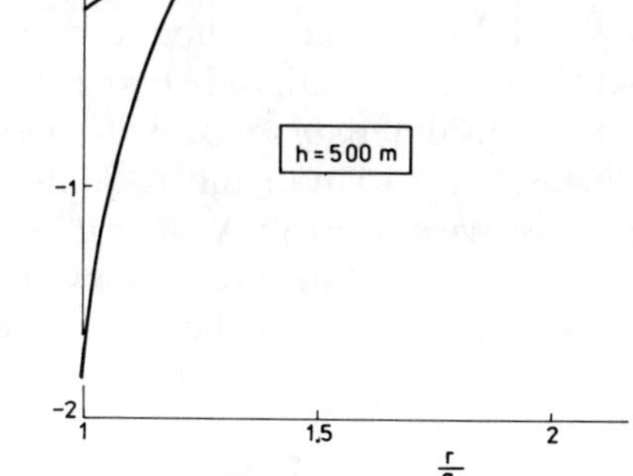

Figure 11.5.4. Evolution in time of irreversible volumetric strain around a well showing dilatancy and compressibility.

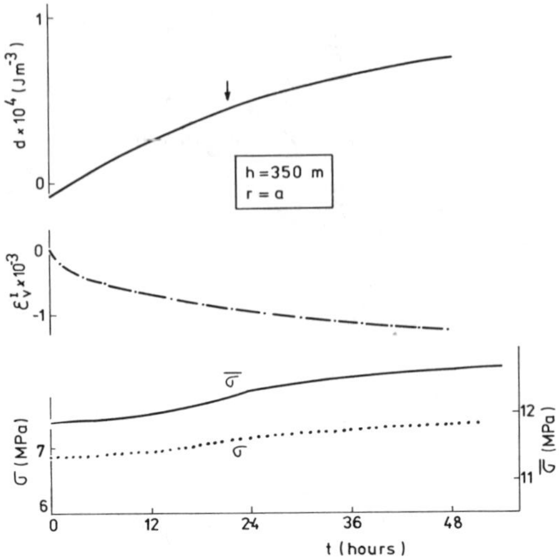

Figure 11.5.5. Variation in time at $r = a$ of damage parameter d, irreversible volumetric strain, mean stress, and equivalent stress. The vertical arrow indicates a failure limit.

continuously decreases with time. At greater distances r during the creep and stress variation period the rock is always in a compressible state, and ε_v^I continuously increases. It is easy to estimate the overall behaviour of a slice $h =$ const. of unit thickness: the greatest part of the volume of this slice is in a compressible state. This conclusion is of significance for the oil industry.

If, for coal (see (8.7.6)), the 'damage parameter' is computed during the creep and slow stress variation period according to the formula (9.3.2), then the variation of d in time is shown in Fig. 11.5.5, for $h = 350$ m and $r = a$. A vertical arrow shows when d reaches the onset of failure $d_f = 4.48 \times 10^3$ J m^{-3}, after about one day. Starting from this moment, the failure is a kind of erosion which takes place around the orifice. This happens, however, only at great depths, and/or generally, after a more or less long time has elapsed since the moment of drilling. From Fig. 11.5.5 it follows also that the volume is always dilatant and that invariants σ and $\bar{\sigma}$ both increase slightly during creep.

Also for *coal*, at $h = 100$ m, it was shown that the most important stress variation takes place at $r = 1.2$ a. In the neighbourhood of the orifice σ and $\bar{\sigma}$ both slightly increase. At the depth considered ($h = 100$ m), the rock is dilatant since for the primary case considered $\sigma_v > \sigma_0$. It is possible that at shallower depths, if $\sigma_v \simeq \sigma_0$, during stress variation the rock may pass from compressibility to dilatancy (seldom *vice versa*). A *global* estimate of the volume behaviour of a slice $h =$ const. of unit thickness shows dilatancy. Thus, if there is fluid in the pores, it will tend to be absorbed by the dilatant portion of the rock.

11.6. A more general primary stress-state (Cristescu (1985*a*, *b*))

In the previous sections the assumption was made that the horizontal stresses are equal in all directions at any depth. Field measurements show, however, that in many locations the horizontal primary stresses depend on orientation. As shown by Lindner and Halpern (1978) (see Chapter 10) in the horizontal plane there is a direction for which σ_h assumes a maximal value, and another direction for which σ_h is minimum. Similar conclusions have also been obtained by other authors (see Saxena *et al.* (1979), Abou-Sayed *et al.* (1978)). For this reason we will further assume that, depending on orientation, at each depth there is a maximal value σ_{h1} and a minimal value σ_{h2} for σ_h, and also that the components σ_{h1}, σ_{h2}, and σ_v are principal stress components. These components will be assumed to be distinct, and $\sigma_{h1} > \sigma_{h2}$.

For the computation of the stress-state around a well we can now use the formulae which are similar to those used for horizontal tunnels (see Chapter 12).

Let us consider a horizontal plane at an arbitrary depth z. When for $z =$ const. we are far enough from the opening, we must obtain the far-field stresses

$$\sigma^S_{xx}|_\infty = \sigma_{h1}, \qquad \sigma^S_{yy}|_\infty = \sigma_{h2}, \qquad \sigma^S_{zz}|_\infty = \sigma_v. \tag{11.6.1}$$

In cylindrical coordinates the components of the primary stresses (far-field stresses) are

$$\begin{aligned} \sigma^S_{rr}|_\infty &= \sigma^P_{rr} = \tfrac{1}{2}(\sigma_{h1} + \sigma_{h2}) + \tfrac{1}{2}(\sigma_{h1} - \sigma_{h2}) \cos 2\theta \\ \sigma^S_{\theta\theta}|_\infty &= \sigma^P_{\theta\theta} = \tfrac{1}{2}(\sigma_{h1} + \sigma_{h2}) - \tfrac{1}{2}(\sigma_{h1} - \sigma_{h2}) \cos 2\theta \\ \sigma^S_{r\theta}|_\infty &= \sigma^P_{r\theta} = -\tfrac{1}{2}(\sigma_{h1} - \sigma_{h2}) \sin 2\theta \\ \sigma^S_{zz}|_\infty &= \sigma^P_{zz} = \sigma_v. \end{aligned} \tag{11.6.2}$$

The relative stresses are

$$\begin{aligned} \sigma^R_{rr} &= p\frac{a^2}{r^2} - \frac{1}{2}(\sigma_{h1} + \sigma_{h2})\frac{a^2}{r^2} - \\ &\quad - \frac{1}{2}(\sigma_{h1} - \sigma_{h2})\left(\frac{4a^2}{r^2} - \frac{3a^4}{r^4}\right)\cos 2\theta \\ \sigma^R_{\theta\theta} &= -p\frac{a^2}{r^2} + \frac{1}{2}(\sigma_{h1} + \sigma_{h2})\frac{\alpha^2}{r^2} - \\ &\quad - \frac{1}{2}(\sigma_{h1} - \sigma_{h2})\frac{3a^4}{r^4}\cos 2\theta \\ \sigma^R_{r\theta} &= -\frac{1}{2}(\sigma_{h1} - \sigma_{h2})\left(\frac{2a^2}{r^2} - \frac{3a^4}{r^4}\right)\sin 2\theta \\ \sigma^R_{zz} &= -2\nu(\sigma_{h1} - \sigma_{h2})\frac{a^2}{r^2}\cos 2\theta, \end{aligned} \tag{11.6.3}$$

and the secondary stresses are

$$\sigma^{S}_{rr} = p\,\frac{a^2}{r^2} + \frac{1}{2}\,(\sigma_{h1} + \sigma_{h2})\left(1 - \frac{a^2}{r^2}\right) +$$

$$+ \frac{1}{2}\,(\sigma_{h1} - \sigma_{h2})\left(1 - \frac{4a^2}{r^2} + \frac{3a^4}{r^4}\right)\cos 2\theta$$

$$\sigma^{S}_{\theta\theta} = -p\,\frac{a^2}{r^2} + \frac{1}{2}\,(\sigma_{h1} + \sigma_{h2})\left(1 + \frac{a^2}{r^2}\right) -$$

$$- \frac{1}{2}\,(\sigma_{h1} - \sigma_{h2})\left(1 + \frac{3a^4}{r^4}\right)\cos 2\theta \qquad (11.6.4)$$

$$\sigma^{S}_{r\theta} = \frac{1}{2}\,(\sigma_{h1} - \sigma_{h2})\left(-1 - \frac{2a^2}{r^2} + \frac{3a^4}{r^4}\right)\sin 2\theta$$

$$\sigma^{S}_{zz} = \sigma_v - \nu(\sigma_{h1} - \sigma_{h2})\,\frac{2a^2}{r^2}\cos 2\theta.$$

The last formulae provide the stress-state in the rock just after the drilling of the well. Let us observe that for such stress-states the mean stress and the equivalent stress are

$$\begin{aligned}\sigma &= \tfrac{1}{3}(\sigma_{rr} + \sigma_{\theta\theta} + \sigma_{zz})\\ \bar{\sigma}^2 &= \sigma^2_{rr} + \sigma^2_{\theta\theta} + \sigma^2_{zz} - \sigma_{rr}\sigma_{\theta\theta} - \sigma_{\theta\theta}\sigma_{zz} - \sigma_{zz}\sigma_{rr} + 3\sigma^2_{r\theta},\end{aligned} \qquad (11.6.5)$$

while for the primary stress-state they are

$$\begin{aligned}\sigma^{P} &= \tfrac{1}{3}(\sigma_{h1} + \sigma_{h2} + \sigma_v)\\ (\bar{\sigma}^{P})^2 &= \left(\frac{\sigma_{h1} + \sigma_{h2}}{2} - \sigma_v\right)^2 + 3\left(\frac{\sigma_{h1} - \sigma_{h2}}{2}\right)^2,\end{aligned} \qquad (11.6.6)$$

and neither σ^{P} nor $\bar{\sigma}^{P}$ depends on the angle θ.

In order to give a numerical example, we will again use the constitutive equation for *granite* given in §8.5. The primary stress-state (11.6.2) has been obtained accepting also the formulae

$$\sigma_v = 0.027\,h, \qquad \sigma_{h2} = 5.4 + 0.012\,h, \qquad \sigma_{h1} = n\sigma_{h2} \qquad (11.6.7)$$

which are used here for the sole purpose of numerical illustration. In (11.6.7), h is given in metres, stresses in MPa, and, in order to compare various possible cases among themselves, the following integer values have been assigned to the coefficient n: $n = 1$; $n = 2$; $n = 4$; $n = 6$.

Figure 11.6.1 shows some of the results obtained for $h = 2$ km. The open circles represent the secondary stress-states on the circumference $r = a$ for $\theta =$

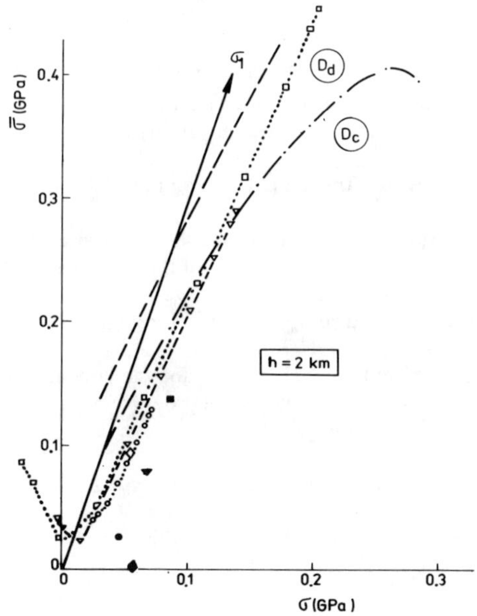

Figure 11.6.1. Primary stress-states (solid figures) and secondary ones (open figures) showing locations of the domains of tensile stress-states around the circumference of a well, and the domains of the stress-states producing dilatancy (states approaching failure).

0°, 10°, . . . , 90° respectively and $n = 2$. The extreme left-hand point corresponds to $\theta = 0°$ and the extreme right-hand point to $\theta = 90°$. The solid circle represents the primary stress-state according to formulae (11.6.7). In a similar way, the stress-states corresponding to $n = 4$ are represented by triangles, and those for $n = 6$ by squares.

The particular case when $n = 1$ and the primary stress-state is hydrostatic, $\sigma_v = \sigma_{h1} = \sigma_{h2}$ with σ_v obtained from (11.6.7), can be represented by a solid rhomb situated on the 0σ-axis. The corresponding secondary stress which is unique (not depending on θ), is represented by an open diamond. This last stress-state has the same mean stress as the corresponding primary stress-state, and it is still a point in the compressibility domain D_c. Thus in this case the rock around the well remains in a compressible state.

If steadily increasing values are assigned to n, the first regions where the rock becomes dilatant correspond to $\theta = 90°$, i.e. to those directions where σ_h is minimum. For further increased values of n the regions of rock dilatancy increase around $\theta = 90°$. For high values of n and/or great depths, in the direction in which the minimal value of σ_h ($\theta = 90°$ in the present example) acts, failure due to significant dilatancy may occur as described in Chapter 9. Failure is also possible in the neighbourhood of the circumference $r = a$ in the direction of maximal σ_h, this time by tensile stresses (in the present example for $\theta = 0°$) (see Fig. 11.6.1). These problems will be presented in detail in the next chapter.

Exercises

1. Deduce formulae (11.2.5).

2. Give a detailed derivation of formulae (11.2.7)—(11.2.11).

3. Using formulae (11.2.12)—(11.2.15) plot the variation of the stresses and strains at various depths. Consider three cases: $\sigma_h > \sigma_v$, $\sigma_h = \sigma_v$, and $\sigma_h < \sigma_v$. Compare the results.

4. Give detailed derivations of formulae (11.3.7) and (11.3.8); prove that (11.3.9)—(11.3.11) hold.

5. Using constitutive constants of your choice for the linear viscoelastic model (possibly those suggested at §11.3), make a graphical representation of the viscoelastic solution (11.3.3)—(11.3.8) for various values of p and σ_h. Discuss the case $p > \sigma_h$, $p = \sigma_h$, and $p < \sigma_h$. Then change k by one or several orders of magnitude and compare the results obtained.

6. Obtain the formulae (11.4.5)—(11.4.7).

7. Repeat the numerical example of §11.4.1, but for the constitutive equation given in §8.6. Use various values of p and h and change k by several orders of magnitude.

8. Find the characteristic lines of the partial differential system of equations (11.1.6), (11.1.7), and (11.4.11).

9. Repeat the arguments given in conjunction with Fig. 11.5.1 and Fig. 11.5.2, but for the constitutive equation for coal given in §8.7.

10. Repeat the arguments given in conjunction with Fig. 11.6.1, but for the constitutive equation given in §8.7; use various values for h.

12

Creep and dilatancy/compressibility of rocks around horizontal tunnels

12.1. Formulation of the problem

The determination of the stress, strain, and displacement distribution around a horizontal tunnel is an important engineering problem. The initial solution was obtained assuming the rock to be linearly elastic. The stress distribution was obtained by analogy with the problem of stress distribution around holes made through elastic plates (see Habib (1973)). Thus the problem was assumed to be a plane stress problem. The first solution seems to have been obtained by G. Kirsch in 1898 (for the history of the problem, see Obert and Duvall (1967)).

Other authors have assumed plane strain states, supposing however that the horizontal and vertical primary stresses are equal (Baklashov and Kartoziya (1975)); incompressible elastic or elastic/plastic models (Krenk (1978), Telles (1983, Chapter 5), Aitaliev *et al.* (1986)), strain softening elasto/plastic model (Yuritzinn *et al.* (1982)), or elastic/perfectly plastic model characterized by the Mohr—Coulomb yield criterion (Detournay and Fairhurst (1982*b*), Detournay (1985), Detournay and Fairhurst (1987)), or various kinds of linear rheological models (Borm (1983), Baoshen (1979)), or viscoplastic models (Nguyen and Berest (1979), Berest and Nguyen (1979), Nguyen *et al.* (1983) and Ottosen (1986)) have also been used. Kaiser (1979), Kaiser *et al.* (1983), and Kaiser and Morgenstern (1982) have studied the closure of the wall of a tunnel in the laboratory. Measurements *in situ* of the closure of tunnel walls have also been reported (Kaiser *et al.* (1983), Berest *et al.* (1980), etc.), as well as the cracking of the lining and the study of the stress distribution around the tunnel (Panet (1969)).

In what follows the problem will be approached employing various constitutive equations, but finally the elastic/viscoplastic constitutive equation will be used in order to describe not only deformation by creep but also dilatancy and/or compressibility of the volume of the rock as well as possible damage and failure. The rock is assumed to be an infinite body in which a cylindrical opening has been excavated. In other words, a plane strain state will be assumed. The present text follows the papers by Cristescu (1985*a, e*), Cristescu *et al.* (1985), and the results by Massier (D. Massier, personal communication).

Let us assume that the tunnel is located at a certain depth where the values of the primary horizontal and vertical stresses σ_h and σ_v, generally distinct, are known. It is also assumed that in a neighbourhood of the tunnel these components are constant and equal to their values at the depth of the tunnel axis. The influence of the earth's surface is disregarded, so the tunnel can be imagined as an infinitely long cylindrical cavity excavated in an infinite space. A cylindrical coordinate system will be chosen with the $0z$-axis coinciding with the symmetry axis of the tunnel (Fig. 12.1.1). On the surface $r = a$ of the tunnel a possible pressure may act, due to various possible causes. Therefore, the boundary conditions are

$$\left.\begin{array}{l} r = a \\ \text{any } z \end{array}\right\} : \quad \sigma^S_{rr} = p, \qquad \sigma^S_{r\theta} = 0, \tag{12.1.1}$$

and

$$\left.\begin{array}{l} r \to \infty \\ \text{any } z \end{array}\right\} : \quad \sigma^S_{xx}\big|_\infty = \sigma^S_{zz}\big|_\infty = \sigma_h, \qquad \sigma^S_{yy}\big|_\infty = \sigma_v, \tag{12.1.2}$$

with $0x$ the horizontal and $0y$ the vertical Cartesian axes.

Conditions (12.1.2) can be written in cylindrical coordinates as

$$\begin{aligned} \sigma^S_{rr}\big|_\infty &= \sigma^P_{rr} = \tfrac{1}{2}(\sigma_h + \sigma_v) + \tfrac{1}{2}(\sigma_h - \sigma_v)\cos 2\theta \\ \sigma^S_{\theta\theta}\big|_\infty &= \sigma^P_{\theta\theta} = \tfrac{1}{2}(\sigma_h + \sigma_v) - \tfrac{1}{2}(\sigma_h - \sigma_v)\cos 2\theta \\ \sigma^S_{r\theta}\big|_\infty &= \sigma^P_{r\theta} = -\tfrac{1}{2}(\sigma_h - \sigma_v)\sin 2\theta; \end{aligned} \tag{12.1.3}$$

see also Carter and Booker (1982).

Let us review the basic equations of the problem. The equilibrium equations written in relative stresses are

$$\begin{aligned} &\frac{\partial \sigma^R_{rr}}{\partial r} + \frac{1}{r}\frac{\partial \sigma^R_{r\theta}}{\partial \theta} + \frac{\partial \sigma^R_{rz}}{\partial z} + \frac{\sigma^R_{rr} - \sigma^R_{\theta\theta}}{r} = 0 \\ &\frac{\partial \sigma^R_{r\theta}}{\partial r} + \frac{1}{r}\frac{\partial \sigma^R_{\theta\theta}}{\partial \theta} + \frac{\partial \sigma^R_{z\theta}}{\partial z} + 2\,\frac{\sigma^R_{r\theta}}{r} = 0 \\ &\frac{\partial \sigma^R_{rz}}{\partial r} + \frac{1}{r}\frac{\partial \sigma^R_{\theta z}}{\partial \theta} + \frac{\partial \sigma^R_{zz}}{\partial z} + \frac{\sigma^R_{rz}}{r} = 0. \end{aligned} \tag{12.1.4}$$

The components of the small strains are

$$\varepsilon_{rr} = \frac{\partial u_r}{\partial r}, \qquad \varepsilon_{r\theta} = \frac{1}{2}\left[\frac{1}{r}\frac{\partial u_r}{\partial \theta} + \frac{\partial u_\theta}{\partial r} - \frac{u_\theta}{r}\right],$$

$$\varepsilon_{rz} = \frac{1}{2}\left(\frac{\partial u_r}{\partial z} + \frac{\partial u_z}{\partial r}\right),$$

$$\varepsilon_{\theta\theta} = \frac{1}{r}\frac{\partial u_\theta}{\partial \theta} + \frac{u_r}{r}, \qquad \varepsilon_{\theta z} = \frac{1}{2}\left(\frac{\partial u_\theta}{\partial z} + \frac{1}{r}\frac{\partial u_z}{\partial \theta}\right), \tag{12.1.5}$$

$$\varepsilon_{zz} = \frac{\partial u_z}{\partial z}.$$

The compatibility conditions in cylindrical coordinates are

$$2\frac{\partial^2(r\varepsilon_{r\theta})}{\partial r\,\partial \theta} - \frac{\partial^2 \varepsilon_{rr}}{\partial \theta^2} - r\frac{\partial^2(r\varepsilon_{\theta\theta})}{\partial r^2} + r\frac{\partial \varepsilon_{rr}}{\partial r} = 0$$

$$2r\frac{\partial^2 \varepsilon_{\theta z}}{\partial \theta\,\partial z} - r^2\frac{\partial^2 \varepsilon_{\theta\theta}}{\partial z^2} - \frac{\partial^2 \varepsilon_{zz}}{\partial \theta^2} + r\left(2\frac{\partial \varepsilon_{rz}}{\partial z} - \frac{\partial \varepsilon_{zz}}{\partial r}\right) = 0$$

$$2\frac{\partial^2 \varepsilon_{rz}}{\partial r\,\partial z} - \frac{\partial^2 \varepsilon_{zz}}{\partial r^2} - \frac{\partial^2 \varepsilon_{rr}}{\partial z^2} = 0 \tag{12.1.6}$$

$$\frac{\partial^2 \varepsilon_{rz}}{\partial \theta\,\partial r} + 2\frac{\partial \varepsilon_{r\theta}}{\partial z} + r\frac{\partial^2 \varepsilon_{r\theta}}{\partial r\,\partial z} - \frac{\partial \varepsilon_{\theta z}}{\partial r} - r\frac{\partial^2 \varepsilon_{\theta z}}{\partial r^2} -$$

$$- \frac{\partial \varepsilon_{rr}}{\partial \theta\,\partial z} - \frac{1}{r}\left(\frac{\partial \varepsilon_{rz}}{\partial \theta} - \varepsilon_{\theta z}\right) = 0$$

$$r\frac{\partial^2 \varepsilon_{r\theta}}{\partial \theta\,\partial z} + \frac{\partial \varepsilon_{\theta z}}{\partial \theta} + r\frac{\partial^2 \varepsilon_{\theta z}}{\partial r\,\partial \theta} - \frac{\partial^2 \varepsilon_{rz}}{\partial \theta^2} + r\frac{\partial \varepsilon_{rr}}{\partial z} - r\frac{\partial^2(r\varepsilon_{\theta\theta})}{\partial r\,\partial z} = 0$$

$$r\frac{\partial^2 \varepsilon_{\theta z}}{\partial r\,\partial z} + \frac{\partial^2 \varepsilon_{rz}}{\partial \theta\,\partial z} - r\frac{\partial^2 \varepsilon_{r\theta}}{\partial z^2} - \frac{\partial^2 \varepsilon_{zz}}{\partial r\,\partial \theta} -$$

$$- \frac{1}{r}\left(r\frac{\partial \varepsilon_{\theta z}}{\partial z} - \frac{\partial \varepsilon_{zz}}{\partial \theta}\right) = 0.$$

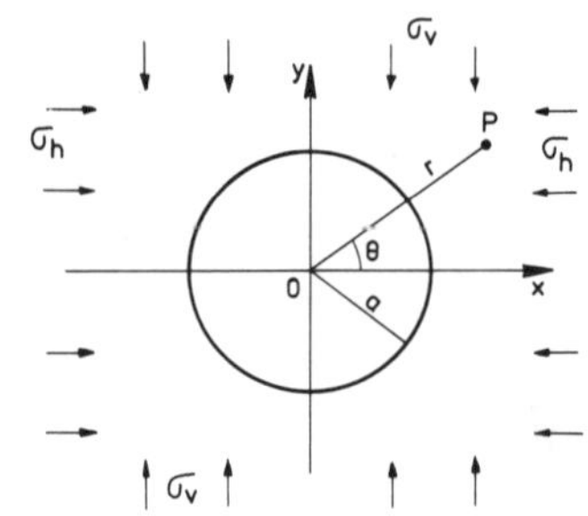

Figure 12.1.1. Scheme of a horizontal tunnel; notation used.

Due to the assumption of a *plane strain* state we have

$$\frac{\partial}{\partial z} = 0, \qquad u_z^{\mathrm{R}} = 0 \tag{12.1.7}$$

and therefore, from (12.1.5) we have, for relative strains

$$\varepsilon_{zz}^{\mathrm{R}} = \varepsilon_{rz}^{\mathrm{R}} = \varepsilon_{\theta z}^{\mathrm{R}} = 0. \tag{12.1.8}$$

12.2. The elastic approach

First let us assume that the rock satisfies a linear elastic constitutive equation written for the relative components in the form of Hooke's law

$$\boldsymbol{\varepsilon}'^{\mathrm{R}} = \frac{1}{2G}\,\boldsymbol{\sigma}'^{\mathrm{R}} \quad \text{and} \quad \sigma^{\mathrm{R}} = 3K\varepsilon^{\mathrm{R}}. \tag{12.2.1}$$

Due to (12.1.8) we also have

$$\sigma_{zz}^{\mathrm{R}} = \nu(\sigma_{rr}^{\mathrm{R}} + \sigma_{\theta\theta}^{\mathrm{R}}), \qquad \sigma_{rz}^{\mathrm{R}} = \sigma_{\theta z}^{\mathrm{R}} = 0. \tag{12.2.2}$$

With these assumptions we return to the plane strain equations discussed in Chapter 5. We can use the Airy function which satisfies the equilibrium conditions and the compatibility condition written in stresses. Starting from the general form of the Airy function and taking into account the boundary conditions of this problem, we choose a function in the form

$$\Phi(r, \theta) = A_1 r^2 + B_1 r^2 \ln r + C_1 \ln r + D_1 +$$
$$+ (A_2 r^2 + B_2 r^4 + C_2 r^{-2} + D_2) \cos 2\theta.$$

However, we set $B_1 = 0$, in order to take into account the boundary conditions in stresses at infinity. The stress components are

$$\sigma_{rr}^{S} = \frac{1}{r}\frac{\partial\Phi}{\partial r} + \frac{1}{r^2}\frac{\partial^2\Phi}{\partial\theta^2}$$

$$= \frac{C_1}{r^2} + 2A_1 + \left(-2A_2 - \frac{4D_2}{r^2} - \frac{6C_2}{r^4}\right)\cos 2\theta$$

$$\sigma_{\theta\theta}^{S} = \frac{\partial^2\Phi}{\partial r^2}$$

$$= 2A_1 - \frac{C_1}{r^2} + \left(2A_2 + 12B_2 r^2 + \frac{6C_2}{r^4}\right)\cos 2\theta \qquad (12.2.3)$$

$$\sigma_{r\theta}^{S} = \frac{1}{r^2}\frac{\partial\Phi}{\partial\theta} - \frac{1}{r}\frac{\partial^2\Phi}{\partial r\,\partial\theta}$$

$$= \left(2A_2 + 6B_2 r^2 - 2\frac{D_2}{r^2} - \frac{6C_2}{r^4}\right)\sin 2\theta.$$

The constants involved in (12.2.3) are determined from the boundary conditions (12.1.1) and (12.1.3), and certainly $B_2 = 0$. Thus we get the *secondary stress-state*

$$\sigma_{rr}^{S} = p\frac{a^2}{r^2} + \frac{1}{2}(\sigma_h + \sigma_v)\left(1 - \frac{a^2}{r^2}\right) +$$

$$+ \frac{1}{2}(\sigma_h - \sigma_v)\left(1 - \frac{4a^2}{r^2} + \frac{3a^4}{r^4}\right)\cos 2\theta$$

$$\sigma_{\theta\theta}^{S} = -p\frac{a^2}{r^2} + \frac{1}{2}(\sigma_h + \sigma_v)\left(1 + \frac{a^2}{r^2}\right) -$$

$$- \frac{1}{2}(\sigma_h - \sigma_v)\left(1 + \frac{3a^4}{r^4}\right)\cos 2\theta \qquad (12.2.4)$$

$$\sigma_{r\theta}^{S} = \frac{1}{2}(\sigma_h - \sigma_v)\left(-1 - \frac{2a^2}{r^2} + \frac{3a^4}{r^4}\right)\sin 2\theta$$

$$\sigma_{zz}^{S} = \sigma_h - \nu(\sigma_h - \sigma_v)\frac{2a^2}{r^2}\cos 2\theta.$$

The *primary* stress-state is defined by (12.1.3) and it is easy to check that

(12.1.3) follows from (12.2.4) for $r \to +\infty$. Now from (12.2.4) and (12.1.3) is obtained the *relative stress-state*

$$\sigma_{rr}^{R} = p\frac{a^2}{r^2} - \frac{1}{2}(\sigma_h + \sigma_v)\frac{a^2}{r^2} - $$

$$- \frac{1}{2}(\sigma_h - \sigma_v)\left(\frac{4a^2}{r^2} - \frac{3a^4}{r^4}\right)\cos 2\theta$$

$$\sigma_{\theta\theta}^{R} = -p\frac{a^2}{r^2} + \frac{1}{2}(\sigma_h + \sigma_v)\frac{a^2}{r^2} - \frac{1}{2}(\sigma_h - \sigma_v)\frac{3a^4}{r^4}\cos 2\theta$$

$$\sigma_{r\theta}^{R} = -\frac{1}{2}(\sigma_h - \sigma_v)\left(\frac{2a^2}{r^2} - \frac{3a^4}{r^4}\right)\sin 2\theta \tag{12.2.5}$$

$$\sigma_{zz}^{R} = -2\nu(\sigma_h - \sigma_v)\frac{a^2}{r^2}\cos 2\theta .$$

The *relative strains* are further obtained from Hooke's law with the stresses (12.2.5):

$$\varepsilon_{rr}^{R} = \frac{1+\nu}{E}\left\{p\frac{a^2}{r^2} - \frac{1}{2}(\sigma_h + \sigma_v)\frac{a^2}{r^2} - \right.$$

$$\left. - \frac{1}{2}(\sigma_h - \sigma_v)\left[4(1-\nu)\frac{a^2}{r^2} - \frac{3a^4}{r^4}\right]\cos 2\theta\right\}$$

$$\varepsilon_{\theta\theta}^{R} = \frac{1+\nu}{E}\left\{-p\frac{a^2}{r^2} + \frac{1}{2}(\sigma_h + \sigma_v)\frac{a^2}{r^2} + \right.$$

$$\left. + \frac{1}{2}(\sigma_h - \sigma_v)\left(4\nu\frac{a^2}{r^2} - \frac{3a^4}{r^4}\right)\cos 2\theta\right\} \tag{12.2.6}$$

$$\varepsilon_{r\theta}^{R} = \frac{1+\nu}{E}\,\frac{1}{2}(\sigma_h - \sigma_v)\left(-\frac{2a^2}{r^2} + \frac{3a^4}{r^4}\right)\sin 2\theta .$$

It is obvious that all these components tend towards zero if $r \to \infty$.

The relative displacements u_r^R and u_θ^R are obtained by integration from $(12.1.5)_{1,4}$, using $(12.2.6)_{1,2}$. We obtain

$$u_r^R = \frac{1+\nu}{E}\left\{-p\,\frac{a^2}{r} + \frac{1}{2}(\sigma_h + \sigma_v)\frac{a^2}{r} + \right.$$

$$\left. + \frac{1}{2}(\sigma_h - \sigma_v)\left[4(1-\nu)\frac{a^2}{r} - \frac{a^4}{r^3}\right]\cos 2\theta\right\} + \varphi'(\theta)$$

$$u_\theta^R = -\frac{1+\nu}{E}\,\frac{1}{2}(\sigma_h - \sigma_v)\left[2(1-2\nu)\frac{a^2}{r} + \frac{a^4}{r^3}\right]\sin 2\theta - \qquad (12.2.7)$$

$$-\varphi(\theta) + \psi(r).$$

These components must satisfy $(12.1.5)_2$ when $(12.2.6)_3$ is also used; thus we obtain

$$\varphi(\theta) = A + a_1 \cos\theta + a_2 \sin\theta \qquad (12.2.8)$$

$$\psi(r) = A + a_3 r.$$

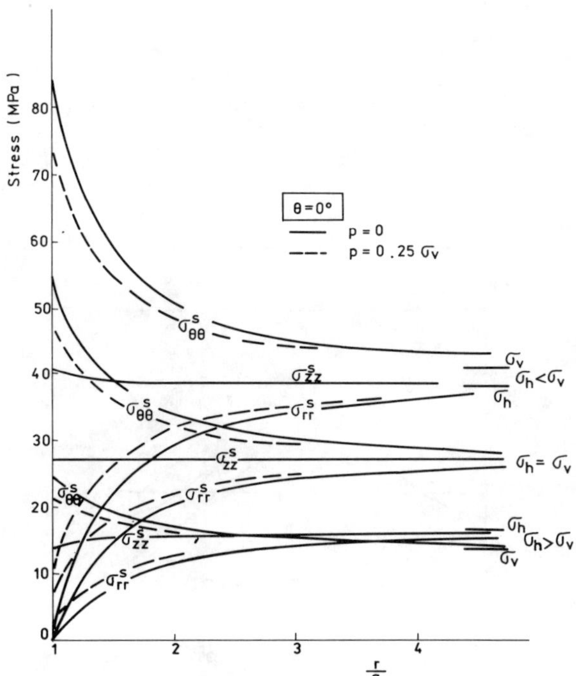

Figure 12.2.1. Secondary stress distribution in the vicinity of a horizontal tunnel for $\theta = 0°$ and three possible primary stress-states (with or without internal pressure).

Since $\mathbf{u} \to \mathbf{0}$ if $r \to \infty$, we get $A = a_1 = a_2 = a_3 = 0$ and thus the *relative displacements* are

$$u_r^{\mathrm{R}} = \frac{1+\nu}{E}\left\{ p\,\frac{a^2}{r} + \frac{1}{2}(\sigma_{\mathrm{h}} + \sigma_{\mathrm{v}})\frac{a^2}{r} + \right.$$

$$\left. + \frac{1}{2}(\sigma_{\mathrm{h}} - \sigma_{\mathrm{v}})\left[4(1-\nu)\frac{a^2}{r} - \frac{a^4}{r^3}\right]\cos 2\theta\right\}$$

$$u_\theta^{\mathrm{R}} = -\frac{1+\nu}{E}\,\frac{1}{2}(\sigma_{\mathrm{h}} - \sigma_{\mathrm{v}})\left[2(1-2\nu)\frac{a^2}{r} + \frac{a^4}{r^3}\right]\sin 2\theta. \tag{12.2.9}$$

These are the displacements with respect to the reference configuration (the configuration before the excavation of the tunnel), i.e. with respect to the primary state. It is obvious that $u_r^{\mathrm{R}} \to 0$ and $u_\theta^{\mathrm{R}} \to 0$ if $r \to \infty$. We recall that $u_z^{\mathrm{R}} = 0$.

The main features of the solution just described are illustrated in several figures drawn for $\nu = 0.2$, $E = 5 \times 10^4$ MPa and $\sigma_{\mathrm{v}} = 0.027h$ and $\sigma_{\mathrm{h}} = 5.4 + 0.0216h$. For instance, in order to illustrate the three possible cases $\sigma_{\mathrm{h}} > \sigma_{\mathrm{v}}$,

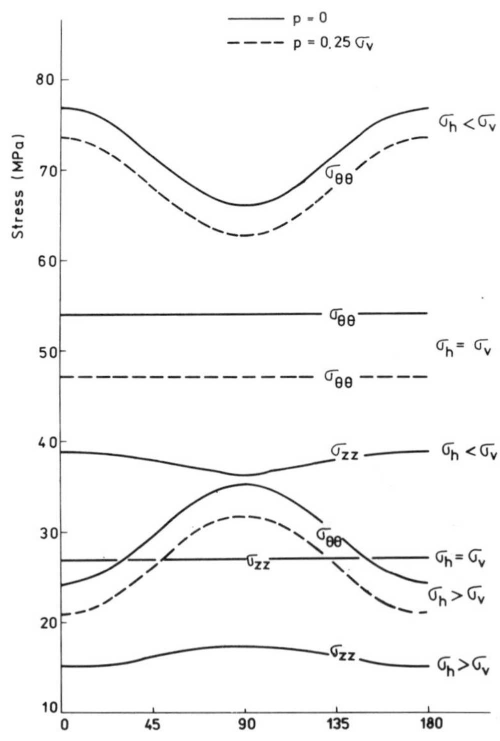

Figure 12.2.2. Variation of the stress components $\sigma_{\theta\theta}$ and σ_{zz} along the tunnel circumference $r = a$ for three possible primary stress-states, with or without internal pressure.

$\sigma_h = \sigma_v$, and $\sigma_h < \sigma_v$, the variation of the components of the secondary stresses at $h = 500$ m, 1000 m, and 1500 m is shown in Fig. 12.2.1 for $\theta = 0°$ and $p = 0$. All components are compressive and the horizontal asymptotes mark the level of the corresponding primary stresses. In order to show a possible influence of the internal pressure, the solutions obtained with $p = 0.25\,\sigma_v$ are represented by broken lines. Similar figures are obtained for $\theta = \frac{1}{2}\pi$.

The variation with θ of the components $\sigma^S_{\theta\theta}$ and σ^S_{zz} along the circumference $r = a$ is shown in Fig. 12.2.2; for $\sigma_h = \sigma_v$ the two components remain constant. The presence of an internal pressure decreases the component $\sigma^S_{\theta\theta}$, but certainly does not affect σ^S_{zz}. If $\sigma_h > \sigma_v$, the maximal value of $\sigma^S_{\theta\theta}$ is obtained for $\theta = \frac{1}{2}\pi$ and $\theta = \frac{3}{2}\pi$ and the minimal one for $\theta = 0°$ and $\theta = \pi$. The situation is reversed if $\sigma_h < \sigma_v$. Similar comments apply for σ^S_{zz}. The variation of $\sigma^S_{r\theta}$ reaches its maximal value for $r = \sqrt{3}a$ and $\theta = (2k + 1)\frac{1}{4}\pi$, with $k = 0, 1, 2, \ldots$

The variation with r of the displacement u^R_r for $\theta = 0$ is shown in Fig. 12.2.3 for both $p = 0$ and $p \neq 0$. The component u^R_r tends quite quickly toward zero when r increases. The component u^R_θ is zero for $\theta = 0$ and $\theta = \frac{1}{2}\pi$, and its variation with r for $\theta = \frac{1}{4}\pi$ is given in Fig. 12.2.4.

Due to excavation, the circumference $r = a$ in the primary state becomes an ellipse after excavation. If $\sigma_h > \sigma_v$, the horizontal diameter of the ellipse is smaller than the vertical one. If $\sigma_h < \sigma_v$, the situation is reversed. Finally, if

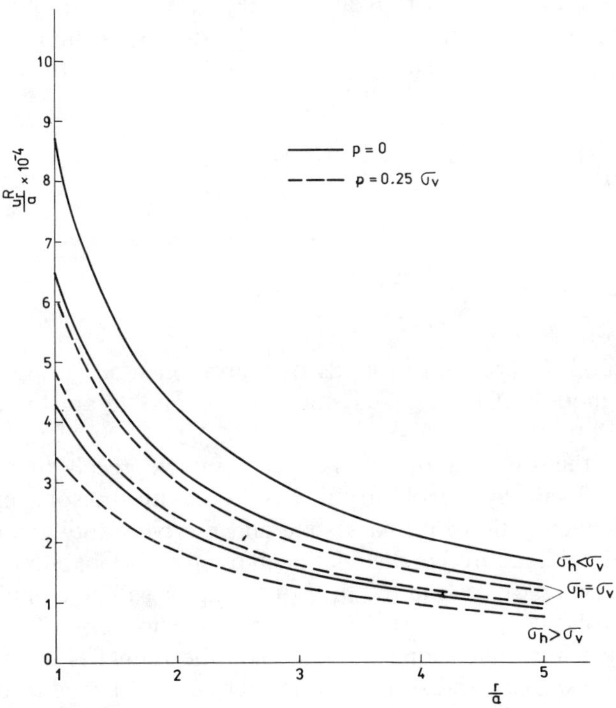

Figure 12.2.3. Distribution of the radial displacement with distance from the tunnel surface for $\theta = 0°$ and three possible primary stress-states (with or without internal pressure).

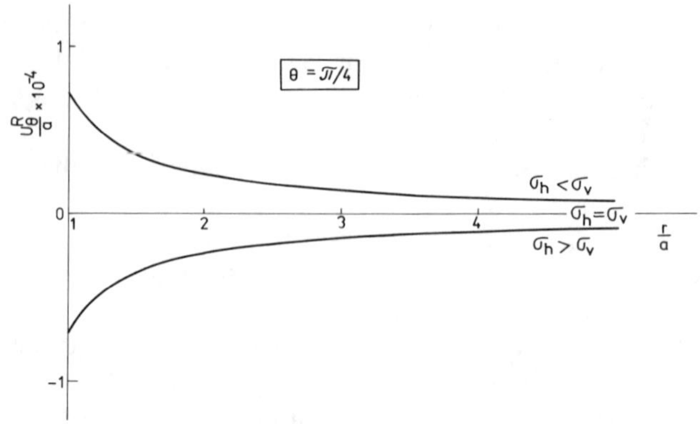

Figure 12.2.4. Circumferential displacement distribution with distance from the tunnel surface, for $\theta = \frac{1}{4}\pi$ and three possible primary stress-states.

$\sigma_h = \sigma_v$, after excavation the opening remains circular with a smaller radius. Generally, the elastic closure of the walls is relatively small for hard rocks.

12.3. Creep around a tunnel according to a linear viscoelastic model

Let us now consider the determination of the stress, strain, and displacement fields assuming that a linear viscoelastic model describes the mechanical properties of some rocks (D. Massier, personal communication). In relative variables we can write

$$\dot{\boldsymbol{\varepsilon}}'^{R} = -k\left(\boldsymbol{\varepsilon}'^{R} - \frac{1}{2G_0}\boldsymbol{\sigma}'^{R}\right) + \frac{1}{2G}\dot{\boldsymbol{\sigma}}'^{R}$$

$$\dot{\varepsilon}^{R} = -k_v\left(\varepsilon^{R} - \frac{1}{3K_0}\sigma^{R}\right) + \frac{1}{3K}\dot{\sigma}^{R}, \tag{12.3.1}$$

where $1/k$ and $1/k_v$ are certain kinds of relaxation times, G_0 and K_0 are the characteristic moduli of the relaxed state, and G and K are the moduli of the instantaneous response (see §5.4).

Concerning the formulation of the problem, we recall that the relations (12.1.1)–(12.1.3) are applicable in this case too, with the sole remark that the parameters characteristic to primary state and the constitutive constants of the model are inter-related through the relationships given in Chapter 10.

A procedure already used in the previous chapter will be employed here too, i.e. we distinguish two time intervals: in a relatively short one $(0, t_0)$ the tunnel is excavated, while in a much longer one (t_0, t_1) the tunnel is in use. In the first time interval the stresses and strains around the tunnel change from the primary state into the initial (i.e. at time t_0) secondary state, according to the instantaneous response of the model (12.3.1):

$$\boldsymbol{\sigma}'^{R_0} = 2G\boldsymbol{\varepsilon}'^{R_0}, \qquad \sigma^{R_0} = 3K\varepsilon^{R_0}. \tag{12.3.2}$$

Due to the boundary conditions and the shape of the domain involved, a plane strain state is again assumed. The relations (12.3.2) are similar to Hooke's law (12.2.1) and therefore the initial relative stress-state is obtained from formulae similar to (12.2.5):

$$\sigma_{rr}^{R_0} = \left(p - \frac{1}{2}(\sigma_h + \sigma_v)\right)\frac{a^2}{r^2} - \frac{1}{2}(\sigma_h - \sigma_v)\left(\frac{4a^2}{r^2} - \frac{3a^4}{r^4}\right)\cos 2\theta$$

$$\sigma_{\theta\theta}^{R_0} = \left(-p + \frac{1}{2}(\sigma_h + \sigma_v)\right)\frac{a^2}{r^2} - \frac{1}{2}(\sigma_h - \sigma_v)\frac{3a^4}{r^4}\cos 2\theta$$

$$\sigma_{r\theta}^{R_0} = \frac{1}{2}(\sigma_h - \sigma_v)\left(-\frac{2a^2}{r^2} + \frac{3a^4}{r^4}\right)\sin 2\theta \qquad (12.3.3)$$

$$\sigma_{zz}^{R_0} = -\frac{3K - 2G}{6K + 2G}\,\frac{1}{2}(\sigma_h - \sigma_v)\frac{4a^2}{r^2}\cos 2\theta$$

$$\sigma^{R_0} = -\frac{3K}{6K + 2G}\,\frac{1}{2}(\sigma_h - \sigma_v)\frac{4a^2}{r^2}\cos 2\theta.$$

The corresponding relative strains are obtained from (12.3.2) and are similar to (12.2.6), i.e.

$$\varepsilon_{rr}^{R_0} = \frac{1}{2G}\left[\left(p - \frac{1}{2}(\sigma_h + \sigma_v)\right)\frac{a^2}{r^2} + \right.$$

$$\left. + \frac{1}{2}(\sigma_h - \sigma_v)\left(-\frac{3K + 4G}{6K + 2G}\,\frac{4a^2}{r^2} + \frac{3a^4}{r^4}\right)\cos 2\theta\right]$$

$$\varepsilon_{\theta\theta}^{R_0} = \frac{1}{2G}\left[\left(-p + \frac{1}{2}(\sigma_h + \sigma_v)\right)\frac{a^2}{r^2} - \right.$$

$$\left. - \frac{1}{2}(\sigma_h - \sigma_v)\left(-\frac{3K - 2G}{6K + 2G}\,\frac{4a^2}{r^2} + \frac{3a^4}{r^4}\right)\cos 2\theta\right] \qquad (12.3.4)$$

$$\varepsilon_{r\theta}^{R_0} = \frac{1}{2G}\,\frac{1}{2}(\sigma_h - \sigma_v)\left(-2\frac{a^2}{r^2} + \frac{3a^4}{r^4}\right)\sin 2\theta$$

$$\varepsilon_{zz}^{R_0} = 0.$$

From the latter formulae the relative displacements at time t_0 are also obtained as

$$u_r^{R_0} = \frac{1}{2G}\left[\left(-p + \frac{1}{2}(\sigma_h + \sigma_v)\right)\frac{a^2}{r} - \right.$$

$$\left. - \frac{1}{2}(\sigma_h - \sigma_v)\left(-\frac{3K+4G}{6K+2G}\frac{4a^2}{r} + \frac{a^4}{r^3}\right)\cos 2\theta\right]$$

$$u_\theta^{R_0} = -\frac{1}{2G}\frac{1}{2}(\sigma_h - \sigma_v)\left(\frac{3G}{6K+2G}\frac{4a^2}{r} + \frac{a^4}{r^3}\right)\sin 2\theta. \tag{12.3.5}$$

In the long time interval (t_0, t_1) which follows, deformation by creep takes place. During this time period the stress can be obtained from an Airy-type function. The procedure makes use of the Laplace transform

$$\tilde{f}(\lambda) = \int_0^\infty f(t')\exp(-\lambda t')\,\mathrm{d}t'$$

with $t' = t/t_*$ a dimensionless parameter. Applying the Laplace transform to the constitutive equation (12.3.1) and using the initial conditions (12.3.2) we get

$$\tilde{\boldsymbol{\varepsilon}}'^{R} = \frac{\dfrac{1}{2G}\dfrac{\lambda}{t_*} + \dfrac{k}{2G_0}}{\dfrac{\lambda}{t_*} + k}\,\tilde{\boldsymbol{\sigma}}'^{R}, \qquad \tilde{\varepsilon}^{R} = \frac{\dfrac{1}{3K}\dfrac{\lambda}{t_*} + \dfrac{k_v}{3K_0}}{\dfrac{\lambda}{t_*} + k_v}\,\tilde{\sigma}^{R}. \tag{12.3.6}$$

Using these relations, the representation

$$\tilde{\sigma}_{rr} = \frac{1}{r}\frac{\partial\tilde{\Phi}}{\partial r} + \frac{1}{r^2}\frac{\partial^2\tilde{\Phi}}{\partial\theta^2}, \qquad \tilde{\sigma}_{\theta\theta} = \frac{\partial^2\tilde{\Phi}}{\partial r^2},$$

$$\tilde{\sigma}_{r\theta} = \frac{1}{r^2}\frac{\partial\tilde{\Phi}}{\partial\theta} - \frac{1}{r}\frac{\partial^2\tilde{\Phi}}{\partial r\,\partial\theta}, \tag{12.3.7}$$

with Φ the Airy function, and the compatibility conditions where the Laplace transform has also been applied, we get for the transform of the Airy function a biharmonic equation

$$\Delta\Delta\tilde{\Phi} = 0. \tag{12.3.8}$$

It follows that the Airy function itself is biharmonic. The boundary conditions (12.1.3) and (12.1.1) suggest choosing the Airy function in the form (12.2.3) in which the coefficients would be the same as for the elastic model (the boundary conditions are the same and are not time-dependent). Thus the relative components σ_{rr}^{R}, $\sigma_{\theta\theta}^{\mathrm{R}}$, $\sigma_{r\theta}^{\mathrm{R}}$ remain constant in the interval (t_0, t_1) and only σ_{zz}^{R} may possibly vary in time:

$$\sigma_{rr}^{\mathrm{S}} = \sigma_{rr}^{\mathrm{S}_0}, \qquad \sigma_{\theta\theta}^{\mathrm{S}} = \sigma_{\theta\theta}^{\mathrm{S}_0}, \qquad \sigma_{r\theta}^{\mathrm{S}} = \sigma_{r\theta}^{\mathrm{S}_0}. \tag{12.3.9}$$

In order to obtain the variation in time of σ_{zz}^{R} and that of the strain components, we start from the constitutive equations

$$\begin{aligned}
\dot{\varepsilon}_{rr}^{\prime\mathrm{R}} &= -k\left[\varepsilon_{rr}^{\prime\mathrm{R}} - \frac{1}{4G_0}\left(\sigma_{rr}^{\mathrm{R}_0} - \sigma_{\theta\theta}^{\mathrm{R}_0} - \sigma_{zz}^{\prime\mathrm{R}}\right)\right] - \frac{1}{4G}\dot{\sigma}_{zz}^{\prime\mathrm{R}}\\
\dot{\varepsilon}_{\theta\theta}^{\prime\mathrm{R}} &= -k\left[\varepsilon_{\theta\theta}^{\prime\mathrm{R}} - \frac{1}{4G_0}\left(\sigma_{\theta\theta}^{\mathrm{R}_0} - \sigma_{rr}^{\mathrm{R}_0} - \sigma_{zz}^{\prime\mathrm{R}}\right)\right] - \frac{1}{4G}\dot{\sigma}_{zz}^{\prime\mathrm{R}}\\
\dot{\varepsilon}_{r\theta}^{\prime\mathrm{R}} &= -k\left[\varepsilon_{r\theta}^{\mathrm{R}} - \frac{1}{2G_0}\sigma_{r\theta}^{\mathrm{R}}\right]\\
-\dot{\varepsilon}_{zz}^{\prime\mathrm{R}} &= \dot{\varepsilon}^{\mathrm{R}} = -k\left[\varepsilon^{\mathrm{R}} + \frac{1}{2G_0}\sigma_{zz}^{\prime\mathrm{R}}\right] - \frac{1}{2G}\dot{\sigma}_{zz}^{\prime\mathrm{R}}\\
\dot{\varepsilon}^{\mathrm{R}} &= -k_{\mathrm{v}}\left[\varepsilon^{\mathrm{R}} - \frac{1}{6G_0}\left(\sigma_{rr}^{\mathrm{R}_0} + \sigma_{\theta\theta}^{\mathrm{R}_0} + \sigma_{zz}^{\prime\mathrm{R}}\right)\right] + \frac{1}{6K}\dot{\sigma}_{zz}^{\prime\mathrm{R}}.
\end{aligned} \tag{12.3.10}$$

From $(12.3.10)_{4,5}$, by separation of variables, we obtain the differential equations for ε and σ_{zz}:

$$\left(\frac{1}{G} + \frac{1}{3K}\right)\ddot{\varepsilon}^{\mathrm{R}} + \left[k\left(\frac{1}{3K} + \frac{1}{G_0}\right) + k_{\mathrm{v}}\left(\frac{1}{3K_0} + \frac{1}{G}\right)\right]\dot{\varepsilon}^{\mathrm{R}} +$$

$$+ kk_{\mathrm{v}}\left(\frac{1}{3K_0} + \frac{1}{G_0}\right)\varepsilon^{\mathrm{R}} = \frac{kk_{\mathrm{v}}}{6K_0G_0}\left(\sigma_{rr}^{\mathrm{R}_0} + \sigma_{\theta\theta}^{\mathrm{R}_0}\right)$$

$$\left(\frac{1}{G} + \frac{1}{3K}\right)\ddot{\sigma}_{zz}^{\prime\mathrm{R}} + \left[k\left(\frac{1}{3K} + \frac{1}{G_0}\right) + k_{\mathrm{v}}\left(\frac{1}{3K_0} + \frac{1}{G}\right)\right]\dot{\sigma}_{zz}^{\prime\mathrm{R}} +$$

$$+ kk_{\mathrm{v}}\left(\frac{1}{3K_0} + \frac{1}{G_0}\right)\sigma_{zz}^{\prime\mathrm{R}} = -\frac{kk_{\mathrm{v}}}{3K_0}\left(\sigma_{rr}^{\mathrm{R}_0} + \sigma_{\theta\theta}^{\mathrm{R}_0}\right)$$

with the solutions of the form

$$\sigma_{zz}^{\prime\mathrm{R}} = a_1\exp(\lambda_1 t) + a_2\exp(\lambda_2 t) - \frac{G_0}{G_0 + 3K_0}\left(\sigma_{rr}^{\mathrm{R}_0} + \sigma_{\theta\theta}^{\mathrm{R}_0}\right)$$

$$\varepsilon^{\mathrm{R}} = b_1\exp(\lambda_1 t) + b_2\exp(\lambda_2 t) + \frac{1}{2G_0 + 6K_0}\left(\sigma_{rr}^{\mathrm{R}_0} + \sigma_{\theta\theta}^{\mathrm{R}_0}\right).$$

Replacing them into $(12.3.10)_{4,5}$ we get the relations between the constants involved as

$$\left(\frac{1}{6K}+\frac{1}{2G}\right)a_1\lambda_1=-(k-k_v)b_1-\left(\frac{k}{2G_0}+\frac{k_v}{6K_0}\right)a_1$$

$$\left(\frac{1}{6K}+\frac{1}{2G}\right)a_2\lambda_2=-(k-k_v)b_2-\left(\frac{k}{2G_0}+\frac{k_v}{6K_0}\right)a_2.$$

Introducing the solutions for σ'^{R}_{rr} and ε^{R} into $(12.3.10)_{1-3}$, we can integrate these equations with the initial conditions (12.3.4) and $(12.3.3)_4$ to obtain

$$\varepsilon^{R}_{rr}=\frac{1}{2}\left[\left(\frac{1}{2G}-\frac{1}{2G_0}\right)\exp(-kt)+\frac{1}{2G_0}\right](\sigma^{R_0}_{rr}-\sigma^{R_0}_{\theta\theta})+$$

$$+\frac{3}{2}\left[B_1\exp(\lambda_1 t)+B_2\exp(\lambda_2 t)+\frac{1}{2G_0+6K_0}\right](\sigma^{R_0}_{rr}+\sigma^{R_0}_{\theta\theta})$$

$$\varepsilon^{R}_{\theta\theta}=-\frac{1}{2}\left[\left(\frac{1}{2G}-\frac{1}{2G_0}\right)\exp(-kt)+\frac{1}{2G_0}\right](\sigma^{R_0}_{rr}-\sigma^{R_0}_{\theta\theta})+$$

$$+\frac{3}{2}\left[B_1\exp(\lambda_1 t)+B_2\exp(\lambda_2 t)+\frac{1}{2G_0+6K_0}\right](\sigma^{R_0}_{rr}+\sigma^{R_0}_{\theta\theta})$$

(12.3.11)

$$\varepsilon^{R}_{r\theta}=\left[\left(\frac{1}{2G}-\frac{1}{2G_0}\right)\exp(-kt)+\frac{1}{2G_0}\right]\sigma^{R_0}_{r\theta}$$

$$\sigma^{R}_{zz}=\left\{\frac{3K-2G}{6K+2G}-\frac{3}{2}A_1[1-\exp(\lambda_1 t)]-\right.$$

$$\left.-\frac{3}{2}A_2[1-\exp(\lambda_2 t)]\right\}(\sigma^{R_0}_{rr}+\sigma^{R_0}_{\theta\theta})$$

where

$$A_1 = \frac{1}{\lambda_1 - \lambda_2}\left\{\left(\frac{G}{G+3K} - \frac{G_0}{G_0+3K_0}\right)\lambda_2 + \right.$$

$$\left. + \frac{3KG}{(G+3K)^2}\left[\left(\frac{G}{G_0} - 1\right)k - \left(\frac{K}{K_0} - 1\right)k_v\right]\right\}$$

$$A_2 = -\frac{1}{\lambda_1 - \lambda_2}\left\{\left(\frac{G}{G+3K} - \frac{G_0}{G_0+3K_0}\right)\lambda_1 + \right.$$

$$\left. + \frac{3KG}{(G+3K)^2}\left[\left(\frac{G}{G_0} - 1\right)k - \left(\frac{K}{K_0} - 1\right)k_v\right]\right\}$$

$$(12.3.12)$$

$$B_1 = -\frac{1}{k-k_v}A_1\left[\left(\frac{1}{2G} + \frac{1}{6K}\right)\lambda_1 + \left(\frac{k}{2G_0} + \frac{k_v}{6K_0}\right)\right]$$

$$B_2 = -\frac{1}{k-k_v}A_2\left[\left(\frac{1}{2G} + \frac{1}{6K}\right)\lambda_2 + \left(\frac{k}{2G_0} + \frac{k_v}{6K_0}\right)\right],$$

while λ_1 and λ_2 are the two real negative distinct solutions of the algebraic equation

$$\left(\frac{1}{G} + \frac{1}{3K}\right)\lambda^2 + \left[k\left(\frac{1}{3K} + \frac{1}{G_0}\right) + k_v\left(\frac{1}{3K_0} + \frac{1}{G}\right)\right]\lambda +$$

$$+ kk_v\left(\frac{1}{G_0} + \frac{1}{3K_0}\right) = 0. \qquad (12.3.13)$$

By integration of equations (12.3.11) with the initial conditions (12.3.5) we obtain the displacements

$$u_r^{\mathrm{R}} = u_r^{\mathrm{R}_0} - \left(\frac{1}{2G} - \frac{1}{2G_0}\right)[1-\exp(-kt)]\left[-\left(p - \frac{\sigma_{\mathrm{h}}+\sigma_{\mathrm{v}}}{2}\right)\frac{a^2}{r} + \right.$$

$$\left. + \frac{\sigma_{\mathrm{h}}-\sigma_{\mathrm{v}}}{2}\left(\frac{2a^2}{r} - \frac{a^4}{r^3}\right)\cos 2\theta\right] - \{B_1[1-\exp(\lambda_1 t)] +$$

$$+ B_2[1-\exp(\lambda_2 t)]\}\frac{\sigma_{\mathrm{h}}-\sigma_{\mathrm{v}}}{2}\frac{6a^2}{r}\cos 2\theta$$

$$(12.3.14)$$

$$u_\theta^{\mathrm{R}} = u_\theta^{\mathrm{R}_0} + \left(\frac{1}{2G} - \frac{1}{2G_0}\right)[1-\exp(-kt)]\frac{1}{2}(\sigma_{\mathrm{h}}-\sigma_{\mathrm{v}})\frac{a^4}{r^3}\sin 2\theta +$$

$$+ \{B_1[1-\exp(\lambda_1 t)] + B_2[1-\exp(\lambda_2 t)]\}\frac{1}{2}(\sigma_{\mathrm{h}}-\sigma_{\mathrm{v}})\frac{6a^2}{r}\sin 2\theta.$$

The ultimate values of these displacements are obtained for $t_1 \to \infty$, i.e.

$$\lim_{t_1\to\infty} u_r^{\mathrm{R}} = \frac{1}{2G_0}\left\{-\left(p - \frac{\sigma_{\mathrm{h}}+\sigma_{\mathrm{v}}}{2}\right)\frac{a^2}{r} + \right.$$

$$\left. + \frac{\sigma_{\mathrm{h}}-\sigma_{\mathrm{v}}}{2}\left[2\frac{3K_0+4G_0}{3K_0+G_0}\frac{a^2}{r} - \frac{a^4}{r^3}\right]\cos 2\theta\right\}$$

$$(12.3.15)$$

$$\lim_{t_1\to\infty} u_\theta^{\mathrm{R}} = -\frac{1}{2G_0}\frac{\sigma_{\mathrm{h}}-\sigma_{\mathrm{v}}}{2}\left[\frac{6G_0}{3K_0+G_0}\frac{a^2}{r} + \frac{a^4}{r^3}\right]\sin 2\theta.$$

It is also easy to check that

$$\lim_{r\to\infty} u_r^{\mathrm{R}} = 0, \qquad \lim_{r\to\infty} u_\theta^{\mathrm{R}} = 0 \tag{12.3.16}$$

as expected. The ultimate strains are obtained from $(12.3.11)_{1-3}$ as

$$\lim_{t_1\to\infty} \varepsilon_{rr}^{\mathrm{R}} = \frac{1}{2G_0}\sigma_{rr}^{\mathrm{R}_0} + \left(\frac{1}{3K_0} - \frac{1}{2G_0}\right)\sigma^{\mathrm{R}_0}$$

$$\lim_{t_1\to\infty} \varepsilon_{\theta\theta}^{\mathrm{R}} = \frac{1}{2G_0}\sigma_{\theta\theta}^{\mathrm{R}_0} + \left(\frac{1}{3K_0} - \frac{1}{2G_0}\right)\sigma^{\mathrm{R}_0} \tag{12.3.17}$$

$$\lim_{t_1\to\infty} \varepsilon_{r\theta}^{\mathrm{R}} = \frac{1}{2G_0}\sigma_{r\theta}^{\mathrm{R}_0}.$$

We now give a numerical example for *andesite* and the following values for the constitutive constants: $G = 10.64$ GPa, $K = 32.76$ GPa, $G_0 = 8.89$ GPa, $K_0 = 14.81$ GPa, $k = 2.26 \times 10^{-7}\ \mathrm{s}^{-1}$, and $k_v = 3 \times 10^{-6}\ \mathrm{s}^{-1}$. The distribution of u_r^R for $\theta = 0$ and $\theta = \frac{1}{2}\pi$ in the case $\sigma_h > \sigma_v$ and $h = 500$ m, is shown in Fig. 12.3.1. For $\theta = 0$ the displacement u_r^R has a sudden increase at $t = 0$, followed by a slow increase which lasts for about 300 days. For $\theta = \frac{1}{2}\pi$ the variation of u_r^R is more complicated: after the initial sudden increase follows a decrease lasting about 50 days and finally again another small increase. The influence of a possible existing internal pressure is also shown by broken lines.

The variation of u_θ^R shown in Fig. 12.3.2 for $\sigma_h > \sigma_v$ shows that u_θ^R is first negative and only after a certain time interval does it become positive.

Let us examine what will be the change in the shape of the tunnel surface during creep (D. Massier, personal communication (1984)). An arbitrary point belonging to the initial circular surface of the tunnel, of coordinates $x = a\cos\theta$, $y = a\sin\theta$, will have at moment t the coordinates

$$\begin{aligned} x^* &= x - (u_r^R \cos\theta - u_\theta^R \sin\theta)_{r=a} \\ y^* &= y - (u_r^R \sin\theta + u_\theta^R \cos\theta)_{r=a}, \end{aligned} \tag{12.3.18}$$

i.e.

$$\begin{aligned} x^* = a\cos\theta \Biggl\{ 1 + \Biggl[\frac{1}{2G_0} + \Bigl(\frac{1}{2G} - \\ - \frac{1}{2G_0} \Bigr) \exp(-kt) \Biggr] (p - \sigma_h) - \frac{\sigma_h - \sigma_v}{2} \frac{3}{3K_0 + G_0} - \\ - \frac{9KG}{3K + G} \frac{\sigma_h - \sigma_v}{2} \frac{1}{\lambda_1 - \lambda_2} \Biggl[\frac{g(\lambda_1)}{\lambda_1} \exp(\lambda_1 t) - \\ - \frac{g(\lambda_2)}{\lambda_2} \exp(\lambda_2 t) \Biggr] \Biggr\} \\ y^* = a\sin\theta \Biggl\{ 1 + \Biggl[\frac{1}{2G_0} + \Bigl(\frac{1}{2G} - \\ - \frac{1}{2G_0} \Bigr) \exp(-kt) \Biggr] (p - \sigma_v) + \frac{\sigma_h - \sigma_v}{2} \frac{3}{3K_0 + G_0} + \\ + \frac{9KG}{3K + G} \frac{\sigma_h - \sigma_v}{2} \frac{1}{\lambda_1 - \lambda_2} \Biggl[\frac{g(\lambda_1)}{\lambda_1} \exp(\lambda_1 t) - \\ - \frac{g(\lambda_2)}{\lambda_2} \exp(\lambda_2 t) \Biggr] \Biggr\}, \end{aligned} \tag{12.3.19}$$

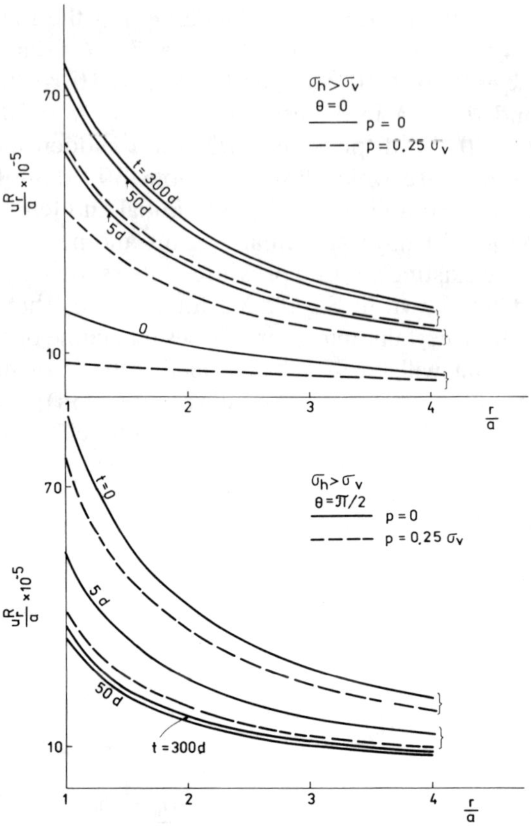

Figure 12.3.1. Evolution in time of the radial displacement in the vicinity of a tunnel for $\theta = 0°$ and $\theta = \frac{1}{2}\pi$ with or without internal pressure and $\sigma_h > \sigma_v$.

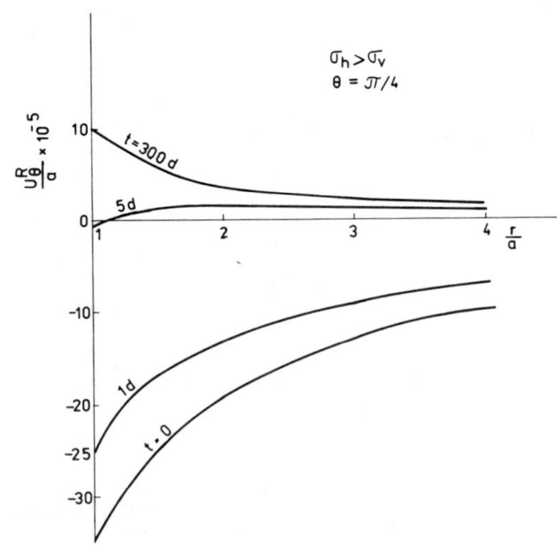

Figure 12.3.2. Evolution in time of the circumferential displacement for $\theta = \frac{1}{4}\pi$ and $\sigma_h > \sigma_v$.

where $g(\lambda)$ is the polynomial

$$g(\lambda) = \frac{1}{KG} \lambda^2 + \left(\frac{k}{G_0 K} + \frac{k_v}{G K_0} \right) \lambda + \frac{1}{G_0 K_0} k k_v. \tag{12.3.20}$$

From (12.3.19) it follows that the initial circular boundary becomes an ellipse whose semiaxes vary in time. Just after excavation, due to the instantaneous response of the rock the semiaxes are

$$\begin{aligned} &a \left\{ 1 + \frac{1}{2G} \left[p - \frac{1}{2} (\sigma_h + \sigma_v) - \frac{3K + 7G}{3K + G} \frac{1}{2} (\sigma_h - \sigma_v) \right] \right\} \\ &a \left\{ 1 + \frac{1}{2G} \left[p - \frac{1}{2} (\sigma_h + \sigma_v) + \frac{3K + 7G}{3K + G} \frac{1}{2} (\sigma_h - \sigma_v) \right] \right\}. \end{aligned} \tag{12.3.21}$$

When $t \to \infty$ the ultimate shape of the ellipse has the semiaxes

$$\begin{aligned} &a \left\{ 1 + \frac{1}{2G_0} \left[p - \frac{1}{2} (\sigma_h + \sigma_v) - \frac{3K_0 + 7G_0}{3K_0 + G_0} \frac{1}{2} (\sigma_h - \sigma_v) \right] \right\} \\ &a \left\{ 1 + \frac{1}{2G_0} \left[p - \frac{1}{2} (\sigma_h + \sigma_v) + \frac{3K_0 + 7G_0}{3K_0 + G_0} \frac{1}{2} (\sigma_h - \sigma_v) \right] \right\}. \end{aligned} \tag{12.3.22}$$

Generally, the time interval during which creep of the rock takes place depends mainly on the values of k and k_v, while the absolute value of the ultimate displacement depends on the values of G_0 and K_0, or more precisely, on how much these moduli differ from G and K respectively. For smaller values of k and k_v the creep lasts a longer time, while for smaller values of G_0 and K_0 as compared to G and K respectively, the ultimate displacements are more significant.

A final remark: the component σ_{zz} varies only slightly during the creep period, generally by less than 8% from its primary value.

12.4. Creep according to an elastic/viscoplastic model

The same problem will be considered, with the constitutive equation of the form

$$\begin{aligned} \dot{\boldsymbol{\varepsilon}} = &\left(\frac{1}{3K} - \frac{1}{2G} \right) \dot{\sigma} \mathbf{1} + \frac{1}{2G} \dot{\boldsymbol{\sigma}} + \\ &+ k(\sigma, \bar{\sigma}, d) \left\langle 1 - \frac{W^{\mathrm{I}}(t)}{H(\sigma, \bar{\sigma})} \right\rangle \frac{\partial H}{\partial \boldsymbol{\sigma}}, \end{aligned} \tag{12.4.1}$$

with

$$W^{\mathrm{I}}(t) = \int_0^t \boldsymbol{\sigma}(\tau) \cdot \dot{\boldsymbol{\varepsilon}}^{\mathrm{I}}(\tau)\,\mathrm{d}\tau \tag{12.4.2}$$

the irreversible stress-work and

$$H(\sigma, \bar{\sigma}) = W^{\mathrm{I}}(t) \tag{12.4.3}$$

the equation of the stabilization boundary. In these equations σ is the mean stress and

$$\bar{\sigma}^2 = \sigma_{rr}^2 + \sigma_{\theta\theta}^2 + \sigma_{zz}^2 - \sigma_{rr}\sigma_{\theta\theta} - \sigma_{\theta\theta}\sigma_{zz} - \sigma_{zz}\sigma_{rr} + 3\sigma_{r\theta}^2 \tag{12.4.4}$$

the equivalent stress.

It is assumed again that in a relatively short time interval $t \in (0, t_0)$ the tunnel (or a certain portion of it) it excavated, and further, that during a much longer time interval $t \in (t_0, t_1)$ the tunnel is in use. By idealization we assume that during the first time interval the response of the rock is 'instantaneous' and in the second time interval various time effects such as creep and slow stress variation may occur. Therefore, if the tunnel is excavated at time t_0, then just after excavation the stresses in the rock are (12.2.4), the strains (12.3.4), and the displacements (12.3.5). Concerning the time interval (t_0, t_1), several variants of the solution based on distinct assumptions are discussed.

Let us first consider a *simplified solution*. This solution is obtained by assuming for the elastic/viscoplastic model the same kind of stress variation as for the linear viscoelastic model: in the time interval $t \in (t_0, t_1)$ the stress components σ_{rr}, $\sigma_{\theta\theta}$, and $\sigma_{r\theta}$ are supposed to remain constant and equal to $(12.2.4)_{1-3}$. Since $\dot{\varepsilon}_{zz} = 0$, i.e. the strain state is a plane strain-state, we have to accept that σ_{zz} is variable in this time interval and therefore satisfies the differential equation

$$\frac{1}{3}\left(\frac{1}{3K} + \frac{1}{G}\right)\dot{\sigma}_{zz}^{\mathrm{R}} = -k\left(1 - \frac{W^{\mathrm{I}}}{H}\right)\frac{\partial H}{\partial \sigma_{zz}^{\mathrm{R}}}\,. \tag{12.4.5}$$

A numerical integration of this equation is possible using iterations and the initial conditions $\sigma_{zz}^{\mathrm{R}_0} = \sigma_{\mathrm{h}}$. An example for granite has shown a quite small (negligible) variation of σ_{zz} and an early fast stabilization of this variation. Thus it can be assumed that all stress components are constant during the creep of the rock surrounding the tunnel, at least as a first estimate of the solution.

In order to write down the formulae describing deformation by creep, first the relationship for the variation in time of $W^{\mathrm{I}}(t)$ when all stress components are constant, is obtained as:

$$1 - \frac{W^{\mathrm{I}}(t)}{H(\sigma, \bar{\sigma})} = \left(1 - \frac{W^{\mathrm{IP}}}{H(\sigma, \bar{\sigma})}\right)\exp\left[\frac{k}{H}\frac{\partial H}{\partial \boldsymbol{\sigma}} \cdot \boldsymbol{\sigma}(t_0 - t)\right]. \tag{12.4.6}$$

The variation of $W^{\mathrm{I}}(t)$ lasts a longer or shorter time depending mainly on the value of the constitutive parameter k. In order to give a numerical example for *granite* (see §8.5), if $k = 2.26 \times 10^{-5}\ \mathrm{s}^{-1}$ the creep lasts only 5 days, for $k = 2.26 \times 10^{-6}\ \mathrm{s}^{-1}$ it lasts 50 days, for $k = 2.26 \times 10^{-7}\ \mathrm{s}^{-1}$ it lasts 500 days, and for $k = 2.26 \times 10^{-8}\ \mathrm{s}^{-1}$ it lasts more than 13 years. As already mentioned, a reasonably correct value for k can be obtained by observing the convergence in time of the walls of the tunnel only.

Further, the formulae for the strain variation are obtained by integration of (12.4.1), using (12.4.6) and the initial conditions

$$t = t_0 : \boldsymbol{\varepsilon}^{\mathrm{R}} = \boldsymbol{\varepsilon}^{\mathrm{R}_0}, \qquad (\boldsymbol{\varepsilon}^{\mathrm{I}})^{\mathrm{R}_0} = 0. \tag{12.4.7}$$

We obtain

$$\boldsymbol{\varepsilon}^{\mathrm{R}} = \boldsymbol{\varepsilon}^{\mathrm{R}_0} + \frac{\left(1 - \dfrac{W^{\mathrm{IP}}}{H}\right)\dfrac{\partial H}{\partial \boldsymbol{\sigma}}}{\dfrac{1}{H}\dfrac{\partial H}{\partial \boldsymbol{\sigma}} \cdot \boldsymbol{\sigma}} \left\{1 - \exp\left[\frac{k}{H}\frac{\partial H}{\partial \boldsymbol{\sigma}} \cdot \boldsymbol{\sigma}(t_0 - t)\right]\right\}. \tag{12.4.8}$$

As a numerical example for *granite*, Fig. 12.4.1 presents for $\sigma_{\mathrm{h}} > \sigma_{\mathrm{v}}$, $\sigma_{\mathrm{v}} = \gamma h$ and $h = 500$ m the distribution with distance of the strain components $\varepsilon_{rr}^{\mathrm{R}}$ and $\varepsilon_{\theta\theta}^{\mathrm{R}}$ at various times. The curves shown for $t = t_0$ correspond to the

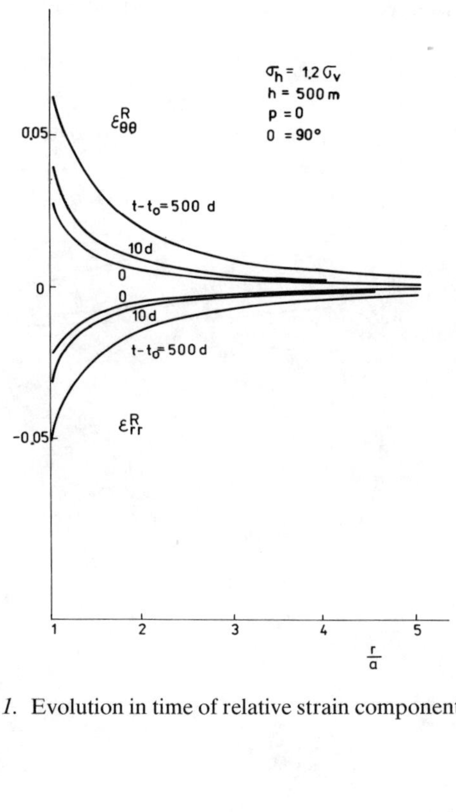

Figure 12.4.1. Evolution in time of relative strain components $\varepsilon_{rr}^{\mathrm{R}}$ and $\varepsilon_{\theta\theta}^{\mathrm{R}}$.

initial strain distribution according to (12.3.4). Stabilization takes place after about 500 days if $k = 2.26 \times 10^{-7}\ s^{-1}$. At a relatively small distance from the opening both components become negligible.

Figure 12.4.2 shows the variation of $\varepsilon^{R}_{r\theta}$ for two cases: $\sigma_h > \sigma_v$ and $\sigma_h < \sigma_v$. The conclusions from this figure are self-explanatory.

A more *exact solution* can be obtained using numerical methods as described in the previous chapter. Let us discuss briefly for the model given for *coal* (see (8.7.6)) the main conclusions obtained for depth $h = 100$ m and $\sigma_h = \sigma_v$. The first conclusion is that the stress σ_{rr} relaxes slightly during the creep period, this relaxation bringing it closer to its initial primary value σ_h. This is shown in Fig. 12.4.3; the dotted line is the initial secondary stress distribution $\sigma^{S_0}_{rr}$ and the full line the stress distribution after 18 days, when the deformation by creep has already stopped.

A second conclusion is that the slow stress variation comes to a halt faster near the free surface $r = a$, so that the deformation by creep lasts a longer time somewhere within the rock. The magnitude of the irreversible rate of deformation during creep is governed by the magnitude of the expression $1 - (W^{I}(t)/H(\sigma, \bar{\sigma}))$. It is easy to see that just after excavation this expression has a maximal value near the free surface $r = a$. At later times the maximum of the rate of deformation is located somewhere near $r = a$, but not exactly on this surface. For instance, at $r = a$ stabilization of the deformation by creep takes place after 5 days, but at $r \simeq 1.1a$, the creep of the rock and the stress variation both last for about 18 days.

A comparison of the two solutions mentioned is shown in Fig. 12.4.4 for the

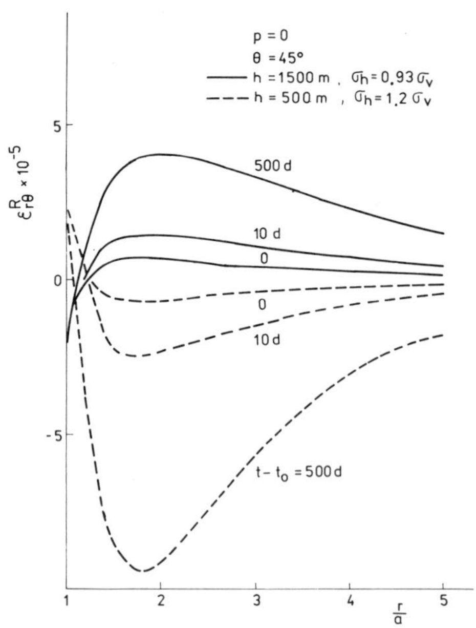

Figure 12.4.2. Evolution in time of the relative strain component $\varepsilon^{R}_{r\theta}$ for two possible depths.

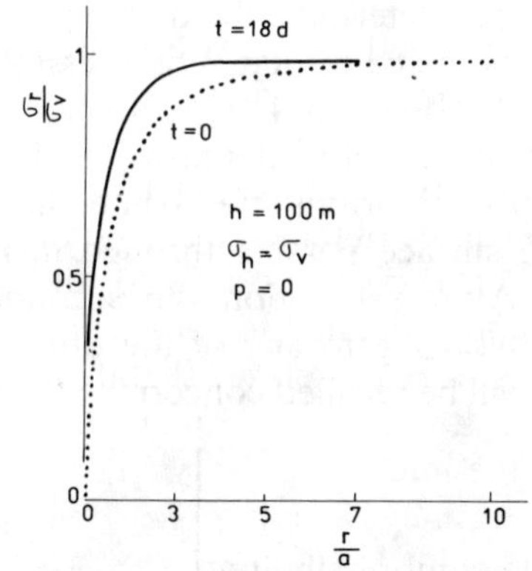

Figure 12.4.3. Radial stress distribution in the neighbourhood of a tunnel showing stress relaxation: initial elastic stress distribution (……), ultimate state (———).

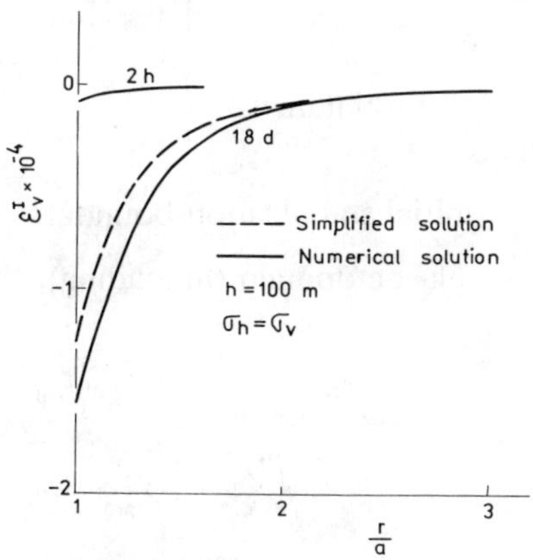

Figure 12.4.4. Evolution in time of the irreversible volumetric strain in the neighbourhood of a tunnel, according to the simplified solution (–––) and numerical solution (———).

variation of ε_v^I. The rock is everywhere dilatant at this depth since at h = 100 m, $\sigma_v > \sigma_0$. As expected, the simplified solution underestimates the magnitude of the deformations.

12.5. Creep, dilatancy/compressibility, damage, and failure around a tunnel

As already mentioned, the elastic/viscoplastic constitutive equation (12.4.1) can describe both volume dilatancy and volume compressibility. It is therefore important to find out under what circumstances and where around the tunnel the rock becomes dilatant, and where compressible. Furthermore, it is of great importance to predict where damage and, finally, failure may be possible and when it may be expected. This subject will be considered in the following (Cristescu (1985*e*)).

First let us see how we can determine the domains where after excavation the rock becomes dilatant, where it becomes compressible, and where an elastic state is possible around the opening. Let us assume that for the rock under consideration a constitutive equation of the form (12.4.1) is determined. Let $P(\sigma^P, \bar{\sigma}^P)$ be the primary equilibrium stress-state shown in a $\sigma 0 \bar{\sigma}$-plane (Fig. 12.5.1). The stabilization surface passing through this point has the equation $H(\sigma, \bar{\sigma}) = H(\sigma^P, \bar{\sigma}^P)$. After excavation the secondary stress-state may be represented by a point belonging to any of the domains shown in this figure. The following definitions will be recalled concerning the secondary stress-state:

$$\left.\begin{array}{l} H(\sigma, \bar{\sigma}) > H(\sigma^P, \bar{\sigma}^P) \\ \dfrac{\partial H}{\partial \sigma} > 0 \quad \text{— irreversible compressibility} \\ \dfrac{\partial H}{\partial \sigma} < 0 \quad \text{— irreversible dilatancy} \\ \dfrac{\partial H}{\partial \sigma} = 0 \quad \text{— compressibility/dilatancy boundary} \end{array}\right\} \text{viscoplastic deformation (loading)} \tag{12.5.1}$$

$$H(\sigma, \bar{\sigma}) = H(\sigma^P, \bar{\sigma}^P) \quad \text{— initial stabilization boundary} \tag{12.5.2}$$

$$H(\sigma, \bar{\sigma}) < H(\sigma^P, \bar{\sigma}^P) \quad \text{— elastic domain (unloading).} \tag{12.5.3}$$

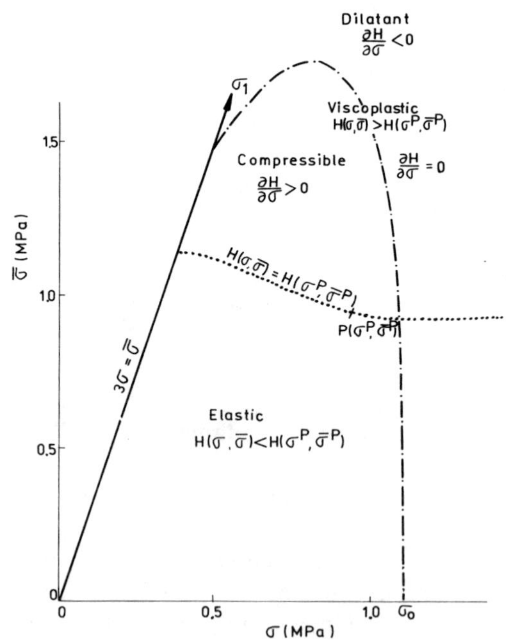

Figure 12.5.1. Various possible stress-states due to excavation. Point P represents the primary stress. Domains above the dotted line represent loading domains (see (12.5.1)); those below it are unloading domains (see (12.5.3)). The border (— · — ·) is the compressibility/dilatancy boundary.

Since the excavation is carried out quickly, we assume that the initial response of the rock is elastic, i.e. that the stresses are (12.2.4), the strains (12.2.6), and the displacements (12.2.7). Introducing (12.2.4) into conditions (12.5.1)–(12.5.3) we can determine the location around the tunnel where the rock is in a dilatant state, etc. For example, all these domains are shown for *coal* (constitutive equation (8.7.4) given in §8.7) and for depth $h = 80$ m in Fig. 12.5.2. Two cases have been considered, $\sigma_h = 0.2\sigma_v$ and $\sigma_h = 0.4\sigma_v$, with $\sigma_v = \gamma h$ in both cases. At the crown the rock is in an elastic state, and it was found that at that depth the failure condition is not satisfied. The domains and boundaries shown in this figure correspond to the domains and boundaries shown in Fig. 12.5.1. Thus in the latter figure, point P represents the primary stress-state for $h = 80$ m and $\sigma_h = 0.4\sigma_v$. In order to show how internal constant pressure may influence the shapes of such domains, these boundaries have also been plotted (using crosses) for the case when $p = 96$ kPa and $\sigma_h = 0.4\sigma_v$. Thus if internal pressure is also present, then the domain of dilatancy is diminished while the elastic domain is enlarged.

We have to point out that the boundaries shown in Fig. 12.5.2 are only the initial boundaries at time $t = t_0$. Since during creep the stresses change slightly (see §12.4), it follows that these boundaries may also change slightly in time. Generally, the domains of dilatancy may either enlarge or shrink slightly in time (see Chapter 11).

The kind of domain of dilatancy which may emerge around a tunnel depends not only on the mechanical properties of the rock (expressed by the constitutive equation) *but also on the primary stress-state.* To illustrate this statement, let us consider the following primary stress-state (see Cristescu (1985*a*)):

$$\sigma_h = n\sigma_v, \qquad \sigma_v = \gamma h, \tag{12.5.4}$$

with $\gamma = 2.7 \times 10^{-5}$, h in metres, and stresses in GPa. Various values will be

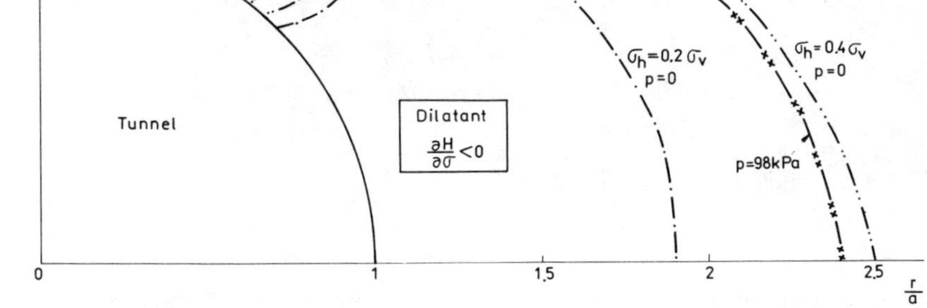

Figure 12.5.2. Various domains around a tunnel: domain of dilatancy, of compressibility, and of elasticity; loading/unloading boundaries (– – – –) and compressibility/dilatancy boundaries (—·—·).

assigned to the coefficient n in order to get various values of the ratio σ_h/σ_v. Let us use the constitutive equation (12.4.1) for the *andesite* described in §8.6. At depth $h = 300$ m and for the following six values of n: 1.5; 0.55; 0.51; 0.5; 0.4; 0.21, the domains (12.5.1)—(12.5.3) are shown in Fig. 12.5.3. For $n = 1.5$ the circumference of the tunnel is surrounded by a circular domain of dilatancy D_d, i.e. both $(12.5.1)_1$ and $(12.5.1)_3$ are satisfied. Outside this domain the rock is in a compressible state according to $(12.5.1)_1$ and $(12.5.1)_2$. The stresses on the boundary between these two domains satisfy $(12.5.1)_4$ and the inequality $(12.5.1)_1$ as well. In both domains the rock deforms by creep. For smaller values of n, the shapes of these domains certainly change. When we consider various decreasing values of n we can find the value for which, for the first time, the compressibility/dilatancy boundary touches the circumference $r = a$ of the tunnel. Thus we find $n = 0.55$. For this value of n the elastic domain defined by (12.5.3) is also present in the neighbourhood of the tunnel, while the boundary (12.5.2) is shown by a dotted line.

If n is decreased even more, the first value of n for which the elastic domain touches the circumference $r = a$ is $n = 0.51$. For even smaller values of n the elastic domain enlarges while the domain of compressibility is no longer in contact with the circumference $r = a$. For very small values of n, only the

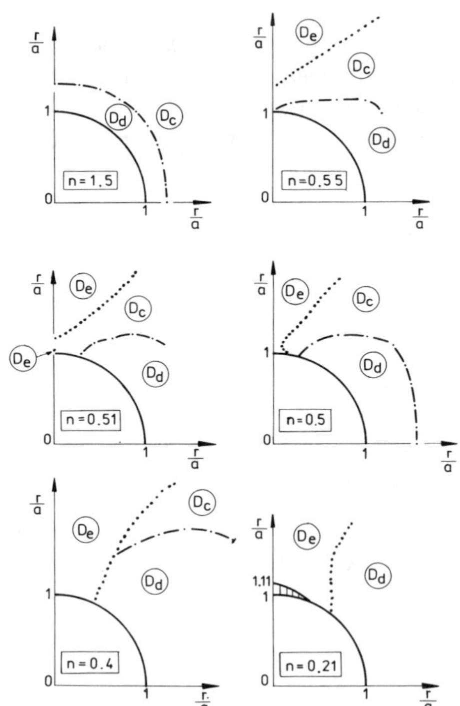

Figure 12.5.3. Domains of dilatancy D_d, of compressibility D_c, and of elasticity D_e around a tunnel for various ratios $\sigma_h/\sigma_v = n$. The hatched area is the volume of the rock which may fall from the crown.

domain of elasticity and that of dilatancy are found around the opening. In the domains D_d and D_c an elastic/viscoplastic deformation takes place for a long period of time. In the domain D_e, unloading takes place with respect to the primary stress-state.

As already mentioned, the boundaries shown in Fig. 12.5.3 may change slightly in time. These boundaries may also change slightly in time if for some reason the ratio n changes; for instance, if another excavation is made in the immediate neighbourhood of the tunnel.

The creep of the rock surrounding the tunnel can be illustrated by representing, at various times, the distribution of the strains with respect to the distance r/a. For instance, Fig. 12.5.4 shows the variations of ε_v^I, ε_{rr} and $\varepsilon_{\theta\theta}$ for *andesite*, $\theta = 0°$, $n = 0.5$, and $k = 5.7 \times 10^{-7}$ s^{-1}. The deformation by creep lasts about 180 days. The irreversible volumetric strain is obtained as explained in §11.4. As a peculiarity of this rock, dilatancy is obtained in the domain $a \leqslant r \leqslant 1.55a$ surrounding the tunnel, and compressibility at greater distances (for $r \geqslant 1.55a$). Compressibility is however, less significant than dilatancy in the immediate vicinity of the tunnel.

Let us consider now the problem of *possible damage and failure* around the tunnel. If for the primary stresses (12.5.4) with $n = 0.21$ we study the distribution of the secondary stresses $\boldsymbol{\sigma}^{S_0}$ around the circumference for the cases $\theta = 0°$, $\theta = \frac{1}{4}\pi$, and $\theta = \frac{1}{2}\pi$, we get the results shown in Fig. 12.5.5. It is obvious that at the crown the stress component $\sigma_{\theta\theta}^{S_0}$ is negative, i.e. it is a tensile stress. The absolute magnitude of this tensile stress can increase with increasing depth. Thus *at the crown, failure due to tensile stresses is possible.*

This can be shown also in a $\sigma 0 \bar{\sigma}$-plane. For *coal* at depth $h = 80$ m, Fig.

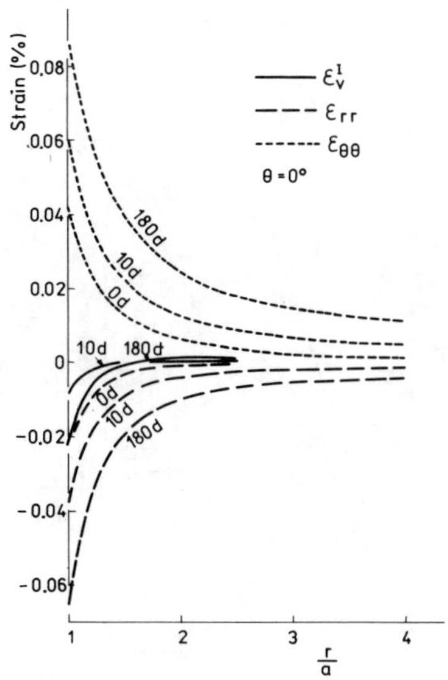

Figure 12.5.4. Evolution in time of the strain components. Close to the tunnel surface the rock is dilatant; at greater distances it is compressible.

12.5.6 shows by solid figures the primary stress-states for four values of the ratio σ_h/σ_v. After excavation of the tunnel the initial secondary stress-state for $r = a$ is distinct in various directions θ. Figure 12.5.6 shows the stress-states after excavation for $\theta = 0°, 10°, \ldots, 90°$, and the four ratios σ_h/σ_v. Thus at the crown the stress-state may become a tensile stress-state with the stresses possibly surpassing the tensile strength of the rock. This is most likely to happen for small values of σ_h/σ_v. In the horizontal direction, equivalent stress $\bar{\sigma}^{S_0}$ may increase greatly. For relatively large values of the ratio σ_h/σ_v the situation may change.

The conclusion is that within a certain domain at the crown *failure due to*

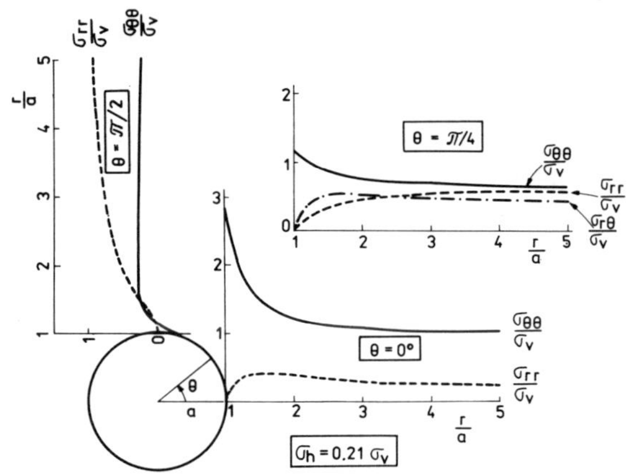

Figure 12.5.5. Initial stress distribution around a tunnel showing possible failure by exceeding the tensile strength at the crown.

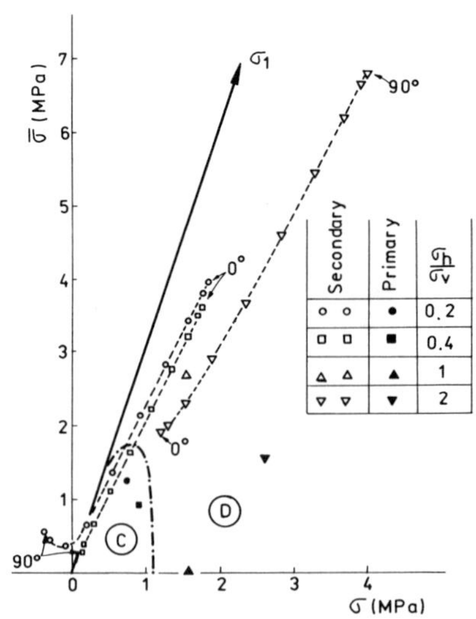

Figure 12.5.6. Primary stresses (solid figures) and secondary stresses (open figures) around a tunnel for various ratios σ_h/σ_v. *C* is the domain of compressibility and *D* the domain of dilatancy.

tensile stresses is possible, mainly at great depths. The exact amount of rock which may fall can be determined quite accurately if the tensile strength of the rock is known. For instance, in the Fig. 12.5.3 the hatched area shows the amount of rock which may fall in a tunnel excavated in *andesite* at the 300 m depth, the tensile strength of the rock being 5.37 MPa and the standard compressive strength 71.6 MPa. Due to various possible faults, joints, and various other kinds of imperfections existing *in situ*, the actual tensile strength may be quite different (generally much smaller) from that determined in the laboratory. Failure which may take place if a principal tensile stress reaches a certain limit is 'instantaneous'.

Another mechanism of rock failure around a tunnel is also possible, as described in Chapter 9, i.e. *damage and failure by dilatancy.* The domains around the circumference $r = a$, where damage by dilatancy is to be expected, are those domains where a significant dilatancy of the volume occurs. Figure 12.5.6 shows that for $\theta = 0°$ if the ratio σ_h/σ_v is small, or for $\theta = 90°$ if this ratio is relatively large, the equivalent secondary stress $\bar{\sigma}^{S_0}$ increases greatly, producing significant dilatancy. Damage by dilatancy occurs *in time* and is *localized* in precisely determined domains. An approximate approach to determining this timing and localization is the following. The initial secondary stress-states are determined from (12.2.4). The variation in time of the irreversible volumetric strain, according to the simplified solution is further obtained from (12.4.8), as

$$\varepsilon_v^I = \frac{\left(1 - \dfrac{W^{IP}}{H}\right)\dfrac{\partial H}{\partial \boldsymbol{\sigma}}}{\dfrac{1}{H}\dfrac{\partial H}{\partial \boldsymbol{\sigma}} \cdot \boldsymbol{\sigma}}\left\{1 - \exp\left[\frac{k}{H}\frac{\partial H}{\partial \boldsymbol{\sigma}} \cdot \boldsymbol{\sigma}(t_0 - t)\right]\right\}. \tag{12.5.5}$$

If we accept that at each material point the mean stress is nearly constant during creep, then it is easy to estimate the expression

$$W_v^I(t) = \int_{t_0}^{t} \sigma(\tau)\dot{\varepsilon}_v^I(\tau)\,\mathrm{d}\tau = \sigma(t_0)[\varepsilon_v^I(t) - \varepsilon_v^I(t_0)] \tag{12.5.6}$$

during creep. From here one can also find the time when the failure condition is satisfied for the first time. Thus, if $W_v^I(\mathrm{cr})$ is a 'critical' value of W_v^I, the corresponding time for the onset of damage can be obtained from

$$t_{cr} = t_0 - \frac{H}{k\dfrac{\partial H}{\partial \boldsymbol{\sigma}} \cdot \boldsymbol{\sigma}} \ln\left\{1 - \frac{\dfrac{1}{H}\dfrac{\partial H}{\partial \boldsymbol{\sigma}} \cdot \boldsymbol{\sigma}}{\left(1 - \dfrac{W^{IP}}{H}\right)\dfrac{\partial H}{\partial \sigma}\,\sigma} W_v^I(\mathrm{cr})\right\}, \tag{12.5.7}$$

t_0 being the moment of excavation. The expression (12.5.6) can also be estimated by numerical procedures.

In order to give an example, let us consider a horizontal tunnel excavated in *andesite* (see §8.6) at $h = 900$ m for $\sigma_h = 0.21\sigma_v$. It is easy to show that the expression (12.5.6) is maximal in the neighbourhood of $\theta = 0°$ and $r = a$. The critical value $W_v^I(\text{cr})$ is reached for the first time at $\theta = 0°$ and $r = a$, 53 hours after excavation. This timing is obtained for the value $k = 5.7 \times 10^{-6}$ s^{-1} of the viscosity coefficient. From this time on, the domain where $W_v^I(t) = W_v^I(\text{cr})$ enlarges, so that after 4 days 16 hours it reaches the points $\theta = 0°$, $r = 1.01a$, and also $\theta = 14°$, $r = a$ (see Fig. 12.5.7). Subsequently, failure proceeds steadily. The computations have to take into account, however, that the shape of the opening is steadily departing from the circular.

The contours shown in Fig. 12.5.7 bear some resemblance to those obtained by Sakurai *et al.* (1985) with an elastic/plastic model based on a Mohr–Coulomb criterion. Rezowalli *et al.* (1984) (see also Zimmerman and King (1985)) presented some preliminary results from a research program involving cross-hole acoustic measurements adjacent to a tunnel mined in a columnar-jointed *basaltic rock* mass; they found that in the horizontal direction both v_P and v_S are lower near the face than in the interior of the rock. This effect is attributed to blast damage, but it may well be due to dilatancy as well.

This kind of damage looks similar to a kind of *erosion*. Since the rock is generally not homogeneous or has weak surfaces (joints, etc.), and since the advanced stage of dilatancy is reached in a whole domain around $\theta = 0°$, it follows that the critical value $W_v^I(\text{cr})$ may be reached somewhere within the rock, for instance around a weak plane (a fault or joint). Adding to this argument the fact that the component σ_{rr} is compressive and quite large at a certain distance from the opening, it follows that failure may occur at a certain distance from the contour $r = a$, and that a significant fragment of rock may suddenly be detached as *violent expulsion*.

Let us give another example to show various possible kinds of damage and failure (and therefore of possible loss of stability) of the rock surrounding a tunnel (or a borehole), with or without an internal pressure. Let us choose a quite unfavourable circumstance. One of the most important factors influencing

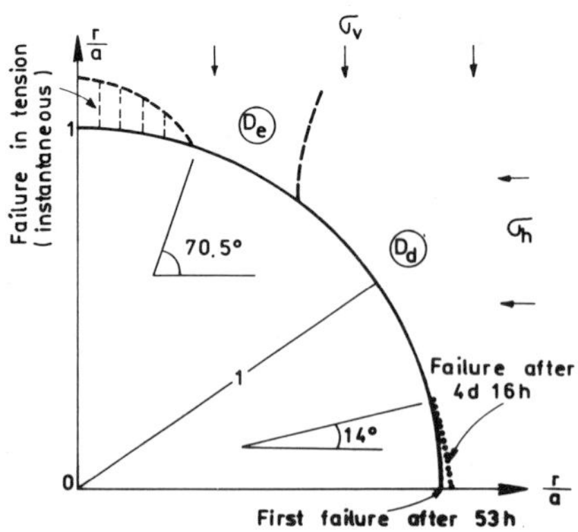

Figure 12.5.7. Domains of dilatancy D_d and unloading D_e around a deep tunnel. Failure is possible either by dilatancy (after long time intervals) or by exceeding tensile strength (instantaneous).

a possible failure has been shown to be the ratio between the horizontal and vertical far-field stresses. Let us consider a case when this ratio is far from unity, for instance $\sigma_h = 0.2\sigma_v$. The depth of the tunnel is certainly another major factor influencing damage, failure, and loss of stability, as well as the mechanical properties of the rock. In order to make an analysis, it is first useful to study the stress distribution around the circumference, occurring immediately after the excavation. For instance, for *coal* (see (8.7.6)) at a depth $h = 350$ m, $\sigma_h = 0.2\sigma_v$ and $\sigma_v = \gamma h$ ($\gamma = 0.0196$) we obtain the stress distribution shown in Fig. 12.5.8 for two cases: $p = 0$ and $p = 0.146\sigma_v$. The primary stress-state is marked by point P and the initial stabilization boundary $H(\boldsymbol{\sigma}) = H(\boldsymbol{\sigma}^P)$ is shown as a broken line. In the neighbourhood of the crown we obtain a tensile stress-state, while along the side walls the stress-states are well into the dilatancy domain (see also Fig. 12.5.6). Again two kinds of failure are possible: at the crown where the maximum principal stress surpasses the tensile strength of the rock (here $\sigma_t = 0.5$ MPa), and along the side walls where evolutive damage is possible. This is shown in Fig. 12.5.9*a* for the case $p = 0$. The domain at the crown where an 'instantaneous' failure due to tensile stress is to be expected is shown surrounded by a thick line.

The evolutive damage of the rock starts at $r = a$ and $\theta = 0°$ after 0.28 days, if the damage threshold is characterized by the value $d_{cr} = 4.48 \times 10^3$ J m^{-3}. This evolutive damage spreads slowly in time and reaches its ultimate position

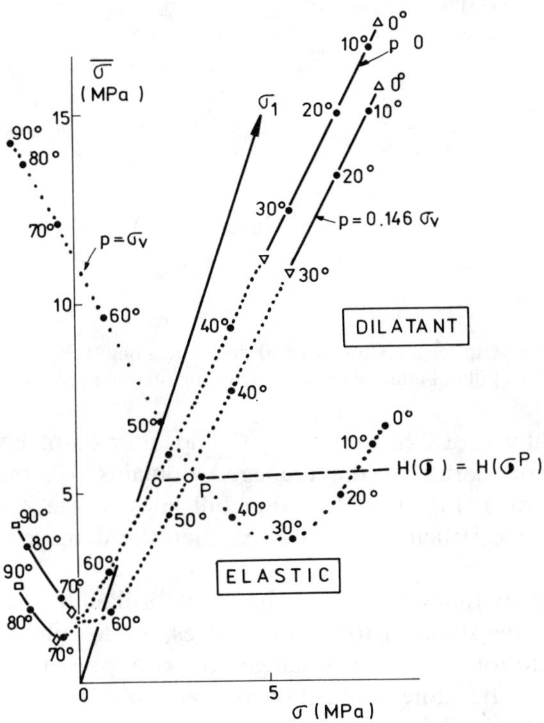

Figure 12.5.8. Stress distribution along the circumference of a tunnel.

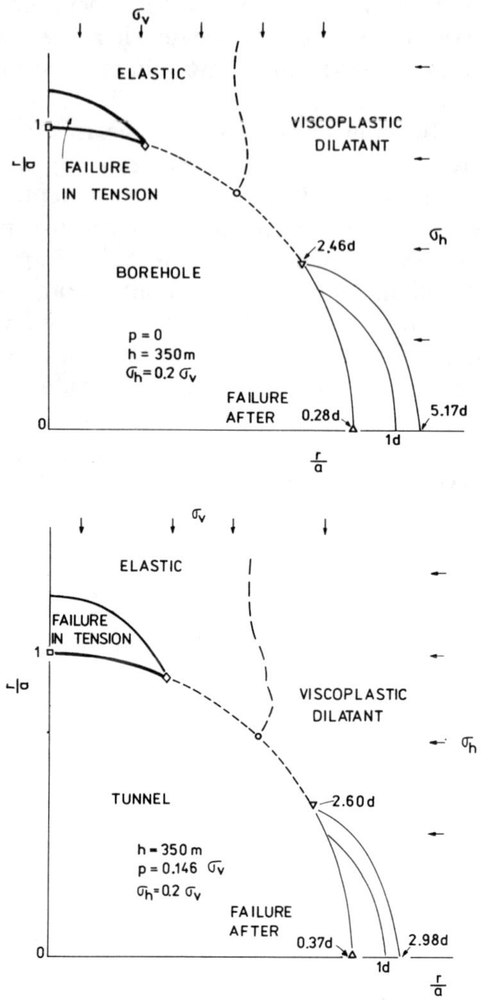

Figure 12.5.9. Failure due to tensile stresses or to dilatancy along the circumference of a tunnel; (*a*) without internal pressure; (*b*) with internal pressure.

after 5.17 days at $r = 1.22a$ and $\theta = 0°$, and after 2.46 days at $r = a$ and $\theta = 33°$. The boundaries of the damaged domains, i.e. the lines of equal damage, are shown in Fig. 12.5.9 as thin full lines. If the rock is non-homogeneous or has some structure weaknesses, then the damaged domains can be much larger.

The portion of the tunnel's contour shown by broken lines is not affected by any failure, unless the stress distribution changes due to failure in other parts of the contour or to other causes (change in pore pressure, in humidity, in temperature, etc.). The same convention is used for marking various lines and points shown in Figs 12.5.8 and 12.5.9. The case when there is an internal constant pressure applied along the contour can be considered in a similar way (see Figs 12.5.8 and 12.5.9*b*); the domain of possible failure at the crown is now larger, while the domain of progressive damage along the walls is smaller.

Figure 12.5.8 shows also the stress distribution which may result if a high pressure is applied inside the opening (as is the case in the hydraulic fracture of a borehole); in the example given, $p = \sigma_v$. In such cases only failure by tensile stresses is possible.

It is interesting to point out that while the vertical diameter of the tunnel decreases suddenly at the moment of excavation and continues to decrease further slowly in time, the horizontal diameter first slightly increases at the moment of excavation, but afterwards decreases slowly in time. This is shown in Fig. 12.5.10; if $p = 0$ the initial instantaneous decrease of u/a is only 5% of the increase which subsequently follows, while for $p = 0.146\sigma_v$ the initial decrease of u/a is 18% less than the maximum increase which follows.

Finally, let us recall that the analysis of evolutive damage depends essentially on the accurate determination of d_{cr} and of the related reference configuration for damage (corresponding to the primary state).

The role of coupled diffusion/deformation phenomena in various facets associated with hydraulic fracturing was analysed by Detournay *et al.* (1986), (1987) within the framework of Biot's theory of poroelasticity (i.e. taking into account the pore pressure in the rock assumed to be elastic). The coupled diffusion/deformation phenomena may well play a significant role during creep of the rock as well.

Tests on small circular openings have been conducted by Kaiser *et al.* (1985) in order to differentiate between yielding and rupture. Tests have been done on *coal* in order to discuss the behaviour of openings in a jointed rock mass during the failure process. The results obtained include that the rupture process is time-dependent; the rupture initiation is accompanied by a two- to three-fold increase in creep closure without a pronounced anisotropic deformation pattern; the least closure is recorded perpendicular to jointing and close to the area where rupture was initiated; the rupture was initiated at stress levels

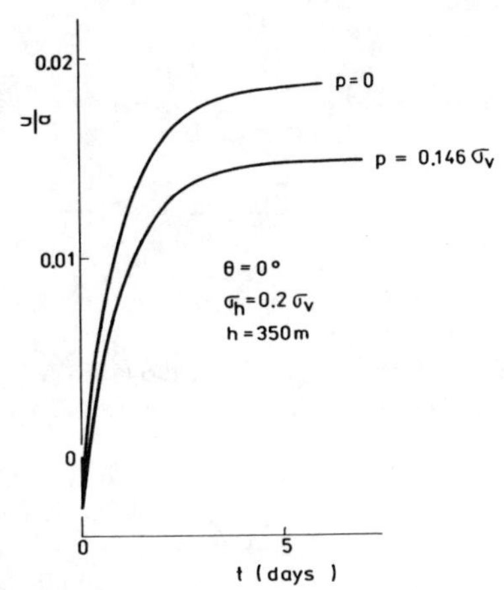

Figure 12.5.10. Variation in time of the displacement of the side wall ($\theta = 0°$).

between 2.0 and 4.2 times the unconfined compressive strength; the stress level corresponding to the tunnel wall rupture initiation point depends on the type and areal distribution of structural weaknesses, while these weaknesses seem to dominate only after significant strain accumulation; rupture is dominated by local weaknesses or heterogeneities.

From the various cases presented above follows the *conclusion* that failure by dilatancy may occur after a certain lapse of time from the moment of excavation (or drilling), depending essentially on the viscosity coefficient, but also on the other mechanical properties of the rock, including the structural weaknesses, as well as on the ratio σ_h/σ_v and the depth. This failure may occur as a failure similar to erosion if the rock is homogeneous, or as a violent expulsion if the rock has structural weaknesses and/or is anisotropic, etc. This may be one of the explanations of the '*rock burst*' phenomenon or of '*outburst*' (a rapid release of fine coal and gas into the working area) which occurs mainly in deep coal mines (see for instance Leighton (1984), MacDonald and Brown (1984)). Other arguments can be added to give a more comprehensive model of outburst, for instance the presence of a gas pressure gradient (Paterson (1986)). Various models have been given to describe the phenomenon of rock burst (see Burgert and Lippmann (1981), Kidybinski (1981), Zubelewicz and Mróz (1983), Petukhov and Liukov (1983), Litwiniszyn (1985), etc.) resulting in a better understanding of various aspects of this phenomenon.

Exercises

1. Show how formulae (12.1.3) can be obtained from (12.1.2).
2. Derive the formulae (12.2.4).
3. Give a detailed derivation of (12.2.9).
4. Find out for what values of r and θ the stress components (12.2.4) take maximal values. Consider the cases $\sigma_h > \sigma_v$ and $\sigma_h < \sigma_v$.
5. Using (12.2.9), show that the circular circumference in the primary state becomes in the secondary state an ellipse of semi-axes

$$a\left\{1+\frac{1}{2G}\left[p-\frac{1}{2}(\sigma_h+\sigma_v)-\frac{3K+7G}{3K+G}\,\frac{1}{2}(\sigma_h-\sigma_v)\right]\right\}$$

$$a\left\{1+\frac{1}{2G}\left[p-\frac{1}{2}(\sigma_h+\sigma_v)+\frac{3K+7G}{3K+G}\,\frac{1}{2}(\sigma_h-\sigma_v)\right]\right\}.$$

6. Starting from (12.2.5) plot the variation of σ^R_{rr}, $\sigma^R_{\theta\theta}$, σ^R_{zz} for $\theta = \frac{1}{2}\pi$, various values of p, and $\sigma_v > \sigma_h$ or $\sigma_h > \sigma_v$.
7. Give detailed derivations of formulae (12.3.11) and (12.3.14).
8. Study the creep of a viscoelastic rock of your choice, around a circular tunnel, following arguments similar to those in §12.3.
9. Derive the formula (12.4.8). Consider a numerical example of your choice (choose another constitutive equation given in the present book, in §8.6 and §8.7, say); consider the cases $\sigma_h > \sigma_v$ and $\sigma_v > \sigma_h$, various depths, and various values of p.
10. Figure 12.5.2 shows domains of dilatancy, compressibility, and elasticity around a tunnel for the constitutive equation for coal (given in §8.7). Make a similar study for some other

constitutive equation of your choice (see §8.6 and §8.7), varying the depth and the ratio between σ_v and σ_h.

11. Make a study similar to the one shown in Fig. 12.5.3 but for another constitutive equation (see §8.6 and §8.7).

12. With reference to Figs 12.5.7 or 12.5.9 determine the domain at the crown of a tunnel where failure is possible by exceeding the tensile strength of the rock. Choose a viscoplastic constitutive equation and various depths. Further, try to find out where a failure threshold by dilatancy is possible.

13

Tunnel support analysis

13.1. Formulation of the problem

After the excavation of a deep tunnel, a support (or lining) is generally applied to limit the closure of the section of the tunnel in a squeezing rock and to ensure the security of the opening. If the rock is not competent, in the absence of the lining, very often failure may occur as a final result of excessive deformation (wall convergence). Sometimes, even in the presence of the lining, structural failure may occur as a result of excessive pressure exerted on the lining by the rock. For an appropriate design of the tunnel support one has to choose the best excavation layout and sequence. Since the convergence of the tunnel takes place slowly in time, and failure may occur either quite soon after excavation or sometimes after a long period of time, possibly destroying also the support, any analysis of the rock—support interaction must be based on the *rheological properties of the rock* and on the mechanical properties of the support. On the other hand, it is well known that the sequence and the excavation procedure and application of the support affect the tunnel performance and its ultimate shape; the excavation layout and support application sequence are therefore essential factors to be considered in tunnel support analysis.

Most authors who have considered tunnel support analysis have used *linear viscoelastic models* to describe the behaviour of the rock. For instance, Baoshen (1979) used a linear standard model and a linear Maxwell model associated with the assumption of volume incompressibility; Popović *et al.* (1979) used a uniaxial linear standard model. Panet (1979*b*) reviewed various rheological models used by several authors (see also Goodman (1980)); Berest and Nguyen (1983*b*), Nguyen *et al.* (1983), *et al.* (1984) used linear viscoplastic models to describe the rock—support interaction. A simplified approach (from the physical point of view) to the rock—support interaction is due to Ardashev *et al.* (1985); the finite-element method is used in conjunction with non-linear stress—strain relations. The finite-element method is used by many authors (Moore and Booker (1982), Rowe *et al.* (1982), Kimura *et al.* (1985), Rodriguez-Roa (1985), Sharma *et al.* (1985), Swoboda (1985), Yufin *et al.* (1985), etc.) in order to study the rock—support interaction. The same method is used in conjunction with elasto/viscoplastic constitutive equation by Sun and Lee (1985) for rocks with orthogonal families of joints. A quite complete time-independent theory of rock—support interaction analysis can be found in Hoek and Brown (1980). The contribution of numerical analysis to the design of shallow tunnels was discussed by Eisenstein (1986). Tunnel stability was studied by Mühlhaus (1985), assuming an elastoplastic model for the rock (satisfying a Mohr—Coulomb yield condition) and considering a uniform internal pressure due to a lining or to rock bolts. Model tests of tunnel excavation, in which the lining and rock bolt were simulated by pieces of paper, were carried out in dry sand by Adachi *et al.* (1982).

In another approach, the concept of the *ground reaction curve*, i.e. the relationship between the ultimate ground displacement and the tunnel support pressure, was introduced without making specific the mechanical properties of the support. This was done either by introducing the concept of equivalent stiffness for an elastic rock (Kaiser (1981)) or a plastic cohesive-frictional dilatant model (Detournay and Fairhurst (1982*a*), Vardoulakis and Detournay (1982), and Detournay and Vardoulakis (1985)) or a Mohr—Coulomb yield criterion (Cividini *et al.* (1985)); this time this curve is called a *characteristic curve*. In some of the examples given by these authors the assumption of volume incompressibility was also accepted. However, as will be shown below, when an elastic/plastic or elastic/viscoplastic model is used for the rock, *the ground reaction curve is not unique* since it depends on the loading history (layout sequence of excavation and support application) and certainly on the mechanical properties of the support.

When considering the mathematical model used for the response of the lining subjected to rock pressure, all authors have assumed a purely linear relationship. In the following we consider both *linear* and *non-linear* relationships for the constitutive equation of the support. The non-linear constitutive equations are used for yieldable supports which are self-adjusting, yielding by sliding between parts of the support, or by compression of wooden strips, or by compression of a stratum of broken rock. Yielding takes place only when the pressure exerted by the rock surpasses a certain limit, thus these supports are self-adjusting to the rock behaviour (creep of rock, pressure exerted by the rock, etc.). For a description of such supports, see for instance Woodruff (1966, Vol. 2).

The constitutive equation which will be used for rock is either a *linear viscoelastic* or an *elastic/viscoplastic* type, since, as already mentioned, time is the major parameter in the rock—support interaction analysis. While the linear viscoelastic constitutive equation allows a much simpler analysis, it cannot describe rock dilatancy, nor possible failure due to dilatancy. With the elastic/viscoplastic constitutive equation, however, one can describe failure due to dilatancy and this constitutive equation is therefore instrumental in the finding of an optimal excavation layout and an optimal sequence procedure in lining application, taking into account the specific properties of the rock, the depth, etc.

If during the operation sequence used, the tunnel surface becomes stress-free at a certain moment (due to the excavation method used: drill and blast, machine excavation, etc.), then one of the main parameters involved in the rock—support interaction analysis is the *time of application of the support* after excavation. If this time interval is too short, then the support may fail due to overloading; if it is too long, then a failure of the rock is possible due to excessive deformation. Generally, the whole *loading history*, i.e. the history of the variation of the pressure exerted by the lining on the tunnel surface, is of great significance for the overall evolution of the behaviour of the rock.

Further, the analysis of rock—support interaction will be presented following a report by Cristescu (1985*f*) and also the papers by Cristescu *et al.* (1987, 1988), and Cristescu (1988*b*).

Let us consider a tunnel of radius a, excavated in a rock which is squeezing (wall convergence), unless a support (lining) is properly installed in due time. The convergence rate can be reduced because of the presence of the lining. Sometimes the convergence by creep is stopped altogether by the lining; in other cases the rock failure is greatly delayed by the presence of the lining. We assume that the lining has an outer radius $b < a$ and an inner radius c. Various kinds of linings will be analysed further. At time t_0, elapsed from the time of excavation (here $t = 0$ is the time when the tunnel surface becomes stress-free, due, for instance, to a blast, etc.) the rock comes in contact with the lining so that t_0 is the time of initiation of the rock—support interaction. For $0 \leqslant t < t_0$ the creep of the rock is not influenced by the lining, while for $t \geqslant t_0$ it is.

The main aspects of the problem to be solved for $t > t_0$ are: the stress evolution in the support, the evolution of the pressure and of the displacement at the rock/lining interface, the ultimate displacement and ultimate pressure at this interface and therefore the ultimate shape of the lining (if a stabilization of the deformation occurs in a certain time interval), under what conditions will stabilization take place, what is the most economical design of the support which would avoid failure, what kind of support would be the most appropriate for the kind of rock considered, the optimal timing t_0, and generally the best procedure and sequence of excavation-lining application, the amount of rock which may break behind the lining, etc.

It must be assumed that the mechanical properties of the material from which the support is made are known. For simplicity, axial symmetry will be assumed throughout.

For methodological reasons we start with the simplest possible case, i.e. linear elastic support and linear viscoelastic rock, and end with the much more involved case: support with non-linear constitutive equation and elastic/visco-plastic rock. The general mathematical formulation of the problem will be given in §13.4.

13.2. Linear elastic support; linear viscoelastic rock

Let us consider a thick tube-shaped lining, of constant thickness $b - c$, made of a linear elastic material—steel or concrete, for instance. If p is the pressure exerted by the rock on the lining and u is the radial displacement of the lining/rock interface, then the constitutive equation for the linear elastic support is (see §5.3)

$$p = \frac{E(b^2 - c^2)}{(1+\nu)[(1-2\nu)b^2 + c^2]} \frac{u - u_0}{b} = q \frac{u - u_0}{b}, \tag{13.2.1}$$

where E and ν are the elastic constants of the material of the lining, q its rigidity, and u_0 the displacement of the rock at the moment when it comes in contact with the lining. Relatively great values of q correspond to *hard* supports, smaller values to *soft* supports.

In order to make the presentation which follows as simple as possible, we make two additional assumptions. First, we assume that the far-field *in situ* stresses are hydrostatic:

$$\sigma_h \simeq \sigma_v = \gamma h \tag{13.2.2}$$

and that it is an equilibrium state for the constitutive equation. Secondly, we assume that the constitutive equation of the rock is linearly viscoelastic of the form (5.4.2). Using these assumptions and recalling (see Chapter 11) that due to excavation the relative mean stress and relative mean strain do not vary, the constitutive equation for the rock written in relative components reduces to

$$\dot{\varepsilon}_\theta = -k\left(\varepsilon_\theta - \frac{1}{2G_0}\sigma_\theta\right) + \frac{1}{2G}\dot{\sigma}_\theta. \tag{13.2.3}$$

Further, with

$$u = r\varepsilon_\theta \tag{13.2.4}$$

and (see (5.3.10) and (11.2.5))

$$\sigma_\theta(t) = -(p(t) - \sigma_v)\frac{a^2}{r^2}, \tag{13.2.5}$$

we get the *differential equation for the rock/support interface* $r = a$ as

$$\dot{u}(t) + \frac{a}{2G}\dot{p}(t) = -k\left[u(t) + \frac{a}{2G_0}(p(t) - \sigma_v)\right]. \tag{13.2.6}$$

This equation describes the convergence of the tunnel surface when pressure $p(t)$, which may be variable, acts on this surface. The pressure $p(t)$ is due to the presence of the lining, and thus depends on the kind of lining.

If by excavation (a blast, say) the pressure at $r = a$ drops instantly from σ_v to zero, then according to (13.2.6) the 'instantaneous' variation of the displacement is governed by

$$\dot{u} + \frac{a}{2G}\dot{p} = 0. \tag{13.2.7}$$

If, however, $p =$ const. and $t \to \infty$, stabilization of the creep occurs. Therefore, the ultimate pressure p_∞ taking values in the interval $0 \leqslant p_\infty \leqslant \sigma_v$ and ultimate displacement u_∞ taking values in the interval $0 \leqslant u_\infty \leqslant (a\sigma_v/2G_0)$ satisfy the relationship

$$p_\infty = \sigma_v - \frac{2G_0}{a} u_\infty \tag{13.2.8}$$

which is called the *ultimate ground reaction curve.* In the present case it is certainly a unique curve. The straight line (13.2.8) is shown in Fig. 13.2.1 as a border line (— · — ·), while the straight line (13.2.7) passing through point $u = 0$, $p = \sigma_v$ (representing the initial state) is shown as a full line. Thus their equation follows from (13.2.6).

If a sudden excavation is carried out, the pressure on the surface $r = a$ drops suddenly to zero and the displacement of the interface $r = a$ becomes $u = (a\sigma_v)/2G$. Further, for $0 \leqslant t < t_0$, the convergence of the walls is governed by (13.2.6) with $p = 0$, i.e.

$$u(t) = \left[\frac{1}{2G_0} + \left(\frac{1}{2G} - \frac{1}{2G_0} \right) \exp(-kt) \right] a\sigma_v, \tag{13.2.9}$$

with the possible ultimate displacement

$$u \Big|_{\substack{t \to \infty \\ p = 0}} = u_\infty^0 = \frac{a\sigma_v}{2G_0} \tag{13.2.10}$$

(see Fig. 13.2.1).

If at time t_0 the support comes in contact with the rock, then from (13.2.1)

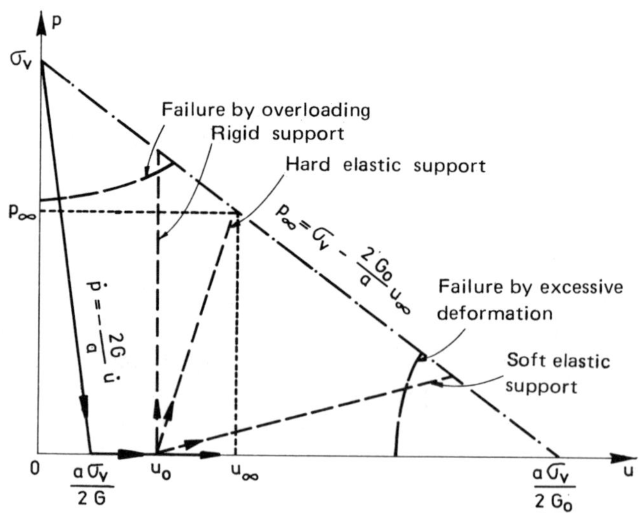

Figure 13.2.1. Possible pressure–displacement relationships on the tunnel surface and rock/lining interface.

and (13.2.6) we get for $t \geqslant t_0$ the displacement of the rock/support interface, influenced by the presence of an elastic support

$$u(t) = \frac{Q}{P} + \left(u_0 - \frac{Q}{P}\right) \exp[P(t_0 - t)] \tag{13.2.11}$$

with

$$P = k \frac{1 + \dfrac{a}{2G_0} \dfrac{q}{b}}{1 + \dfrac{a}{2G} \dfrac{q}{b}}, \qquad Q = \frac{ka}{2G_0} \frac{\sigma_v + \dfrac{q}{b} u_0}{1 + \dfrac{a}{2G} \dfrac{q}{b}}, \tag{13.2.12}$$

if the initial conditions

$$t = t_0 : u = u_0, \qquad p = 0$$

are also used. The ultimate displacement and pressure are

$$u_\infty = a \frac{b\sigma_v + qu_0}{2G_0 b + aq}, \qquad p_\infty = q \frac{a\sigma_v - 2G_0 u_0}{2G_0 b + aq}. \tag{13.2.13}$$

These formulae show how various parameters involved in the problem influence the ultimate displacement and ultimate pressure.

For the *ideal rigid support* installed at time t_0

$$t \geqslant t_0 : u = u_0 = \text{const.} \tag{13.2.14}$$

we get from (13.2.6) for the increase of the pressure on the support

$$p(t) = \left(\sigma_v - \frac{2G_0}{a} u_0\right) \left\{1 - \exp\left[\frac{Gk}{G_0} (t_0 - t)\right]\right\}. \tag{13.2.15}$$

If t_0 is relatively small, then (13.2.15) predicts very high pressures, under which the support may be expected to collapse. The domain of such high pressures producing failure of the support due to overloading is shown schematically in Fig. 13.2.1 by the upper broken line.

The boundary in the u–p plane marking the failure by overloading can be determined by introducing an appropriate failure or yield condition for the support, and knowing the stress distribution in the support $(5.3.7)_{1-3}$. For instance, for a metallic lining a Mises *yield condition*

$$(\sigma_r - \sigma_\theta)^2 + (\sigma_\theta - \sigma_z)^2 + (\sigma_z - \sigma_r)^2 = 2\sigma_Y^2$$

can be assumed for the support. Replacing in this condition the stress distribution in the support as given by $(5.3.7)_{1-3}$ we get the condition for the support to remain elastic as

$$p < \sigma_Y \frac{b^2 - c^2}{2b^2(\nu^2 - \nu + 1)^{1/2}}, \tag{13.2.16}$$

with σ_Y and ν the tensile yield stress and the Poisson ratio of the material of the support.

For a *concrete* or *shotcrete* support, in order to predict failure, one can accept, for instance, a Nadai (1950) type of failure condition

$$(\sigma_r - \sigma_\theta)^2 + (\sigma_\theta - \sigma_z)^2 + (\sigma_z - \sigma_r)^2 = 2[a_0(\sigma_r + \sigma_\theta + \sigma_z) + a_1]^2, \qquad (13.2.17)$$

where

$$a_0 = \frac{\sigma_c - \sigma_t}{\sigma_c + \sigma_t}, \qquad a_1 = \frac{2\sigma_c\sigma_t}{\sigma_c + \sigma_t}, \qquad (13.2.18)$$

with σ_c the uniaxial compressive strength and σ_t the absolute value of the uniaxial tensile strength of the concrete or shotcrete. For such a support the rock pressure must not surpass the limit value (we assume $\nu > 0.1$, say)

$$p < \frac{a_1(b^2 - c^2)}{2b^2[a_0(1 + \nu) - (1 - \nu + \nu^2)^{1/2}]}. \qquad (13.2.19)$$

Conversely, if the pressure p can be estimated or prescribed using other arguments, then the previous inequality can be used to determine the minimal thickness of the wall of the lining:

$$c^2 < b^2\left[1 + \frac{2p[(1 - \nu + \nu^2)^{1/2} - a_0(1 + \nu)]}{a_1}\right]. \qquad (13.2.20)$$

The above formula can be used, for instance, to show how the thickness of the wall of the lining must be increased with increasing depth (increasing p). For example, Fig. 13.2.2 shows the variation of the maximum pressure with the radius of the tunnel for three kinds of *concrete* (curves a, b, c) and two of *shotcrete* (curves d and e). The material constants are given in Table 13.1 and $\sigma_c = 10\sigma_t$. The first conclusion obtained is that the Poisson ratio of the lining material has a very important influence: smaller values of ν increase the maximum pressure. An increased value of σ_c increases the maximum pressure.

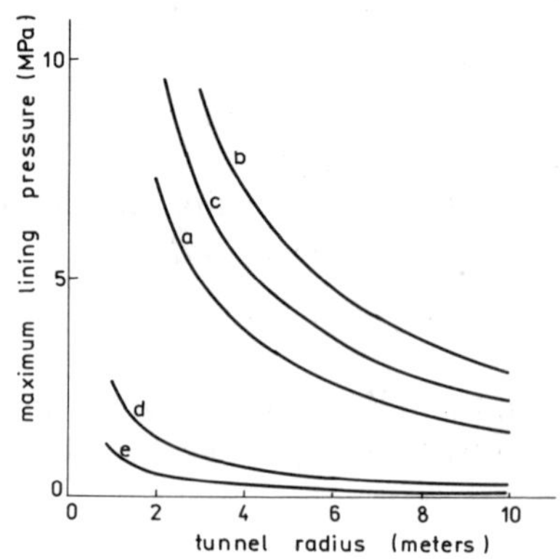

Figure 13.2.2. Variation of maximum lining pressure with tunnel radius, for various linings.

Table 13.1 Lining characteristics

Lining material	ν	σ_c (MPa)	Wall thickness (cm)	Curve in Fig. 13.2.2
Concrete	0.25	35	30	a
Concrete	0.20	35	30	b
Concrete	0.25	49	30	c
Shotcrete	0.25	35	5	d
Shotcrete	0.25	14	5	e

Further, Fig. 13.2.3 shows how the lining thickness varies with the tunnel radius for various values of pressure at the rock/lining interface. The example shown in this figure corresponds to curve a in Fig. 13.2.2. Thus, if the rock/lining interface pressure is prescribed, one can determine for each tunnel radius the appropriate lining thickness, once the material from which the lining is to be made has been chosen.

The boundary of the domain where rock failure may occur due to excessive deformation cannot be obtained using the present constitutive equation. This subject will be approached with the help of an elastic/viscoplastic constitutive equation and of a failure criterion of the form presented in Chapter 9.

Knowing the elastic properties of the support (i.e. knowing q), the time t_0 must be chosen in such a way that the straight line (13.2.1) does not cross either of the two boundaries mentioned above. Generally, if all coefficients from (13.2.6) are made precise, all conceivable displacements u and pressures p are represented by points belonging to the triangle shown in Fig. 13.2.1 (bounded by (13.2.7), (13.2.8) and $p = 0$); points outside this triangle have no physical meaning.

In order to give an example of how the above formulae can be used, let us

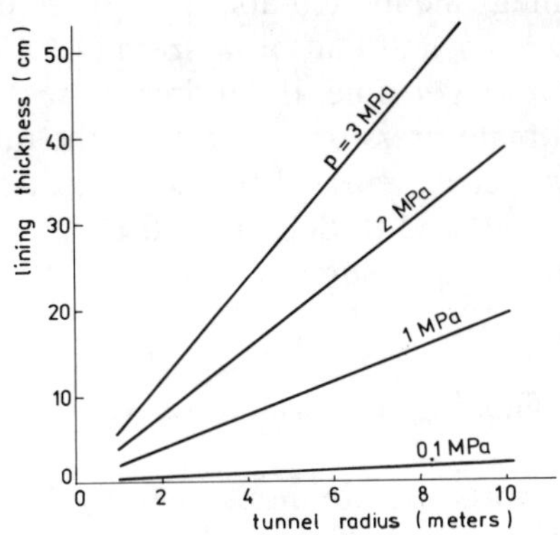

Figure 13.2.3. Lining thickness as function of tunnel radius for various pressure p.

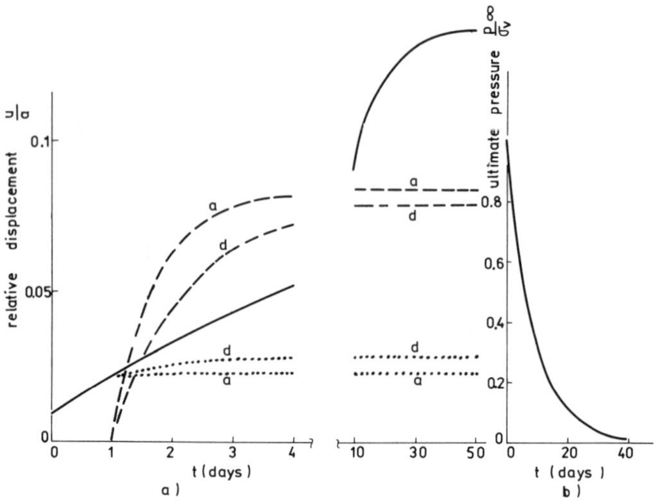

Figure 13.2.4. (*a*) Variation in time of radial displacement in the absence of the support (———), of the radial displacement in the presence of the support (……), and of the pressure on the support (— — — —). (*b*) Ultimate pressure as function of the time of application of the support.

consider a circular tunnel of radius $a = 200$ cm, excavated in a soft rock (*coal*) at a depth of 280 m. The constitutive constants for this rock are

$$G = 300.8 \text{ MPa}, \qquad G_0 = 20 \text{ MPa}, \qquad k = 1.182 \times 10^{-6} \text{ s}^{-1}. \tag{13.2.21}$$

The slow closure of the walls in the absence of any support is shown in Fig. 13.2.4*a* by a full line. This closure by creep lasts more than two months. If a support is applied at various moments, i.e. various values are assigned to the parameter t_0, then this closure is slowed down, as shown by the various dotted lines in Fig. 13.2.4*a*. For instance, if a support is applied after one day, the closure is shown by dotted curves a and d (the same notation as in Table 13.1 and Fig. 13.2.2); a shotcrete lining, being softer, leads to smaller pressures but larger displacements. Fig. 13.2.4*a* shows by broken lines the variations of the pressure for the two linings mentioned above. In these examples the value of b is obtained from $b = a - u_0$ and that of u_0 from (13.2.9) for $t_0 = 1$ day. If the lining is installed after a longer time, the ultimate pressure is certainly smaller. The decrease of the ultimate pressure p_∞/σ_v with the timing t_0 is shown in Fig. 13.2.4*b*. For $t_0 = 0$, we get $p_\infty = \sigma_v$. The decrease of p_∞ when t_0 increases is very fast; this figure can be used to determine the correct time of installation of the support when the ultimate pressure is chosen.

13.3. Non-linear self-adjusting supports; linear viscoelastic rock

With linear elastic supports we do not generally have too much flexibility regarding the timing of the application of the support; the elastic supports are quite stiff, and as a consequence the support has be applied late and at an exact timing. Other kinds of supports may therefore be preferred for heavy rocks when significant deformation by creep is to be expected, i.e. the ground moves toward the excavation and squeezes the supports. Such supports are used mainly for temporary mine openings, for the support of access openings, generally where the primary ground pressures are very high. In such cases a more efficient support is one that can be applied earlier (small values of t_0), but the ultimate pressure on this support must not surpass a certain limit (yieldable ground support or self-adjusting supports). If a rigid support were to be installed in such cases, it would be fractured as the ground squeeze continues.

Let us consider the following kind of support. We assume that in the time interval $0 \leqslant t < t_0$ we have as yet no support, and that wall convergence takes place according to the formula (13.2.9). Further, in the time interval $t_0 \leqslant t \leqslant t_1$ due to the presence of a linear elastic support, convergence is obtained with formula (13.2.11). A support will be called an *elastic/constant pressure support* if it satisfies the following conditions:

$$\begin{aligned} &t_0 \leqslant t \leqslant t_1 : p(t) = q\,\frac{u(t) - u_0}{b} \\ &t \geqslant t_1 : p = p_1 = p(t_1) = \text{const.} \end{aligned} \tag{13.3.1}$$

Thus for $t \geqslant t_1$, convergence is obtained from

$$u(t) = -\frac{a}{2G_0}(p_1 - \sigma_v) + \left[u_1 + \frac{a}{2G_0}(p_1 - \sigma_v)\right]\exp(k(t_1 - t)) \tag{13.3.2}$$

if at $t = t_1$ we have $u = u_1$ and $p = p_1$ (u_1 and p_1 satisfy (13.2.11)). The ultimate displacement u_∞ is obtained either from (13.3.2) or from (13.2.8) for $p_1 = p_\infty$. Such a support would be a very promising one if the magnitude of the pressure p_1 could be controlled *a priori*. There are several kinds of such supports which keep the support pressure ‘nearly’ constant.

13.3.1. First example

Let us consider a tunnel of radius a, along which at a spacing L, circular yieldable steel shaft rings are installed. These circular shaft rings are made from several parts with overlapped ends held together by means of U-bolt clamps. If the ground pressure reaches a certain magnitude, the sliding joint yields, so that the ground pressure is kept under control. Such yieldable steel shaft rings are used where the primary ground pressures are very high.

Let us denote by α the angle between the friction surfaces (of the overlapping between the shaft ring components) and the symmetry axis, by N_T the tensile force in each bolt, by N_n the corresponding normal force on the friction surface, and by T the friction force situated in the cross-section (see Fig. 13.3.1,

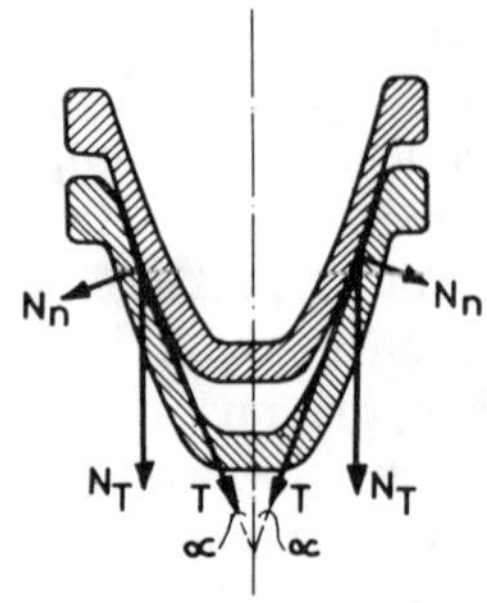

Figure 13.3.1. Cross-section through yieldable steel arc joint.

where a cross-section through the two overlapping shaft rings is shown). Thus we can write

$$N_T = N_n \sin \alpha + T \cos \alpha. \tag{13.3.3}$$

Further, if M is the tightening torque of a nut, we have

$$M = gdN_T \tag{13.3.4}$$

with g a constant (between 0.15 and 0.20) and d the diameter of the bolt. If the number of U-bolt clamps is n, the circumferential force in the shaft ring due to friction is

$$F_f = 2n\mu N_n, \tag{13.3.5}$$

with μ the friction coefficient, either dynamic or static. This friction coefficient is determined in the laboratory, using the formula

$$\mu = \frac{F_f dg \sin \alpha}{2nM - gdF_f \cos \alpha}, \tag{13.3.6}$$

the static coefficients μ_s corresponding to the maximal value of F_f (just before incipient sliding), and the dynamic coefficient μ_d to the minimal value F_f when the sliding is already taking place.

Let N be the circumferential force in the shaft ring, ℓ its width, R the radius of curvature of the shaft ring (at the centre of gravity of the cross-section), E the Young's modulus of the steel, and A the area of the cross-section. From the strength-of-materials formulae we have

$$N = p\ell R, \qquad u - u_0 = \frac{p\ell R^2}{AE}. \tag{13.3.7}$$

If p_m is the mean ground pressure, we can also write

$$\ell p = Lp_m. \tag{13.3.8}$$

With these formulae we obtain

$$p_m = \frac{EA}{LR^2}(u - u_0), \tag{13.3.9}$$

and the convergence of the walls in the presence of the support is obtained from a formula of the form (13.2.11) where, however,

$$P = k \frac{1 + \dfrac{a}{2G_0}\dfrac{EA}{LR^2}}{1 + \dfrac{a}{2G}\dfrac{EA}{LR^2}}, \qquad Q = \frac{ka}{2G_0}\, \frac{\sigma_v + u_0 \dfrac{EA}{LR^2}}{1 + \dfrac{a}{2G}\dfrac{EA}{LR^2}}. \tag{13.3.10}$$

When the ground pressure p_m increases, there is a certain moment t_1 when sliding at the joints starts. This happens when $N = F_{fs}$, where F_{fs} is the static circumferential friction force, i.e. when

$$p_{mi} = \frac{2n\mu_s M}{LRgd(\sin\alpha + \mu_s\cos\alpha)}, \tag{13.3.11}$$

where the subscript i stands for the 'initial' moment of sliding. Due to the sliding the pressure p decreases to its 'final' value when the ground pressure on the support is balanced by the reaction of the support (during sliding)

$$p_{mf} = \frac{2n\mu_d M}{LRgd(\sin\alpha + \mu_d\cos\alpha)}, \tag{13.3.12}$$

where μ_d is the dynamic friction coefficient.

The sliding process is assumed to be instantaneous. Thus the sudden decrease of the pressure and the sudden increase of the displacement are related by (see (13.2.7) and Fig. 13.3.2)

$$u_i - u_f = -\frac{a}{2G}(p_{mi} - p_{mf}). \tag{13.3.13}$$

u_{f1} and p_{mf1} can be determined from the last two equations, i.e. the values of u_f and p_{mf} at the end of the first sliding cycle, when the sliding ceases.

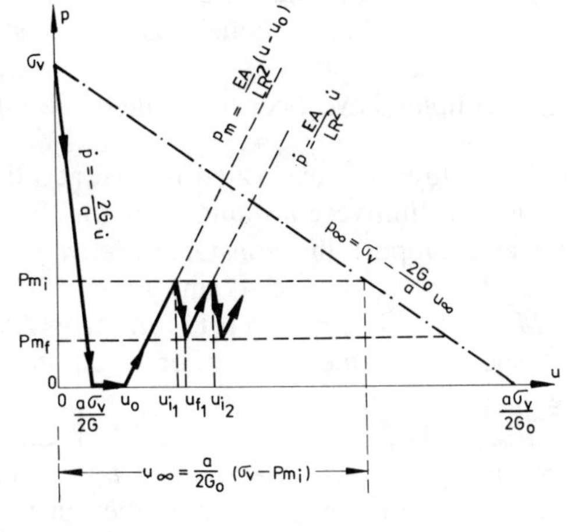

Figure 13.3.2. Variation of pressure and radial displacement in a yieldable steel shaft ring.

Further, in the second cycle, the convergence by creep of the rock continues with the initial data

$$t = t_1 : u = u_{f1}, \qquad p = p_{mf1}, \tag{13.3.14}$$

according to

$$u(t) = \frac{Q}{P} + \left[u_{f1} - \frac{Q}{P} \right] \exp[P(t_1 - t)], \tag{13.3.15}$$

while the pressure is obtained from

$$p(t) - p_{mf} = \frac{EA}{LR^2} [u(t) - u_{f1}]. \tag{13.3.16}$$

Furthermore, there is a time t_2 when condition (13.3.11) is satisfied again and a new sliding cycle takes place, etc.

Since the total displacement is

$$u_\infty = \frac{a}{2G_0} (\sigma_v - p_{mi}), \tag{13.3.17}$$

it is easy to compute the total number of cycles, the time interval in which each cycle takes place, etc.

Generally, the amplitude of the oscillation of the pressure $p_{mi} - p_{mf}$ is quite small and therefore such a yieldable support can be considered to be an elastic/nearly constant pressure support. In fact, the pressure is not only oscillating but also steadily increasing, since with an increase of u, R decreases and p_{mi} increases; furthermore, the sliding conditions are steadily changing.

The magnitude of the sliding pressure p_{mi} is governed by all the factors involved in formula (13.3.11). Increasing the tightening torque M means raising the pressure p_{mi}. The amplitude of the oscillations which follow are governed by the difference between the static and dynamic values of the friction coefficient. In this way it can be explained why the yieldable shaft rings adjust themselves until neighbouring rings carry nearly equal loads and stabilize the ground motion.

Several numerical examples have been computed. For illustration, we give two of them here.

In the first example the tunnel is excavated in coal at a depth of $h = 350$ m ($\sigma_v = 6.86$ MPa). The constitutive constants used for the rock are (13.2.21), while for the tunnel and support the constants are: $a = 168$ cm, $\alpha = 66°$, $A = 30$ cm^2, $L = 70$ cm, $R = 155.4$ cm, $E = 196$ GPa, $\mu_s = 0.475$, $\mu_d = 0.436$, $n = 2$, $M = 14.7$ kN cm, $g = 0.15$, and $d = 2.5$ cm. We obtain for the instantaneous elastic displacement just after excavation $u_e = (a\sigma_v)/2G = 1.94$ cm. Choosing $t_0 = 2$ h, it follows that $u_0 = 2.1436$ cm and further, $p_{mi} = 0.062$ MPa, $p_{mf} = 0.058$ MPa, $u_{i1} = 2.1614$ cm, $u_{f1} = 2.1627$ cm, $u_{i2} - u_{f1} = 0.0178$ cm, $t_{i1} = 8318$ s, and $t_{i2} - t_{f1} = 1117$ s. Furthermore, if the friction conditions do not change steadily, then in the slip—creep cycles which follow, p_{mi}, p_{mf}, u_i, and u_f remain the same. Since the ultimate displacement is $u_\infty = 28.55$ cm, it follows that there are 1391 cycles before stabilization occurs. The first cycle takes place in 1117 s, the second in 1130 s, the third in 1143 s, and so on. This time interval steadily increases according to (13.3.15). The final time of stabilization tends to infinity.

In another case considered, the same constants have been used but $M =$ 29.4 kN cm. Furthermore, $\mu_s = 0.3599$, $\mu_d = 0.3346$, $p_{mi} = 0.09788$ MPa, $p_{mf} = 0.09190$ MPa, $u_{i1} = 2.1717$ cm, $u_{f1} = 2.1734$ cm, $u_{i2} - u_{f1} = 0.02814$ cm, $t_{i1} = 8977$ s, $t_{i2} - t_{f1} = 1788$ s, and $u_\infty = 28.55$ cm; there are 885 cycles and the time interval for each cycle increases: the second cycle takes place in 1806 s, the third in 1839 s, etc.

13.3.2. Second example

Let us now consider a lining made of circular segments of reinforced concrete panels (Fig. 13.3.3), with wooden strips inserted between the longitudinal joints of the panels. We assume again that in the time interval $0 \leqslant t < t_0$ elapsed from the moment of excavation, a convergence of the walls takes place due to creep, and it is only at time t_0 that the rock comes in contact with the lining. For $t \geqslant t_0$ the lining is stressed due to ground pressure and the creep of the rock is influenced in its turn by the presence of the lining.

In order to give a simplified mathematical description of the rock–lining interaction we make two basic assumptions. First, the reinforced concrete panels are much more rigid than the wood and for this reason it will be assumed that the total possible deformation of the lining is essentially due to the compressibility of the wood planks. Secondly, it will be assumed that under the rock pressure, the lining deforms but always remains almost circular. Therefore, the change in length of the lining circumference is assumed to be entirely due to the wood compressibility.

We start the problem by studying the *compressibility of the wood plank*. It is easy to obtain by experiment the relationship describing this compressibility. For instance, Fig. 13.3.4 shows by filled circles the results of the compressibility of a dry fir plank of initial thickness 4 cm compressed perpendicularly to the fibres, i.e. w has the meaning of reduction in thickness. The first conclusion is

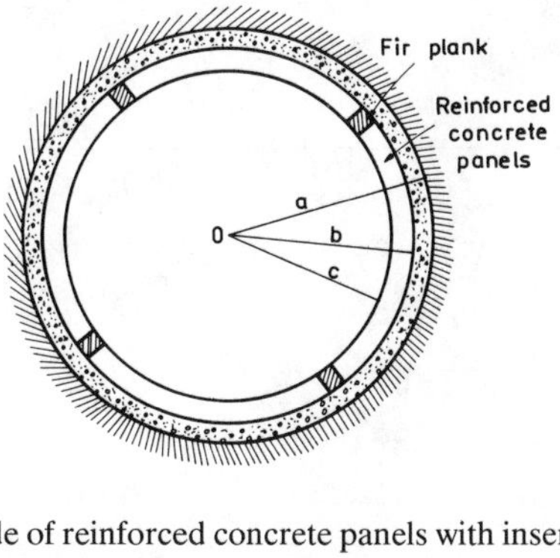

Figure 13.3.3. Lining made of reinforced concrete panels with inserted wooden strips.

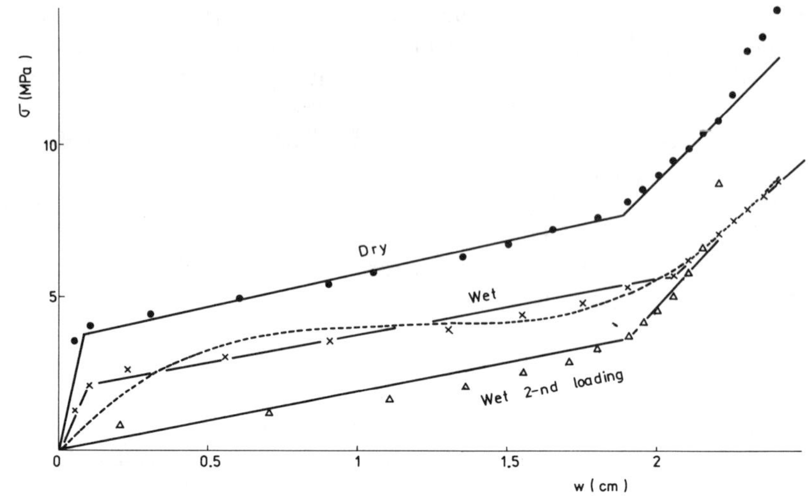

Figure 13.3.4. Stress—displacement relationship for dry or wet wood planks showing a significant volume compressibility.

that the compressibility of the wood is remarkable: the final thickness of the plank is less than half the initial thickness. To keep the figure clear, the unloading curves have not been plotted, but the unloading is practically rigid, i.e. the wood plank compressibility is irreversible. Table 13.2 gives, for a dry fir plank, the dimensions and densities before and after the test.

It is interesting to observe that if the wood plank is wet (24 hours of immersion in water before the test) then the stress—strain curve is lowered significantly (crosses in Fig. 13.3.4), and also that after the test the greatest part of the deformation recovers quite quickly in time. If the wood plank is again immersed in water for 24 hours and afterward compressed a second time, the recorded results are shown in Fig. 13.3.4 by triangles; therefore, even though after the first test the thickness of the plank greatly recovered in time, the wood no longer has the same mechanical properties as the virgin plank. Thus humidity, and the earlier compression to which the plank was subjected,

Table 13.2 Characteristic of a dry fir plank before and after test

	Before test	After test
Length (cm) (along fibre)	14.84	14.85
Width (cm)	15.04	16.10
Thickness (cm)	3.97	1.62
Density ($g\ cm^{-3}$)	0.422	0.963

strongly influence the mechanical properties of the plank when subjected to a second test, in the sense that it becomes much softer. All these properties have a significant influence on the rock—support interaction, as will be shown below.

The above-mentioned tests show that the fir plank is highly compressible and that this compressibility is irreversible (plastic). Since during lining deformation the fir plank is subjected only to a loading process, this loading can be described, for instance, by the empirical formula

$$\sigma = Aw^3 + Bw^2 + Cw, \tag{13.3.18}$$

with A, B, and C material constants to be determined from experimental data. For instance, for a wet plank, using the least-squares method we get $A = 2.503$ MPa cm^{-3}, $B = -8.750$ MPa cm^{-2}, and $C = 10.284$ MPa cm^{-1}. The corresponding stress-displacement relation is shown in Fig. 13.3.4 by a broken line.

The relationship (13.3.18) is cumbersome and difficult to handle in the analysis of a rock—support interaction. It is easier to handle the experimental data of Fig. 13.3.4, if these data are approximated by three straight lines:

$$\begin{aligned} &\sigma = Sw && \text{if } 0 \leqslant \sigma \leqslant \sigma_1 \\ &\sigma - \sigma_1 = S_1(w - w_1) && \text{if } \sigma_1 \leqslant \sigma \leqslant \sigma_2 \\ &\sigma - \sigma_2 = S_2(w - w_2) && \text{if } \sigma_2 \leqslant \sigma, \end{aligned} \tag{13.3.19}$$

with the values for the constants given in Table 13.3. These straight lines are shown in Fig. 13.3.4 by full lines.

Let us assume that along the circumference of the lining, n planks are inserted. The relationship between the compressibility w of a single plank and the radial displacement u of the rock/support interface can be written as

$$w = \frac{2\pi(u - u_0)}{n}. \tag{13.3.20}$$

A simplified relationship between the circumferential stress σ_θ in the support and the pressure p on its outer surface will be used:

$$p = \frac{b - c}{R}\sigma_\theta, \tag{13.3.21}$$

where R is the radius of the support circumference up to the centre of gravity of the cross-section. Using (13.3.20) and (13.3.21) in (13.3.18), we get

$$p = \frac{b - c}{R}\left[A\left(\frac{2\pi(u - u_0)}{n}\right)^3 + B\left(\frac{2\pi(u - u_0)}{n}\right)^2 + C\,\frac{2\pi(u - u_0)}{n}\right] \tag{13.3.22}$$

for the constitutive equation of the support.

Table 13.3 Values of constants involved in the constitutive equation (13.3.19)

	S (MPa cm^{-1})	S_1 (MPa cm^{-1})	S_2 (MPa cm^{-1})	σ_1 (MPa)	σ_2 (MPa)	w_1 (cm)	w_2 (cm)
Dry plank	47.5	2.11	10.50	3.8	7.8	0.08	1.89
Wet plank	22	1.85	8.67	2.1	5.7	0.1	2.05
Wet 2nd loading	—	1.93	10.67	0	3.7	0	1.9

In a similar way we get for the simplified model (13.3.19)

$$p = q\,\frac{u - u_0}{R} \qquad \text{if } 0 \leqslant p \leqslant p_1$$

$$p = p_1 + q_1\,\frac{u - u_0}{R} - \frac{b - c}{R}\,S_1 w_1 \quad \text{if } p_1 \leqslant p \leqslant p_2 \tag{13.3.23}$$

$$p = p_2 + q_2\,\frac{u - u_0}{R} - \frac{b - c}{R}\,S_2 w_2 \quad \text{if } p_2 \leqslant p,$$

with

$$q = \frac{2\pi(b - c)}{n}\,S, \qquad p_1 = \frac{b - c}{R}\,\sigma_1, \qquad \frac{n w_1}{2\pi} = u_1 - u_0, \tag{13.3.24}$$

and similar formulae for q_1, q_2, p_2, and w_2.

If (13.3.23) is introduced in the differential equation (13.2.6), one can describe the convergence of the rock/support interface by formulae of the type (13.2.11), but with other meaning for q involved in the expressions of the coefficients P and Q:

$$P = k\,\frac{1 + \dfrac{a}{2G_0}\,\dfrac{q}{R}}{1 + \dfrac{a}{2G}\,\dfrac{q}{R}}, \qquad Q = \frac{ka}{2G_0}\,\frac{\sigma_v + \dfrac{q}{R}\,u_0}{1 + \dfrac{a}{2G}\,\dfrac{q}{R}}, \tag{13.3.25}$$

used for the case when u varies in the interval $u_0 \leqslant u \leqslant u_1$, and the time in the interval $t_0 \leqslant t \leqslant t_1$ with

$$t_1 - t_0 = \frac{1}{P}\,\ln\frac{u_0 - \dfrac{Q}{P}}{u_1 - \dfrac{Q}{P}}. \tag{13.3.26}$$

Further, in the next interval we have

$$P = k \frac{1 + \dfrac{aq_1}{2G_0 R}}{1 + \dfrac{aq_1}{2GR}},$$

$$Q = \frac{ka}{2G_0} \frac{\dfrac{q_1 u_0}{R} - p_1 + \dfrac{b-c}{R} S_1 w_1 + \sigma_v}{1 + \dfrac{a}{2G} \dfrac{q_1}{R}}, \tag{13.3.27}$$

etc.

Some numerical examples are shown in Fig. 13.3.5 for $a = 170$ cm, $b = 166$ cm, $c = 150$ cm, $R = 156$ cm, and $h = 350$ m. Several cases have been considered: wet planks when $n = 4$ or $n = 10$, and wet planks subjected to a preliminary compression, again for $n = 4$ and $n = 10$; both the non-linear constitutive equation (13.3.22) and the piecewise linearized version (13.3.23) have been used. In all cases the starting point was the choosing of a desired value for the ultimate pressure p_∞, followed by a back-analysis in order to determine t_0. Therefore, u_∞ is also prescribed by (13.2.8). In the case of wet planks, from $(13.3.23)_2$ and (13.2.8) it follows that

$$u_0 = \frac{R}{q_1} \left[\left(\frac{q_1}{R} + \frac{2G_0}{a} \right) u_\infty - \sigma_v + p_1 - \frac{b-c}{R} S_1 w_1 \right]. \tag{13.3.28}$$

If (13.3.22) is used, u_0 follows from (13.3.22) and (13.2.8) by a trial-and-error

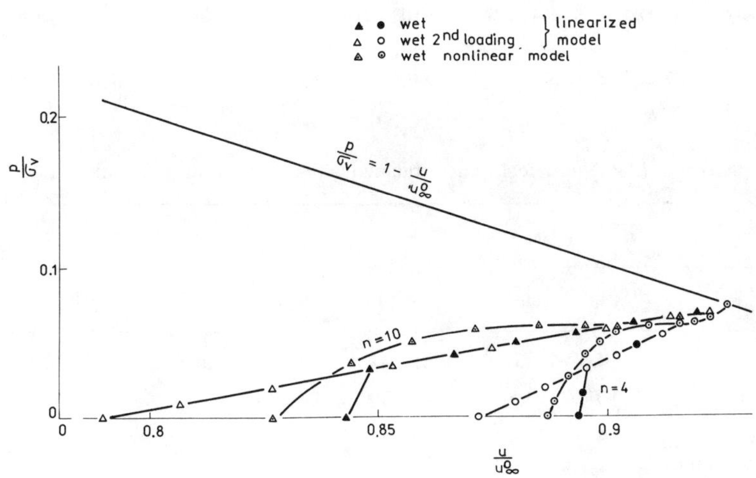

Figure 13.3.5. Variation of pressure and radial displacement at the rock/lining interface for 4 and 10 wood planks inserted along the circumference.

procedure. Further, the time of application of the lining is obtained from (13.2.9) as

$$t_0 = \frac{1}{k} \ln \frac{\dfrac{1}{2G} - \dfrac{1}{2G_0}}{\dfrac{u_0}{a\sigma_v} - \dfrac{1}{2G_0}}, \tag{13.3.29}$$

while u_1 follows from $(13.3.24)_3$, p_1 from (13.3.23), t_1 from (13.3.26), and t_∞ from a similar formula. In the case of wet planks used after a preliminary loading (second loading), the procedure is similar: u_0 follows from $(13.3.23)_2$ (when $p_1 = 0$ and $w_1 = 0$), t_0 from (13.3.29), and t_∞ as before.

The numerical values obtained are given in Table 13.4 for $p_\infty = 0.5$ MPa, $u_\infty = 27.03$ cm, and $u_\infty^0 = 29.155$ cm.

In the case of constitutive relationship (13.3.22) we get for $n = 4$ the values $u_0 = 25.86$ cm and $t_0 = 20.67$ days, while for $n = 10$ it follows that $u_0 = 24.109$ cm and $t_0 = 16.50$ days.

From the above example it is obvious that a more efficient lining is obtained for a greater number of wet planks, or of planks of greater thickness when the time t_0 takes the smallest possible value. It follows also that an even more favourable solution is the use of planks which have been previously compressed before being used in the lining and which are wet.

In the examples given, the influence of a possible temporary additional lining sometimes installed in the interval $0 \leqslant t \leqslant t_0$ (but for a certain shorter prescribed time interval only) has been disregarded in order to keep the examples reasonably simple. Such a temporary lining could also, however, be considered in a more complete rock-support analysis, if necessary. If between the rock and the lining there is a stratum of broken rock, the influence of the compressibility of this stratum could also be included in the rock-support analysis.

One can conclude that by the choice of an appropriate time t_0 and of the

Table 13.4 Numerical values obtained with the constitutive equation (13.3.19)

Plank	n	u_0	t_0	u_1	p_1	t_1	t
		(cm)	(days)	(cm)	(MPa)	(days)	(days)
Wet	4	26.046	21.244	26.110	0.2257	21.734	76.633
Wet	10	24.569	17.439	24.729	0.2257	17.471	84.830
Wet 2nd loading	4	25.422	19.453	u_0	0	t_0	57.637
Wet 2nd loading	10	23.010	14.572	u_0	0	t_0	56.715

number n, and by wetting the planks, the rigidity of the support can be reduced to the desired value.

13.4. Non-linear self-adjusting support; elastic/viscoplastic rock

13.4.1. Formulation of the problem

Let us assume now that the rock under consideration satisfies an elastic/viscoplastic constitutive equation of the kind introduced in Chapter 8. For the component $\dot{\varepsilon}_\theta$ this constitutive equation reads

$$\dot{\varepsilon}_\theta = \left(\frac{1}{3K} - \frac{1}{2G} \right) \dot{\sigma} + \frac{1}{2G}\,\dot{\sigma}_\theta + k \left\langle 1 - \frac{W^{\mathrm{I}}}{H} \right\rangle \frac{\partial H}{\partial \sigma_\theta}. \qquad (13.4.1)$$

Thus any 'instantaneous' response of the rock satisfies (with (13.2.4))

$$\frac{\dot{u}}{a} = \left(\frac{1}{3K} - \frac{1}{2G} \right) \dot{\sigma} + \frac{1}{2G}\,\dot{\sigma}_\theta. \qquad (13.4.2)$$

In other words, if the variation of the stresses and the initial data are prescribed, the instantaneous increase of the rock/support interface displacement can be determined. For the hydrostatic far-field *in situ* stresses (13.2.2), the stresses in the rock after a fast excavation are

$$\sigma_r = -(\sigma_{\mathrm{v}} - p)\frac{a^2}{r^2} + \sigma_{\mathrm{v}} \qquad \sigma_z = \sigma_{\mathrm{v}}, \qquad \sigma_\theta = (\sigma_{\mathrm{v}} - p)\frac{a^2}{r^2} + \sigma_{\mathrm{v}} \qquad (13.4.3)$$

and therefore, at the rock/lining interface $r = a$

$$\sigma_r = p, \qquad \sigma_\theta = 2\sigma_{\mathrm{v}} - p, \qquad \sigma_z = \sigma_{\mathrm{v}}. \qquad (13.4.4)$$

Introducing (13.4.4) into (13.4.2) we obtain the relationship relating the instantaneous variation of pressure and that of the displacement at the interface

$$\frac{\dot{u}}{a} = -\frac{\dot{p}}{2G}\,. \qquad (13.4.5)$$

This relationship holds for any sudden stress variation for which $\dot{\sigma}_r = -\dot{\sigma}_\theta$ and $\dot{\sigma}_z = 0$ are satisfied (at least approximately).

According to (13.4.1) any ultimate equilibrium state must satisfy

$$H(\sigma_\infty, \bar{\sigma}_\infty) = W^{\mathrm{I}}(t)|_{t \to \infty}. \qquad (13.4.6)$$

Thus while the instantaneous response relationship is unique, but certainly depends on the initial data, the locus of ultimate states is not, since it depends on the loading history as well. This will be analysed in detail below, since the optimal solution for the support as well as the prediction of a possible failure occurring after a certain time depend essentially on the loading history.

The formulation of the *initial condition* for the problem under consideration is now more involved than that for the linear viscoelastic model. Two cases can be distinguished, depending on whether $\sigma^P = \sigma_v < \sigma_0$ or $\sigma_v > \sigma_0$ (the meaning of σ_0 for the mathematical model used is given in §4.3.2 or §8.2). For each of these two cases two subcases are possible: the primary stresses (the far-field stresses) at the depth considered satisfy the stability equality

$$H(\sigma^P, 0) = W^{IP}, \tag{13.4.7}$$

or the stability inequality

$$H(\sigma^P, 0) < W^{IP}, \tag{13.4.8}$$

since in all cases $\bar{\sigma}^P = 0$. Equality (13.4.7) corresponds to the case when the primary stress-state is an equilibrium state for the constitutive equation (13.4.1), while the case (13.4.8) corresponds to an elastic one, i.e. to the case when during the natural loading history taking place before excavation the rock was subjected to higher stresses so that the actual primary stresses belong to the elastic domain (following a certain unloading). This unloading may also be due to some other previous neighbouring excavations (above the location chosen for future excavation, for instance).

The four possible cases are shown in Fig. 13.4.1*a–f*. The primary stress satisfying (13.4.7) is shown as point P in Fig. 13.4.1*b*, the initial shape of the corresponding ultimate ground reaction curve as a border line (— · — ·), and that of the 'instantaneous' response at the moment of excavation is shown in Fig. 13.4.1*a* as a full line.

In a similar way the case when inequality (13.4.8) is satisfied is shown in Fig. 13.4.1*f* for both $\sigma_v < \sigma_0$ (point P_1) and $\sigma_v > \sigma_0$ (point P_2). A possible corresponding (primary) stabilization boundary is shown as a broken line. If the

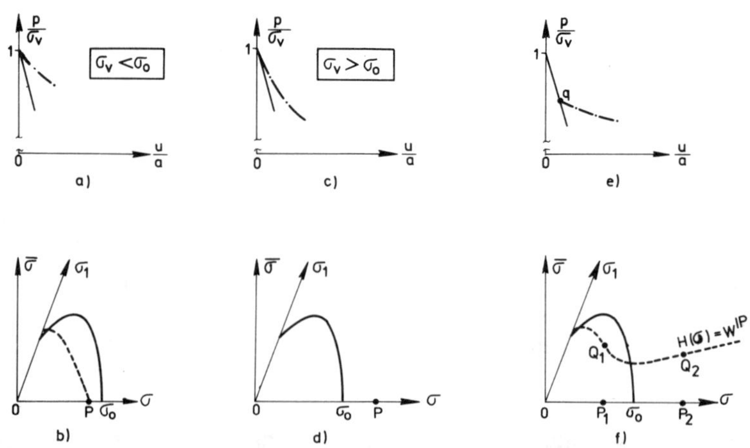

Figure 13.4.1. Formulation of the initial conditions for elastic/viscoplastic constitutive equations, depending on various possible primary stresses.

secondary stress-states (stresses after excavation) are represented as points situated above the point Q_1 which is located on the boundary (or Q_2, respectively) then the ultimate ground reaction curve (— · — ·) and the instantaneous response curve (full line) are shown in Fig. 13.4.1*e*; these two curves intersect each other at a point corresponding to the stress-state in Q. The ultimate ground reaction curve is certainly not unique, but all these curves pass through point q in Fig. 13.4.1*e*. If, however, the secondary stress-state is represented by a point situated under the curve $H(\boldsymbol{\sigma}) = W^{IP}$ passing through point Q, then the rock will remain elastic after excavation.

An intermediate case is shown in Fig. 13.4.1*d*. In this case the primary state is elastic and an equilibrium state, but the smallest increase of $\bar{\sigma}$ leads to an elastic/viscoplastic state. Since the passing from the elastic hydrostatic state to the elastic/viscoplastic state is very smooth, the ultimate ground reaction curves have an initial slope (at point $p = \sigma_v$ and $u = 0$) coinciding with the slope of the instantaneous response curve (Fig. 13.4.1*c*).

13.4.2. Simplified solutions

Let us assume now that after excavation the pressure on the circumference $r = a$ suddenly decreases to a certain non-zero value. The corresponding value of the displacement is obtained from (13.4.5). The straight line (13.4.5) passing through the initial state $u = 0$, $p = \sigma_v$, is shown in Fig. 13.4.2 by a heavy full line AB. If, further, the pressure $0 \leqslant p < \sigma_v$ is somehow kept constant for a long period of time, then the displacement of the rock/lining interface will slowly increase in time due to creep. The simplified solutions are obtained assuming that under constant pressure p the stresses in the rock are (13.4.3) (i.e. a possible stress variation which may slowly change their initial values is

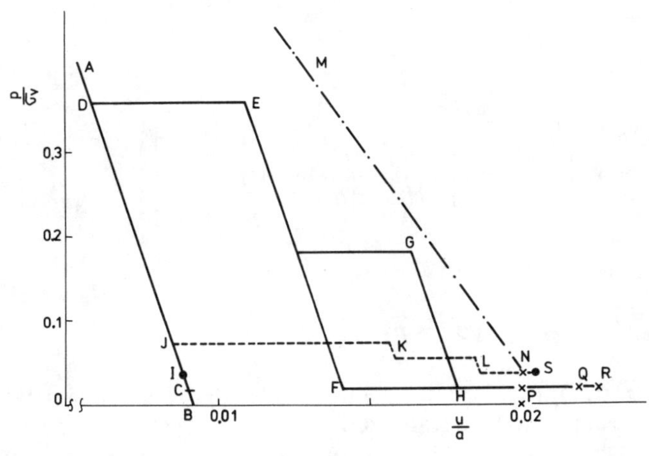

Figure 13.4.2. Various ultimate displacements obtained for several histories of pressure variation acting on the tunnel surface. Crosses mark incipient failure, and the border line (— · — ·) the ultimate ground reaction curve corresponding to a history of variation of the pressure with a single incremental decrease before the ultimate constant pressure level is reached.

disregarded). In order to obtain a slow increase of u we first recall that under a constant stress-state, $W^{\mathrm{I}}(t)$ varies in time according to the relationship (Chapter 8)

$$1 - \frac{W^{\mathrm{I}}(t)}{H(\sigma, \bar{\sigma})} = \left(1 - \frac{W^{\mathrm{IP}}}{H(\sigma, \bar{\sigma})}\right) \exp\left[\frac{k}{H}\frac{\partial H}{\partial \boldsymbol{\sigma}} \cdot \boldsymbol{\sigma}(t_{\mathrm{c}} - t)\right], \qquad (13.4.9)$$

where t_{c} is the time when deformation by creep begins (in the particular case considered here, we have assumed $t_{\mathrm{c}} \simeq 0$). The time of stabilization is obtained from (13.4.9) when $W^{\mathrm{I}}(t)$ reaches a conventional value which is extremely close to $H(\sigma, \bar{\sigma})$. The case when stabilization does not occur will be discussed later on.

Multiplying (13.4.1) by a, we get the *differential equation describing the motion of the lining/rock interface*

$$\dot{u} = -\frac{a}{2G}\dot{p} + ak\left\langle 1 - \frac{W^{\mathrm{I}}}{H}\right\rangle \frac{\partial H}{\partial \sigma_\theta}. \qquad (13.4.10)$$

Further, with (13.4.9) and the initial data

$$t = t_{\mathrm{c}} : u = -\frac{p - \sigma_{\mathrm{v}}}{2G} a, \qquad (13.4.11)$$

we get for $p = $ const. the variation of u in time (wall convergence)

$$\frac{u(t)}{a} = -\frac{p - \sigma_{\mathrm{v}}}{2G} + \frac{H\left(1 - \dfrac{W^{\mathrm{IP}}}{H}\right)\dfrac{\partial H}{\partial \sigma_\theta}}{\dfrac{\partial H}{\partial \boldsymbol{\sigma}} \cdot \boldsymbol{\sigma}} \times$$

$$\times \left\{1 - \exp\left[\frac{k}{H}\frac{\partial H}{\partial \boldsymbol{\sigma}} \cdot \boldsymbol{\sigma}(t_{\mathrm{c}} - t)\right]\right\}, \qquad (13.4.12)$$

where $H(\sigma, \bar{\sigma})$ is computed for

$$\sigma = \sigma_{\mathrm{v}}, \qquad \bar{\sigma} = \sqrt{3}(\sigma_{\mathrm{v}} - p), \qquad (13.4.13)$$

and $W^{\mathrm{IP}} = H(\boldsymbol{\sigma}(0))$, i.e. it is the value of H for the primary stress-state $\sigma_r = \sigma_\theta = \sigma_z = \sigma_{\mathrm{v}}$ assumed to be a stabilized state.

From (13.4.12) for various constant values of p and $t \to \infty$ we get the ultimate values u_∞ when stabilization of deformation by creep occurs. With this procedure the curve shown (— · — ·) in Fig. 13.4.2 is obtained in the p—u plane. This curve is the geometrical locus of all ultimate states (u_∞, p_∞) where stabilization of creep takes place. If in (13.4.12) for p = const. = p_∞, we make $t \to \infty$, we obtain

$$\frac{u_\infty}{a} = -\frac{p_\infty - \sigma_v}{2G} + \left.\frac{H\left(1 - \dfrac{W^{IP}}{H}\right)\dfrac{\partial H}{\partial \sigma_\theta}}{\dfrac{\partial H}{\partial \boldsymbol{\sigma}} \cdot \boldsymbol{\sigma}}\right|_{p = p_\infty} \tag{13.4.14}$$

which is just the *equation of the ultimate ground reaction curve* or the rock/support interface stabilization curve corresponding to the particular loading history considered. It is obvious from (13.4.14) that generally, the u_∞—p_∞ curve is not a straight line and that it passes through the point $u = 0$, $p = \sigma_v$.

The curve u_∞—p_∞ is *not unique* since the ultimate state (u_∞, p_∞) is also *loading-history-dependent* (due to $W^I(t)$). It is easy to show that if during the creep period the pressure p is reduced by a certain quantity before again being reduced to its final value p_∞, then the final stabilization will take place for greater values of u_∞, even if the ultimate pressure p_∞ is the same in the two loading histories (produced in one or two steps). If we reach the same ultimate pressure p_∞ by a creep process in which the pressure was decreased successively several times, then the ultimate displacement u_∞ has an even greater value.

Let us now consider the important case when the value of the pressure p is relatively low. In this case, at the end of the creep period instead of stabilization, the threshold of a *failure process may be reached due to excessive deformation by dilatancy.* The time of the failure threshold is obtained from

$$\mathrm{d}W_v^I = k\left\langle 1 - \frac{W^I}{H}\right\rangle \frac{\partial H}{\partial \sigma}\,\sigma\,\mathrm{d}t, \tag{13.4.15}$$

if (13.4.9) is also taken into account. With the initial condition $t = t_c : W_v^I(t_c) = W_v^{IP}$, we obtain

$$W_v^I(t) - W_v^{IP} = \frac{\left(1 - \dfrac{W^{IP}}{H}\right)\dfrac{\partial H}{\partial \sigma}\,\sigma}{\dfrac{1}{H}\dfrac{\partial H}{\partial \boldsymbol{\sigma}} \cdot \boldsymbol{\sigma}} \times$$

$$\times \left\{1 - \exp\left[\frac{k}{H}\frac{\partial H}{\partial \boldsymbol{\sigma}} \cdot \boldsymbol{\sigma}(t_c - t)\right]\right\}. \tag{13.4.16}$$

Therefore the moment when the failure threshold is reached is obtained if expression $W_v^I(t) - W_v^{IP}$ is replaced by $W_{v(cr)}^I$ (which has been defined in Chapter 9),

$$t_{cr} = t_c - \frac{H}{k \dfrac{\partial H}{\partial \boldsymbol{\sigma}} \cdot \boldsymbol{\sigma}} \ln \left\{ 1 - \frac{\dfrac{1}{H} \dfrac{\partial H}{\partial \boldsymbol{\sigma}} \cdot \boldsymbol{\sigma}}{\left(1 - \dfrac{W^{IP}}{H}\right) \dfrac{\partial H}{\partial \sigma} \sigma} W^{I}_{v(cr)} \right\} . \quad (13.4.17)$$

Let us observe here that $W^{IP}_v = W^{IP}$, since the primary stress-state is hydrostatic, and that $-d_f = W^I_{v(cr)} \simeq W^I_v(t_{cr}) - W^{IP}_v$.

The points marking the failure threshold obtained during deformation by creep under relatively small values of p are shown by crosses lying along the line NP in Fig. 13.4.2.

If the same final pressure p_c is reached by a much more involved loading history, DEFQ say, then the failure threshold will take place later on and for larger displacement, at point Q. In a similar way the loading history DEGHR produces a failure threshold at R.

On the other hand, it may happen that a deformation by creep under constant pressure starting from point I may lead to the failure threshold at point N, while the loading history JKLS ending at the same pressure level may lead, after a longer time, to stabilization at point S.

In order to determine when the failure threshold is reached for the various loading histories mentioned above, we must determine when $W^I_v(t)$ surpasses the critical value $W^I_{v(cr)}$. For the loading history DEFQ for instance, a formula of the form

$$W^I_v(t) - W^{IP}_v = \frac{\left(1 - \dfrac{W^{IP}}{H}\right) \dfrac{\partial H}{\partial \sigma} \sigma}{\dfrac{1}{H} \dfrac{\partial H}{\partial \boldsymbol{\sigma}} \cdot \boldsymbol{\sigma}} \times$$

$$\times \left\{ 1 - \exp\left[\frac{k}{H} \frac{\partial H}{\partial \boldsymbol{\sigma}} \cdot \boldsymbol{\sigma}(-t_F) \right] \right\} \Bigg|_{p = p_D} +$$

$$+ \frac{\left(1 - \dfrac{W^I(E)}{H}\right) \dfrac{\partial H}{\partial \sigma} \sigma}{\dfrac{1}{H} \dfrac{\partial H}{\partial \boldsymbol{\sigma}} \cdot \boldsymbol{\sigma}} \times$$

$$\times \left\{ 1 - \exp\left[\frac{k}{H} \frac{\partial H}{\partial \boldsymbol{\sigma}} \cdot \boldsymbol{\sigma}(t_F - t) \right] \right\} \Bigg|_{p = p_F} \quad (13.4.18)$$

has been used, where $W^I(E)$ is the value of $W^I(t)$ at E and also at F.

Thus by *repeatedly reducing the pressure acting on the rock surface where the final lining will be installed, the failure threshold can possibly be avoided at the expense of additional deformation* by creep of the rock. Furthermore, if by the method of excavation the pressure p is not allowed to drop to zero but is kept for a while nearly constant at a certain non-zero level, the attainment of the failure threshold can be avoided or at least delayed. Similar results are obtained if a tunnel of smaller radius is first excavated (see, for instance, the so-called 'New Austrian Method'; Bieniawski (1984), Teodorescu (1977)), and further, if the radius is enlarged once or in several successive steps, or if several tunnels of smaller radii are successively bored in order to get the larger final shape of the desired tunnel, while simultaneously strengthening step by step the contour by rock bolts, shotcrete, or other kinds of supports. No section of excavated tunnel surface should be left unsupported even temporarily (Bieniawski (1984)). In this case, the stress at what would be the final rock/lining interface decreases steadily by unloading (in the sense of a decrease of p) while the deformation increases; as a consequence it is possible to find a law for the decrease of p (excavation layout and support application sequence) so that either the failure threshold can be avoided altogether, or failure by dilatancy alters a smaller region around the opening.

Figure 13.4.2 has been obtained for a certain kind of coal for which (σ_v = 5.49 MPa):

$$H(\sigma, \bar{\sigma}) := \frac{a_0}{a_2 \dfrac{\sigma}{\sigma_*} + a_1} \left(\frac{\bar{\sigma}}{\sigma_*} \right)^2 + b_0 \frac{\bar{\sigma}}{\sigma_*} + \\ + \begin{cases} c_0 \sin \left(\omega \dfrac{\sigma}{\sigma_*} + \varphi \right) + c_1 & \text{if } 0 \leqslant \sigma \leqslant \sigma_0 \\ c_0 + c_1 & \text{if } \sigma_0 \leqslant \sigma, \end{cases} \tag{13.4.19}$$

with $a_0 = 7.65 \times 10^{-4}$ MPa, $a_1 = 0.55$, $a_2 = 8.1599 \times 10^{-3}$, $b_0 = 0.001$ MPa, $c_0 = 4.957 \times 10^{-4}$ MPa, $c_1 = 4.8955 \times 10^{-4}$, $\omega = 171.927°$, $\varphi = -99.068°$, $\sigma_0 = 1.0996$ MPa, $\sigma_* = 1$ MPa, $K = 769.6$ MPa, $G = 300.8$ MPa, $k = 6 \times 10^{-6}\ \text{s}^{-1}$ and $d_f = 4.48 \times 10^3\ \text{J m}^{-3}$. For this example the failure threshold for creep under constant p, starting from B, occurs after 2.47 days, from C after 3.1 days, and from I after 6.3 days. However, the failure threshold for the loading history DEFQ occurs after 1 + 2.7 days (1 day along DE and 2.7 days along FQ) and along DEGHR after 1 + 1 + 2.5 days, while stabilization at S occurs after 1 + 1 + 11.1 days.

Let us now consider the problem of sudden excavation, i.e. the case when the tunnel surface $r = a$ becomes stress-free for a certain time interval. We assume that just after excavation the pressure drops quite suddenly to zero at the tunnel surface $r = a$, and is kept zero for the time interval $t \in [0, t_0]$, after which an elastic support is installed. Thus t_0 is the installation time or, more precisely, the time when the rock comes in contact with the lining. It will be assumed that the constitutive equation of the support is of the form (13.3.1). If the installation time t_0 is conveniently selected, then u_0 follows from (13.4.12) with $p = 0$ and $t_c = 0$.

Further, (13.4.10) must be integrated, taking into account $(13.3.1)_1$ and the initial data

$$t = t_0 : u = u_0, \quad p = p_0 = 0. \tag{13.4.20}$$

Substituting $(13.3.1)_1$ into (13.4.10) we obtain

$$\left(\frac{b}{q} + \frac{a}{2G}\right)\dot{p} = ak\left\langle 1 - \frac{W^{\mathrm{I}}(t)}{H}\right\rangle \frac{\partial H}{\partial \sigma_\theta}. \tag{13.4.21}$$

If we eliminate the time between (13.4.15) and (13.4.21), we get

$$\left(\frac{b}{qa} + \frac{1}{2G}\right)\frac{\dfrac{\partial H}{\partial \sigma}\sigma}{\dfrac{\partial H}{\partial \sigma_\theta}}\,\mathrm{d}p = \mathrm{d}W_{\mathrm{v}}^{\mathrm{I}}, \tag{13.4.22}$$

and with a similar procedure, starting from

$$\mathrm{d}W^{\mathrm{I}} = k\left\langle 1 - \frac{W^{\mathrm{I}}(t)}{H}\right\rangle \frac{\partial H}{\partial \boldsymbol{\sigma}} \cdot \boldsymbol{\sigma}\,\mathrm{d}t, \tag{13.4.23}$$

we get

$$\left(\frac{b}{qa} + \frac{1}{2G}\right)\frac{\dfrac{\partial H}{\partial \boldsymbol{\sigma}}\cdot\boldsymbol{\sigma}}{\dfrac{\partial H}{\partial \sigma_\theta}}\,\mathrm{d}p = \mathrm{d}W^{\mathrm{I}}. \tag{13.4.24}$$

In both equations (13.4.22) and (13.4.24) the left-hand side depends on p alone, if we take into account (13.4.13) as well. These equations will be integrated for the particular case (13.4.19) (i.e. coal) with the initial data

$$t = t_0 : p = p_0 = 0, \qquad W^{\mathrm{I}} = W^{\mathrm{I}}(t_0), \qquad W_{\mathrm{v}}^{\mathrm{I}} = W_{\mathrm{v}}^{\mathrm{I}}(t_0). \tag{13.4.25}$$

We first compute the expression

$$\frac{\dfrac{\partial H}{\partial \sigma}\sigma}{\dfrac{\partial H}{\partial \sigma_\theta}} = \frac{\alpha(\sigma_{\mathrm{v}} - p)^2 + \gamma\sigma_{\mathrm{v}}}{\dfrac{\alpha}{3\sigma_{\mathrm{v}}}(\sigma_{\mathrm{v}} - p)^2 + \beta(\sigma_{\mathrm{v}} - p) + \dfrac{b_0}{\sigma_*}\dfrac{\sqrt{3}}{2} + \dfrac{\gamma}{3}}$$

$$= 3\sigma_{\mathrm{v}} + \frac{A}{p - p_1} + \frac{B}{p - p_2},$$

where

$$\alpha = -\frac{3a_0 a_2 \sigma_v}{\sigma_*^3 \left(a_2 \dfrac{\sigma_v}{\sigma_*} + a_1\right)^2}, \qquad \beta = \frac{3a_0}{\sigma_*(a_2\sigma_v + a_1\sigma_*)},$$

$$\gamma = \begin{cases} c_0 \dfrac{\omega}{\sigma_*} \cos\left(\omega \dfrac{\sigma_v}{\sigma_*} + \varphi\right) & \text{if } 0 \leqslant \sigma_v \leqslant \sigma_0 \\ 0 \quad \text{if } \sigma_v \geqslant \sigma_0 \end{cases}$$

$$A = \frac{9\sigma_v^2}{\alpha} \; \frac{\dfrac{1}{2}\left\{\beta\sigma_v\left(2 + \dfrac{3\beta}{\alpha}\right) + \beta\left[\dfrac{9\beta^2}{\alpha^2}\sigma_v^2 - \left(\dfrac{b_0}{\sigma_*}\dfrac{\sqrt{3}}{2} + \dfrac{\gamma}{3}\right)\dfrac{12\sigma_v}{\alpha}\right]^{1/2}\right\} - \beta\sigma_v - \dfrac{b_0}{\sigma_*}\dfrac{\sqrt{3}}{2}}{\left[\dfrac{9\beta^2}{\alpha^2}\sigma_v^2 - \left(\dfrac{b_0}{\sigma_*}\dfrac{\sqrt{3}}{2} + \dfrac{\gamma}{3}\right)\dfrac{12\sigma_v}{\alpha}\right]^{1/2}}$$

$$B = \frac{9\beta\sigma_v^2}{\alpha} - A \tag{13.4.26}$$

$$\left.\begin{matrix} p_1 \\ p_2 \end{matrix}\right\} = \frac{1}{2}\left\{\sigma_v\left(2 + \frac{3\beta}{\alpha}\right) \pm \left[\frac{9\beta^2}{\alpha^2}\sigma_v^2 - \left(\frac{b_0}{\sigma_*}\frac{\sqrt{3}}{2} + \frac{\gamma}{3}\right)\frac{12\sigma_v}{\alpha}\right]^{1/2}\right\}.$$

Now we can integrate (13.4.22) with (13.4.25) to get

$$\left(\frac{b}{qa} + \frac{1}{2G}\right)\left\{3\sigma_v(p(t) - p_0) + A \ln \frac{p(t) - p_1}{p_0 - p_1} + \right.$$

$$\left. + B \ln \frac{p(t) - p_2}{p_0 - p_2}\right\} = W_v^{\mathrm{I}}(t) - W_v^{\mathrm{I}}(t_0). \tag{13.4.27}$$

In a similar way, from (13.4.24) and (13.4.25) we obtain

$$\left(\frac{b}{qa} + \frac{1}{2G}\right)\left\{3\sigma_v\left(1 + \frac{2\beta}{\alpha}\right)(p(t) - p_0) + C \ln \frac{p(t) - p_1}{p_0 - p_1} + \right.$$

$$\left. + D \ln \frac{p(t) - p_2}{p_0 - p_2}\right\} = W^{\mathrm{I}}(t) - W^{\mathrm{I}}(t_0), \tag{13.4.28}$$

where

$$C = \frac{Q + Pp_1}{p_1 - p_2}, \qquad D = P - C$$

$$P = \frac{3\sigma_v}{\alpha}\left(3\beta\sigma_v + \frac{6\beta^2\sigma_v}{\alpha} - \frac{\sqrt{3}\,b_0}{\sigma_*}\right) \tag{13.4.29}$$

$$Q = -\frac{3\sigma_v^2}{\alpha}\left(\frac{\sqrt{3}\,b_0}{2\sigma_*} + 3\beta\sigma_v + \frac{6\beta^2\sigma_v}{\alpha} + \right.$$

$$\left. + \frac{3\sqrt{3}\,\beta b_0}{\alpha\sigma_*} + \frac{2\beta\gamma}{\alpha}\right).$$

The equations (13.4.27) and (13.4.28) allow us to obtain the values of W_v^I and W^I corresponding to various values of p along the straight line

$$p = q\,\frac{u - u_0}{b}, \tag{13.4.30}$$

which is the constitutive equation of the support as long as it is elastic. When W_v^I surpasses the critical value $W_{v(cr)}^I$, the failure threshold by dilatancy is reached. If, however, during the period when p is increased, the condition $W^I(t) \simeq H(\sigma, \bar{\sigma})$ is at a certain moment approximately satisfied, a stabilization of the creep occurs. The value of p corresponding to a possible failure by dilatancy is obtained from (13.4.27), and that which may correspond to a possible stabilization of deformation by creep, from (13.4.28). Further, in the analysis of the rock—support interaction we can assign various values to u_0, i.e. to the time t_0, in order to examine the influence of the time of installation of the support on a possible stabilization of the deformation or on a possible failure.

The time of stabilization or the time when the failure threshold is reached can be obtained only by numerical procedures, for instance by substituting $W^I(t)$ from (13.4.28) into (13.4.23) which thus becomes a differential equation in p and t.

In order to provide a numerical example, we have again considered the viscoplastic potential (13.4.19), and the constitutive equation of the form (13.3.9) for the support. The results are shown in Fig. 13.4.3. The border line (— · — ·) is the same as that in Fig. 13.4.2. Stabilization along various straight lines (13.4.9) written for various values of u_0 occurs along the broken line shown in Fig. 13.4.3 which does not coincide with the border line. Thus, for the present loading history, stabilization occurs at smaller displacements than in the previous loading history considered (creep under constant pressure). The states corresponding to a failure threshold are practically the same for the two loading histories discussed.

Finally, let us consider a support of the type examined in the first example in §13.3. Therefore, after excavation the pressure decreases to zero, it is then kept

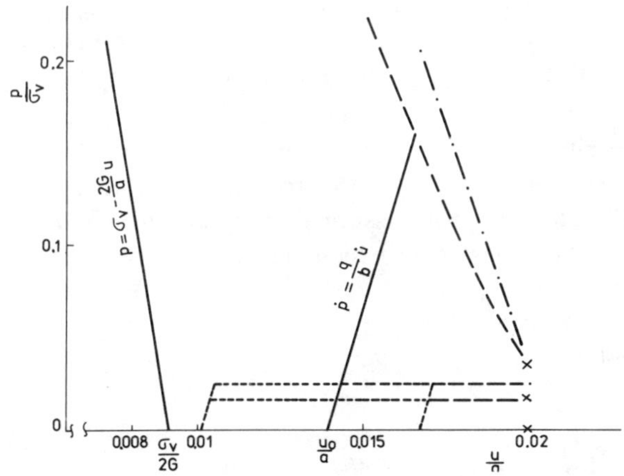

Figure 13.4.3. Variation of pressure and radial displacement at the rock/lining interface for a yieldable steel shaft ring and various loading histories. Crosses mark incipient failure; the border line (— · — ·) and the broken line (——–) are two possible ultimate ground reaction curves.

at zero for a certain time interval, and further, it begins to increase if at $t = t_0$ a support is installed. We assumed that in the time interval $t \in [t_0, t_1)$, the support is elastic, but for $t \geqslant t_1$ the pressure is kept nearly constant (see (13.3.1)), i.e.

$$\begin{aligned} p_m &= \frac{EA}{LR^2}(u - u_0) \quad \text{for } t_0 \leqslant t \leqslant t_1 \\ p_{mf} &\leqslant p \leqslant p_{mi} \quad \text{for } t \geqslant t_1. \end{aligned} \tag{13.4.31}$$

For instance, for $t_0 = 2$ h, $p_{mi}/\sigma_v = 0.0161$ and $p_{mf}/\sigma_v = 0.0143$, we reach the failure threshold at slightly higher displacements than those obtained in the previously considered loading history (the plot of p—u relationship is shown as the lower broken line in Fig. 13.4.3). For a higher tightening torque of the nuts ($p_{mi}/\sigma_v = 0.025$, $p_{mf}/\sigma_v = 0.0232$) we get the upper broken line, and again the failure threshold is reached at slightly greater displacements.

Similar results are obtained for $t_0 = 1$ day (broken lines).

13.4.3. General solutions

In a similar way to that described in Chapter 11 and in Appendix 2, one can obtain a general solution of the problem by numerical integration of the system (11.1.6), (11.1.7), and (11.4.11) with the appropriate boundary and initial conditions. Let us recall that the solution described in the previous section, though useful, is only a simplified solution since it was *a priori* assumed that during creep all the stress components are constant. Furthermore, it is not a rigorously obtained solution since, for instance, the compatibility equation (11.1.7) was not taken into account.

In the following the mathematical formulation of the problem is quite general. Thus, it will be assumed that due to excavation occurring at time $t = 0$, along the circumference $r = a$ the stress σ_r suddenly decreases from σ_v to a certain value p with $0 \leqslant p < \sigma_v$. Further on, for $t > 0$, due to the presence of a lining of constant pressure, the component σ_r is kept constant ($\sigma_r = p =$ const.) along the circumference $r = a$, while the other stress components may vary slowly in time, simultaneously with a rock deformation by creep.

Since a 'fast' excavation was assumed, the '*initial*' *conditions* for the slow variations of stresses and strains which occur for $t > 0$, are just the 'elastic' solution discussed in §11.2:

$$\begin{aligned}
\sigma_r &= \sigma_r^{S_0} = (p - \sigma_h)\frac{a^2}{r^2} + \sigma_h \\
\sigma_\theta &= \sigma_\theta^{S_0} = -(p - \sigma_h)\frac{a^2}{r^2} + \sigma_h \\
\sigma_z &= \sigma_z^{S_0} = \sigma_v \\
\varepsilon_r &= \varepsilon_r^{R_0} = \frac{p - \sigma_h}{2G}\frac{a^2}{r^2} \\
\varepsilon_\theta &= \varepsilon_\theta^{R_0} = -\frac{p - \sigma_h}{2G}\frac{a^2}{r^2} \\
\varepsilon_z &= \varepsilon_z^{R_0} = 0 \\
u &= u^{R_0} = -\frac{p - \sigma_h}{2G}\frac{a^2}{r}, \qquad w^{R_0} = 0.
\end{aligned} \tag{13.4.32}$$

In the above formulae, p may possibly be zero.

The *boundary conditions* are formulated on the tunnel surface

$$r = a, \qquad t > 0 : \sigma_r = p, \tag{13.4.33}$$

and at far distances r_∞ (generally $r_\infty = 10a$ or $20a$, say)

$$\left.\begin{aligned} r &= r_\infty \\ t &\geqslant 0 \end{aligned}\right. : \quad \begin{aligned} \boldsymbol{\sigma} &= \boldsymbol{\sigma}^{S_0} \quad \text{(according to (13.4.32))} \\ \boldsymbol{\varepsilon} &= \boldsymbol{\varepsilon}^{R_0} \\ \mathbf{u} &= \mathbf{u}^{R_0}. \end{aligned} \tag{13.4.34}$$

The system of five partial differential equations with five unknown functions σ_r, σ_θ, σ_z, ε_r, ε_θ, and two independent variables r and t, is integrated by a numerical procedure using the finite-difference method as described in Appendix 2. The main objective is to find the ultimate ground reaction for the programme of variation of σ_r as described above, and simultaneously to find the solution, i.e. the variation of all unknown functions for $t > 0$ and $a \leqslant r < r_\infty$.

The numerical example is given for a tunnel excavated in coal (elastic/viscoplastic constitutive equation with the potential (8.7.6)) at the depth $h =$ 350 m. The ultimate ground reaction curve obtained with the simplified solution (constant stresses during creep) is shown in Fig. 13.4.4a as a non-linear curve

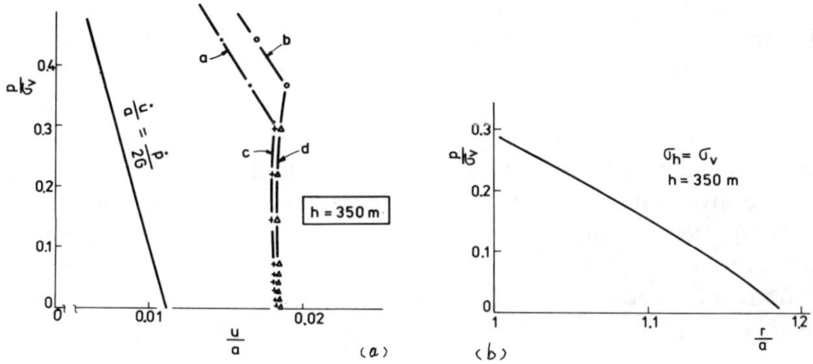

Figure 13.4.4. (*a*) Ultimate ground reaction curves and the curves marking the onset of failure, obtained by two methods for a tunnel excavated in coal. (*b*) The amount of damaged rock behind the lining.

(— · — ·). For smaller values of the pressure p, due to significant deformation, failure due to dilatancy is possible. Using as criterion for the onset of failure by dilatancy the value $d_f = 4.48 \times 10^3 \text{ J m}^{-3}$ of the damage parameter, and the formula (13.4.16), we find that the curve in the p—u-plane marking the onset of failure by dilatancy is the curve c (×—×—).

Using now the numerical solution and the same law of variation of σ_r at $r = a$, we obtain the ultimate ground reaction curve shown as curve b in Fig. 13.4.4*a* (○—○—). Therefore, the ultimate displacement is more significant than the one obtained with the simplified solution (due to the increase of the component σ_θ during creep above $\sigma_\theta^{S_0}$, in the neighbourhood of the circumference $r = a$). For relatively small values of p, due to excessive deformation, the onset of failure is again reached but for slightly greater displacements than obtained with the simplified solution (curve d in Fig. 13.4.4*a*, △—△—). The shape of the ground reaction curve is determined by checking when the condition $\langle 1 - W^I/H \rangle = 0$ is satisfied at various pressure levels, while the onset of failure is determined when the irreversible volumetric stress-work

$$W_v^I(t) - W_v^{IP} = \int_0^t \sigma(\tau)\dot{\varepsilon}_v^I(\tau)\,d\tau$$

$$= \int_0^t \sigma(\tau)\dot{\varepsilon}_v(\tau)\,d\tau + \frac{(\sigma^{S_0})^2 - \sigma^2(t)}{2K}, \tag{13.4.35}$$

reaches its critical value $W_v^I(t) - W_f^{IP} = -d_f$. Here $\sigma^{S_0} = \sigma^P$ since $\sigma_h = \sigma_v$. The values of all the variables from the right-hand side in formula (13.4.35) are obtained using the numerical method described above. It follows from Fig. 13.4.4*a* that while the ground reaction curves obtained with the two methods are distinct, the onset of failure curves obtained with these two methods are practically the same.

Let us recall that both curves, i.e. the ultimate ground reaction and the curve marking the onset of failure, are not unique since both depend on the loading history (i.e. on the law of variation of σ_r on $r = a$, and therefore, on the layout sequence of excavation and support application). Any other law of variation of σ_r on $r = a$, for instance those presented in the previous section, can also be considered in conjunction with the numerical method, but the procedure is much more involved. That is why for a preliminary qualitative analysis one can use the simplified solution, which may be revealing for the determination and optimization of the appropriate layout sequence of excavation and support application and predicts practically the same onset of failure. For practical purposes one can try to find various kinds of 'corrections' of the simplified solution.

Though in the example considered, for $p/\sigma_v < 0.3$ a failure behind the lining of constant pressure takes place, one may ask how important is the volume of the rock which will be ultimately damaged by microcracking, and how great will be the additional loading on the lining? The amount of this failed rock, obtained using the simplified solution, is shown in Fig. 13.4.4*b*. For relatively small values of p, the amount of damaged rock may be significant mainly if there are also some weak planes, other structural weaknesses, etc.

How fast the damage by dilatancy spreads into the rock is shown in Fig. 13.4.5 for three levels of pressure, as obtained with the simplified solution. In the absence of the lining ($p = 0$), the amount of the damaged rock is quite significant: damage starts at $t = 0.78$ days after excavation and lasts more than five days, though the rapid deterioration takes place mainly in the first two days. The presence of a constant pressure certainly reduces the amount of damaged rock behind the lining.

The variation in time of the displacement of the rock/support interface is shown in Fig. 13.4.6 for various values of p, as obtained using the numerical solution; triangles at the end of the curves mark the onset of failure, while a circle means stabilization. During the creep of the rock the stresses vary slightly; for instance Fig. 13.4.7 shows the slight increase of the component σ_θ for various constant values of p and $r = a$. The first points on these curves (at $t = 0$) correspond to $\sigma_\theta^{S_0}$, the last points to the onset of failure.

Similar computations have been performed for a depth of 260 m, using the same constitutive equation. The results are shown in Fig. 13.4.8. The ground

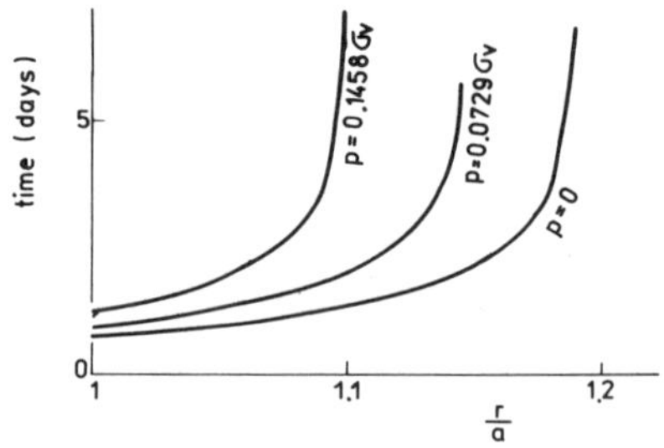

Figure 13.4.5. Spreading of the domain of damage into the rock surrounding the tunnel surface.

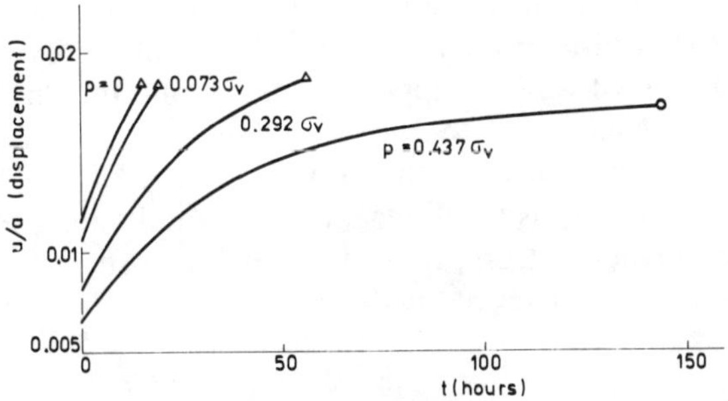

Figure 13.4.6. Variation in time of the displacement of the rock/support interface for various applied constant pressures; triangles indicate onset of failure, the circles indicate stabilization.

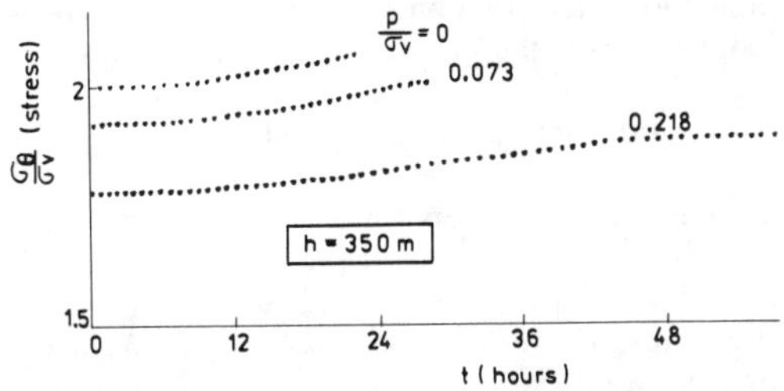

Figure 13.4.7. Slow increase of component σ_θ during creep, for various constant applied pressures.

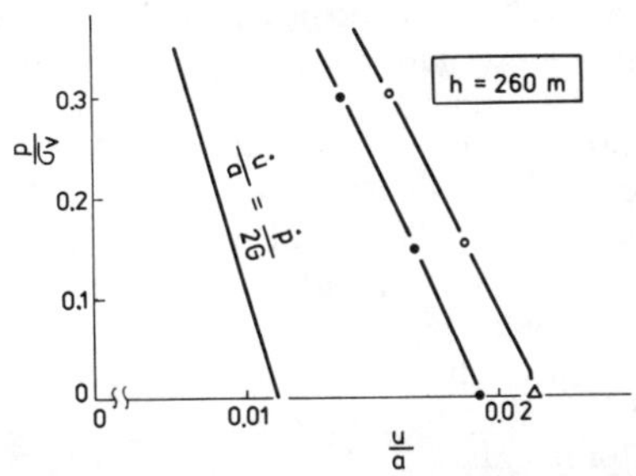

Figure 13.4.8. Ultimate ground reaction curves for tunnel excavated in coal at $h = 260$ m; onset of failure is incipient only for $p = 0$ (numerical solution).

reaction curve obtained with the simplified method is shown as (• — • —), while that obtained with the numerical method as (○—○—). For $p = 0$, the triangle in Fig. 13.4.8 marks the onset of failure according to formula (13.4.35). Thus for this kind of *coal*, for depths greater than 250 m, say, a failure due to dilatancy is to be expected even for a hydrostatic far-field stress distribution $\sigma_h = \sigma_v$. If $\sigma_h \neq \sigma_v$, such a failure may be expected even at smaller depths.

Conclusions

Though the above analysis has made use of several simplifying assumptions (axial symmetry, hydrostatic far-field stresses, etc.), several conclusions of general significance can be formulated. If the rock is a squeezing one (visco-plastic), in order to avoid failure by excessive deformation (excessive dilatancy), the appropriate design of tunnel support set-up is one in which the excavation layout and sequence would never reduce simultaneously to zero the pressure all along the entire circumference of the future surface of the tunnel. In other words, the final pressure p_∞ is to be reached by steadily decreasing step by step the initial stress (far-field stresses) to the final pressure p_∞, possibly by a decrease carried out successively along the contour.

Exercises

1. Prove formula (13.2.7).
2. Give a detailed derivation of the solution (13.2.9)—(13.2.13). By assigning various values to the parameters involved, show how the ultimate displacement and pressure depend on these parameters.
3. Give a detailed derivation of formula (13.2.15) and plot the p—t_0 curve for various values of G_0.
4. Using (13.2.19), find the failure boundary for various mechanical properties of the support and its size.
5. Using (13.2.20) and setting $p = \sigma_v$, find the increase in the wall thickness with depth.
6. Give a detailed derivation of the formulae (13.3.3)—(13.3.16) or establish similar ones for another arch geometry or a more elaborate distribution of frictional forces. Give a numerical example for the quantities involved in Fig. 13.3.2, by using various values for M. Determine the dynamic and static friction coefficients in the laboratory. Study the influence of the spacing L.
7. Study the dependence of the friction coefficients on the geometry of the cross-section of the yieldable shaft ring (i.e. for various values of α, d, n). Make laboratory experiments and check these results with the prediction of theoretical formulae.
8. Compress a piece of wooden plank of your choice. Find a compressibility relationship for that particular wood. Use it in an analysis of the type (13.3.18)—(13.3.27).
9. Repeat the test of Exercise 8, but use wet wooden planks.
10. Give other examples of non-linear support. For instance, consider a layer of broken rock between the ground and the outside surface of a circular cast-in-place concrete lining. Find by laboratory experiment a volumetric compressibility relationship for a stratum of broken rock. Use it in a non-linear lining analysis.
11. As in Problem 10, but for mortar.
12. Consider the first example from §13.3. Repeat the computation of the numerical example, but for $M = 29.4$ kN cm, $\mu_s = 0.3599$, and $\mu_d = 0.3346$. Then increase n (for instance $n = 3$). Reconsider the whole problem for any other similar support of your choice.
13. Give a detailed derivation of the formulae (13.4.12) and (13.4.14).
14. Show by numerical examples how the ground reaction curve depends on the loading history for constitutive equation (13.4.1). Consider various laws for the decrease of p.
15. Give all details necessary to obtain formulae (13.4.27)—(13.4.28).
16. Repeat the numerical examples shown in Fig. 13.4.3, but using as the lining model the one analysed in example §13.3.2.

Appendix 1

A short introduction to fracture mechanics
Prof. dr. Dorel Homentcovschi

A1.1. Introduction

As shown in previous chapters, randomly oriented microcracks are always present in most rocks. At the beginning of the deformation due to a compressive loading, these cracks are growing in a more or less random manner. However, at an advanced stage of compressive (uniaxial or triaxial) loading, the growth of the cracks is no longer random; there is a tendency for the microcracks to orient their growth according to the direction of the maximum compressive applied stress. In this Appendix an analytical motivation of this crack growth process is given on the basis of fracture mechanics (in the framework of linear plane elasticity theory).

Brace and Bombolakis (1963) and others have found that in compressive tests a flat crack does not grow in its own plane, but tends to orient itself in the direction of the main compressive applied stress. A complete mathematical theory of this phenomenon is due to S. Nemat-Nasser and H. Horii (see Nemat-Nasser and Horii (1982), Horii and Nemat-Nasser (1985, 1986)). In this Appendix a fracture theory is presented which provides the necessary information for the study of microcrack growth in rocks. It is assumed that the reader is familiar with the Kolosow–Muskhelishvili representation in plane elasticity. The development given here avoids the use of singular integral equations; instead, the problem is reduced to the solution of an infinite system of linear equations. Numerical work using this approach is still in progress, so the examples given here are taken from those obtained by Nemat-Nasser and Horii.

In the present Appendix, compressive stresses are considered to be negative.

A1.2. The fundamental relations of the plane theory of elasticity

The plane state of elastic equilibrium of a solid body is a model corresponding to two possible physical problems. The first one corresponds to the plane strain problem occurring when wide cylindrical bodies (with stress-free ends) are subjected to forces normal to the generatrices and constant along them. The second variant corresponds to the generalized plane stress-state encountered in thin plates subjected to the action of some forces situated in their median plane. We consider a Cartesian system of coordinates $0xy\hat{z}$, chosen so that the $x0y$ plane coincides either with the cross-section plane in the case of plane strain-state or with the median plane of the plate in the case of the generalized plane stress-state. What is common to the two physical problems is the fact that the elastic state of the body can be completely described in terms of three stress components σ_{xx}, σ_{yy}, σ_{xy} and two displacement components u, v, all being functions of two spatial variables x, y only. In both cases the mathematical problem is reduced to solving a biharmonic equation subject to certain boundary conditions.

In the plane theory of elasticity, the stress and the displacement components can be expressed in terms of two analytical functions $\Phi(z)$, $\Psi(z)$ of the complex variable $z = x + iy$ via the Kolosow and Muskhelishvili formulae (see Muskhelishvili (1953) and Homentcovschi (1986)):

$$\sigma_{xx} + \sigma_{yy} = 2\{\Phi(z) + \overline{\Phi(z)}\} \tag{A1.2.1}$$

$$\sigma_{yy} - \sigma_{xx} + 2i\sigma_{xy} = 2\{\bar{z}\Phi'(z) + \Psi(z)\} \tag{A1.2.2}$$

$$\begin{aligned} &2G(u + iv) = K\varphi(z) - z\overline{\Phi(z)} - \overline{\Psi(z)} \\ &\varphi'(z) = \Phi(z), \qquad \psi'(z) = \Psi(z) \end{aligned} \tag{A1.2.3}$$

where $G = E/(2(1 + \nu))$, E is Young's modulus, ν is the Poisson ratio and $K = 3 - 4\nu$ in the case of plane strain-state while $K = (3 - \nu)/(1 + \nu)$ in the case of generalized plane stress-state.

Let $\widehat{AB}$ be an arbitrary arch in the interior of the domain $\mathscr{D}$ occupied by the elastic body (Fig. A1.1). Let $X_n\,\mathrm{d}s$ and $Y_n\,\mathrm{d}s$ denote the components of the stress vector $\mathbf{F}_n\,\mathrm{d}s$ acting on the element $\mathrm{d}s$ of the arch $\widehat{AB}$. We have

$$i\int_0^s [X_n + iY_n]\,\mathrm{d}s = \varphi(t) + t\,\overline{\Phi(t)} + \overline{\psi(t)} + \text{const.} \tag{A1.2.4}$$

with t the complex variable on the arch $\widehat{AB}$ and s the curvilinear coordinate of point B with respect to the point A.

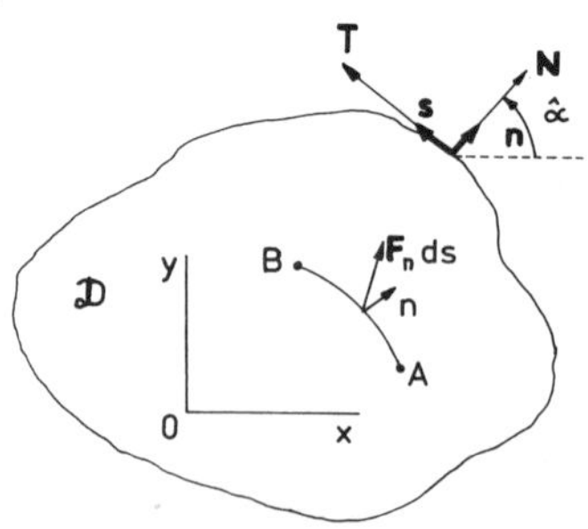

Figure A1.1 Notation in the domain $\mathscr{D}$, definition of the traction vector.

Let $\hat{\alpha}$ be the angle between the normal to $\widehat{AB}$ and the $0x$-axis. Then

$$\frac{\mathrm{d}t}{\mathrm{d}s} = \exp\left\{i\left(\hat{\alpha} + \frac{1}{2}\pi\right)\right\} = i\exp(i\hat{\alpha}), \qquad \frac{\mathrm{d}s}{\mathrm{d}t} = -i\exp(-i\hat{\alpha}). \quad \text{(A1.2.5)}$$

By differentiation of the relation (A1.2.4) we obtain

$$i(X_n + iY_n) = \frac{\mathrm{d}}{\mathrm{d}t}\{\varphi(t) + t\overline{\Phi(t)} + \overline{\psi(t)}\}\frac{\mathrm{d}t}{\mathrm{d}s}$$

and further

$$\varphi'(t) + \overline{\Phi(t)} + t\frac{\mathrm{d}\overline{\Phi(t)}}{\mathrm{d}\bar{t}}\frac{\mathrm{d}\bar{t}}{\mathrm{d}t} + \frac{\mathrm{d}\overline{\psi(t)}}{\mathrm{d}\bar{t}}\frac{\mathrm{d}\bar{t}}{\mathrm{d}t}$$

$$= (X_n + iY_n)\exp(-i\hat{\alpha}). \quad \text{(A1.2.6)}$$

We also have

$$(X_n + iY_n)\exp(-i\hat{\alpha}) = (X_n\cos\hat{\alpha} + Y_n\sin\hat{\alpha}) + i(Y_n\cos\hat{\alpha} - X_n\sin\hat{\alpha})$$
$$= N + iT$$

with N and T the projection of the stress vector on the normal and tangent to the arch $\widehat{AB}$, respectively. Finally, the relation (A1.2.6) becomes

$$N + iT = \Phi(t) + \overline{\Phi(t)} + \frac{\mathrm{d}\bar{t}}{\mathrm{d}t}\{t\overline{\Phi'(t)} + \overline{\Psi(t)}\}. \quad \text{(A1.2.7)}$$

The resultant of the stress vector $X_R + iY_R$ acting on the arch $\widehat{AB}$ (in the sense of the normal **n**) can be written with (A1.2.7) as

$$X_R + iY_R = \int_0^s (X_n + iY_n)\,\mathrm{d}s = -i\int_0^t (X_n + iY_n)\exp(-i\hat{\alpha})\,\mathrm{d}t$$

$$= -i\int_0^t (N + iT)\,\mathrm{d}t. \quad \text{(A1.2.8)}$$

In a similar way, by differentiation of (A1.2.3) along the arch $\widehat{AB}$ it follows that

$$2G\frac{\mathrm{d}}{\mathrm{d}t}(u + iv) = K\Phi(t) - \overline{\Phi(t)} - \frac{\mathrm{d}\bar{t}}{\mathrm{d}t}\{t\overline{\Phi'(t)} + \overline{\Psi(t)}\}. \quad \text{(A1.2.9)}$$

The left-hand side of this relation can also be expressed in terms of intrinsic variables (independent of the system of coordinates). Thus

$$\frac{\mathrm{d}}{\mathrm{d}t}(u+iv)=\frac{\mathrm{d}}{\mathrm{d}s}(u+iv)\frac{\mathrm{d}s}{\mathrm{d}t}=\left(\frac{\mathrm{d}u}{\mathrm{d}s}+i\frac{\mathrm{d}v}{\mathrm{d}s}\right)(-i)\exp(-i\hat{\alpha})$$

$$=-i\left\{\frac{\mathrm{d}}{\mathrm{d}s}[(u+iv)\exp(-i\hat{\alpha})]+\right.$$

$$\left.+[(u+iv)\exp(-i\hat{\alpha})]\,i\frac{\mathrm{d}\hat{\alpha}}{\mathrm{d}s}\right\}.$$

Finally we have

$$\frac{\mathrm{d}}{\mathrm{d}t}(u+iv)=-i\frac{\mathrm{d}}{\mathrm{d}s}(v_{\mathrm{n}}+iv_{\mathrm{t}})+(v_{\mathrm{n}}+iv_{\mathrm{t}})\tilde{k},$$

with $\tilde{k}$ the curvature of the arch $\widehat{AB}$ at the point considered, and v_{n} and v_{t} the projections of the displacement vector on the normal to the arch $\widehat{AB}$ and on the tangent, respectively. In the case when $\overline{AB}$ is a segment of a straight line, $\tilde{k}=0$, and therefore

$$\frac{\mathrm{d}}{\mathrm{d}t}(u+iv)=-i\frac{\mathrm{d}}{\mathrm{d}s}(v_{\mathrm{n}}+iv_{\mathrm{t}}).$$

If the multiply connected domain $\mathscr{D}$ is bounded by the outer boundary $\mathscr{C}_0$ and the inner boundaries $\mathscr{C}_1,\ldots,\mathscr{C}_m$, in order to have uniform displacements in $\mathscr{D}$ the functions $\Phi(z)$ and $\Psi(z)$ must be of the form

$$\Phi(z)=-\frac{1}{2\pi(1+K)}\sum_{j=1}^{m}\frac{X_{Rj}+iY_{Rj}}{z-z_j}+\Phi^*(z)$$

$$\Psi(z)=\frac{K}{2\pi(1+K)}\sum_{j=1}^{m}\frac{X_{Rj}-iY_{Rj}}{z-z_j}+\Psi^*(z).$$

Here, $X_{Rj}+iY_{Rj}$ denotes the stress resultant on the boundary $\mathscr{C}_j$ ($j=1,\ldots,m$), z_j is a point in the interior of $\mathscr{C}_j$, while $\Phi^*(z)$, $\Psi^*(z)$ are holomorphic functions in $\mathscr{D}$ with holomorphic primitives in the domain (therefore analytic and uniform).

In the case in which the curve $\mathscr{C}_0$ is unbounded, the behaviour of the functions $\Phi(z)$ and $\Psi(z)$ at infinity must be prescribed. The condition of uniformity for the displacements leads to the following representations in the neighbourhood of the point at infinity:

$$\Phi(z) = -\frac{X_{RT} + iY_{RT}}{2\pi(1 + K)z} + B + iC + \frac{a_2}{z^2} + \frac{a_3}{z^3} + \dots$$

$$\Psi(z) = \frac{K(X_{RT} - iY_{RT})}{2\pi(1 + K)z} + B' + iC' + \frac{b_2}{z^2} + \frac{b_3}{z^3} + \dots .$$

Here

$$X_{RT} + iY_{RT} = \sum_{j=1}^{m} (X_{Rj} + iY_{Rj})$$

is the resultant of the external forces applied on the boundary of the domain, and the constants B, B', C' are determined by the uniform state of the stress field at infinity, while C describes the rotation at infinity.

If the far field stresses of the elastic body are defined by the principal stress components p_∞ and q_∞, it follows that

$$B = \frac{p_\infty + q_\infty}{4}, \qquad B' + iC' = -\frac{p_\infty - q_\infty}{2} \exp(-2i\gamma)$$

with γ the angle between the $0x$-axis and the direction of p_∞.

The far-field stresses generate on the element of arch of orientation $\hat{\alpha}$ the tractions obtained from (A1.2.7) as

$$(N + iT)_\infty = \frac{p_\infty + q_\infty}{2} + \frac{p_\infty - q_\infty}{2} \exp(2i(\gamma - \hat{\alpha})). \tag{A1.2.10}$$

A1.3. Integral representation of the functions $\Phi(z)$ and $\Psi(z)$

In what follows we are giving some integral representations of the functions $\Phi(z)$ and $\Psi(z)$, which involve physical quantities from (A1.2.7) and (A1.2.9). Summing these relations we obtain

$$\Phi(t) = \frac{2G}{1 + K} \frac{\mathrm{d}}{\mathrm{d}t} (u + iv) + \frac{1}{1 + K} (N + iT). \tag{A1.3.1}$$

If the arch $\widehat{AB}$ is a discontinuity line, the relation (A1.3.1) holds on each of its sides:

$$\begin{aligned} \Phi^+(t) &= \frac{2G}{1 + K} \frac{\mathrm{d}}{\mathrm{d}t} (u + iv)^+ + \frac{1}{1 + K} (N + iT)^+ \\ \Phi^-(t) &= \frac{2G}{1 + K} \frac{\mathrm{d}}{\mathrm{d}t} (u + iv)^- + \frac{1}{1 + K} (N + iT)^- \end{aligned} \tag{A1.3.2}$$

where $\Phi^+(t)$ is the limiting value of the function $\Phi(z)$ when $z \to t \in \widehat{AB}$ from the left-hand side of the arch $\widehat{AB}$ (from the opposite side of **n**) while $\Phi^-(t)$ is the limiting value obtained from the right-hand side of the function at point t. From the sum and the difference of the two relations (A1.3.2) we get

$$\begin{aligned}\Phi^+(t) + \Phi^-(t) &= \frac{2}{1+K} f'(t) + \frac{2}{1+K} p(t) \\ \Phi^+(t) - \Phi^-(t) &= iQ(t),\end{aligned} \tag{A1.3.3}$$

where the Savruk (1981) notation is used:

$$\begin{aligned}&(N+iT)^{\pm} = p(t) \pm q(t) \\ &2G \frac{\mathrm{d}}{\mathrm{d}t}(u+iv)^{\pm} = f'(t) \mp \frac{1+K}{2i} g'(t) \\ &g'(t) - \frac{2i}{1+K} q(t) = Q(t).\end{aligned} \tag{A1.3.4}$$

From the second relation (A1.3.3) we can obtain integral representation of the functions $\Phi(z)$. Thus taking into account the relations given by Plemelj (see Homentcovschi (1986)) we have

$$\Phi(z) = \frac{1}{2\pi}\int_{\widehat{AB}} \frac{Q(t)}{t-z}\,\mathrm{d}t + \Phi_0(z) = \Phi_{\widehat{AB}}(z) + \Phi_0(z). \tag{A1.3.5}$$

The function $\Phi_0(z)$ which is holomorphic on the arch $\widehat{AB}$ characterizes the influence of other discontinuity lines situated inside the domain $\mathscr{D}$, the influence of the boundaries, and/or that of the far field elastic state (in the case of unbounded domains).

Taking into account the relation (A1.3.1), the formula (A1.2.7) yields

$$\bar{t}\,\Phi'(t) + \Psi(t) = (N - iT)\frac{\mathrm{d}\bar{t}}{\mathrm{d}t} - (\Phi(t) + \overline{\Phi(t)})\frac{\mathrm{d}\bar{t}}{\mathrm{d}t} \tag{A1.3.6}$$

which holds on each side of a possible discontinuity line $\widehat{AB}$. Taking into account (A1.3.3) and (A1.3.4) from (A1.3.6) it follows that

$$[\bar{t}\,\Phi'(t) + \Psi(t)]^+ - [\bar{t}\,\Phi'(t) + \Psi(t)]^- = i\{\overline{Q(t)} - Q(t) - 2i\overline{q(t)}\}\frac{\mathrm{d}\bar{t}}{\mathrm{d}t}.$$

The jump of the function $\Phi'(t)$ when crossing the arch $\widehat{AB}$ is obtained by differentiation of the second relation (A1.3.3) and hence

$$\Psi^+(t) - \Psi^-(t) = -i\frac{\mathrm{d}}{\mathrm{d}t}(\bar{t}Q(t)) + i[\overline{Q(t)} - 2i\overline{q(t)}]\frac{\mathrm{d}\bar{t}}{\mathrm{d}t}.$$

The function $\Psi(z)$ holomorphic in $\mathscr{D}$ has the representation

$$\Psi(z) = -\frac{1}{2\pi}\int_{\widehat{AB}} \frac{\mathrm{d}(\bar{t}Q(t))}{t-z} + \frac{1}{2\pi}\int_{\widehat{AB}} \frac{\overline{Q(t)} - 2i\overline{q(t)}}{t-z}\,\mathrm{d}\bar{t} + \Psi_0(z). \tag{A1.3.7}$$

In the case in which $\widehat{AB}$ is a closed curve, the first integral from (A1.3.7) can be transformed by integration by parts to

$$\Psi(z) = -\frac{1}{2\pi}\int_{\widehat{AB}} \frac{\bar{t}Q(t)}{(t-z)^2}\,dt + \frac{1}{2\pi}\int_{\widehat{AB}} \frac{\overline{Q(t)} - 2i\overline{q(t)}}{t-z}\,d\bar{t} + \Psi_0(z)$$

$$= \Psi_{\widehat{AB}}(z) + \Psi_0(z) \tag{A1.3.8}$$

where the function $\Psi_0(z)$ plays the same role as the function $\Phi_0(z)$. The relationship (A1.3.8) holds also for more general cases, when the arch $\widehat{AB}$ is open and the function $Q(t)$ has integrable singularities at the extremities of the arch.

Let us observe that in the relation (A1.2.7) and (A1.2.9) the function $\Psi(z)$ enters through

$$S(z, \bar{z}) = \bar{z}\Phi'(z) + \Psi(z)$$

only. With the representations (A1.3.5) and (A1.3.8) we can write

$$S(z, \bar{z}) = \frac{1}{2\pi}\int_{\widehat{AB}} \frac{(\bar{z} - \bar{t})Q(t)}{(t-z)^2}\,dt + \frac{1}{2\pi}\int_{\widehat{AB}} \frac{\overline{Q(t)} - 2i\overline{q(t)}}{t-z}\,d\bar{t} +$$

$$+ \bar{z}\Phi_0'(z) + \Psi_0(z) \equiv S_{\widehat{AB}}(z, \bar{z}) + S_0(z, \bar{z}). \tag{A1.3.9}$$

The relations (A1.2.1)–(A1.2.3), (A1.3.5), (A1.3.7) allow us to express the stresses and displacements in the elastic media in terms of the jumps of the stresses and displacements across the discontinuity line $\widehat{AB}$. The advantage of the representation (A1.3.5), (A1.3.8), due to Savruk (1981), is that the densities involved in the Cauchy-type integrals are functions with obvious mechanical meaning. Moreover, in many physical problems these (or some of their combinations) are just the mechanical parameters which are of interest and are to be determined by solving the problem under consideration.

A1.4. The main boundary-value problems

Let us assume now that the arch $\widehat{AB}$ is a crack in the elastic medium. In this case the condition $(A1.3.4)_1$ or $(A1.3.4)_2$ (or some combination of them) is the boundary condition of the problem. The difference between these two relations, corresponding respectively to the two sides '+' and '−' of the crack, has been used in order to establish the integral representation of the analytical functions $\Phi(z)$ and $\Psi(z)$. Adding the two conditions $(A1.3.4)_1$ and $(A1.3.4)_2$ corresponding to the two sides of the arch $\widehat{AB}$, we obtain

$$(N + iT)^+ + (N + iT)^- = 2p(t) \tag{A1.4.1}$$

$$G\left\{\frac{d}{dt}(u + iv)^+ + \frac{d}{dt}(u + iv)^-\right\} = f'(t). \tag{A1.4.2}$$

Using here the representation of the functions Φ and Ψ given in the previous section, we get the integral equation of the problem.

In the case in which the stresses are prescribed on the two sides of the crack, we have to solve the *first boundary-value problem* of plane elasticity; the functions $p(t)$ and $q(t)$ are prescribed, while the functions $f'(t)$ and $g'(t)$ must be determined. The integral representation (A1.3.5) and (A1.3.7) involves the unknown function $Q(t)$ (which replaces $f'(t)$) besides the prescribed density $q(t)$. The function $Q(t)$ must satisfy a normalization condition which follows from the requirement that the displacements be single-valued: $g(a) = g(b) = 0$ (a and b are the complex points corresponding to the tips A and B of the crack). From the relation $(A1.3.4)_3$, it follows that

$$\int_a^b Q(t)\,\mathrm{d}t = -\frac{2i}{1+K}\int_a^b q(t)\,\mathrm{d}t.$$

From (A1.2.8) and $(A1.3.4)_1$ we also obtain

$$2\int_a^b q(t)\,\mathrm{d}t = \int_a^b \{(N+iT)^+ - (N+iT)^-\}\,\mathrm{d}t = i(\hat{X}_\mathrm{R} + i\hat{Y}_\mathrm{R}). \quad \text{(A1.4.3)}$$

$\hat{X}_\mathrm{R}$ and $\hat{Y}_\mathrm{R}$ are the components of the resultant of all the forces which are acting on the two sides of the crack on the $0x$- and $0y$-axis, respectively. Finally the single-valued condition of the displacements becomes

$$\int_a^b Q(t)\,\mathrm{d}t = \frac{1}{1+K}(\hat{X}_\mathrm{R} + i\hat{Y}_\mathrm{R}). \quad \text{(A1.4.4)}$$

In the case of the first boundary-value problem, the relation (A1.4.1) yields the integral equation for the determination of the function $Q(t)$ (and therefore of the function $g'(t)$) while the relation (A1.4.2) further determines the function $f'(t)$.

The second main boundary-value problem consists in the determination of the elastic state in the case when the displacements along the arch $\widehat{AB}$ are prescribed; in addition to the conditions at infinity and on the other boundaries. It is assumed that the functions $f'(t)$ and $g'(t)$ are known while $p(t)$ and $q(t)$ are to be determined. The function $q(t)$ is determined from the integral equation which is derived from relation (A1.4.2), and further the function $p(t)$ follows directly from relation (A1.4.1). The general theory (Muskhelishvili (1953)) shows that in this case it is still necessary to prescribe (by the formulation of the problem) the resultant $\hat{X}_\mathrm{R} + i\hat{Y}_\mathrm{R}$ of the stresses on the two sides of the crack; according to (A1.4.3) this can be written as

$$\int_a^b q(t)\,\mathrm{d}t = \frac{i}{2}(\hat{X}_\mathrm{R} + i\hat{Y}_\mathrm{R}). \quad \text{(A1.4.5)}$$

Note that, taking into account the definition $(A1.3.4)_3$ of the function $Q(t)$, the relation (A1.4.5) reduces to (A1.4.4) with the corresponding interpretation of the parameters involved.

The first and second boundary-value problems generally occur in the case when the elastic body is subjected to tensile stresses at infinity (or on the outer boundary). If at large distances the elastic body is subjected to compression ($p_\infty < 0$, $q_\infty < 0$) the crack is closed and the boundary conditions of the problem are of a *mixed type*:

$$\begin{gathered} v_n^+ = v_n^-, \qquad T^+ = T^-, \qquad N^+ = N^- \\ T^+ = -\tau_c + \mu^* N^+ \end{gathered} \tag{A1.4.6}$$

with τ_c the cohesion and μ^* the friction coefficient between the two sides of the crack; both are positive constants assumed to be known. Relation $(A1.4.6)_1$ states that the crack is closed; $(A1.4.6)_4$ expresses the friction between the two sides of the crack. From (A1.4.6) we also have

$$(N + iT)^+ - (N + iT)^- = 0, \quad \text{i.e. } q(t) = 0. \tag{A1.4.7}$$

In the case of rectilinear crack, $\tilde{k} = 0$ and then $(A1.4.6)_1$ yields

$$\operatorname{Im}\left\{\frac{\mathrm{d}}{\mathrm{d}t}(u + iv)^+ - \frac{\mathrm{d}}{\mathrm{d}t}(u + iv)^-\right\} = 0$$

and therefore

$$\operatorname{Re}\{g'(t)\} = 0, \qquad \operatorname{Re}\{Q(t)\} = 0. \tag{A1.4.8}$$

The equation for the determination of $Q(t)$ (a pure imaginary function) is obtained from $(A1.4.6)_4$:

$$\operatorname{Re}\{(\mu^* + i)(N + iT)^+\} - \tau_c = 0. \tag{A1.4.9}$$

A1.5. The influence functions corresponding to the elementary crack

In what follows we assume that the crack $\overline{AB}$ is rectilinear with the tips at a and b. Since the cracks we are studying are constituted by several rectilinear segments (disjoint or not), or can be approximated by such kinds of elements, the rectilinear finite crack $\overline{ab}$ will be referred as the *elementary crack*.

In the relation (A1.3.5) we consider the change of variables

$$\begin{aligned} t &= \frac{b-a}{2}\xi + \frac{b+a}{2} \qquad (t \to \xi) \\ z &= \frac{b-a}{2}Z + \frac{b+a}{2} \qquad (z \to Z). \end{aligned} \tag{A1.5.1}$$

It then follows that

$$\Phi_{\overline{AB}}(z) = \frac{1}{2\pi}\int_{-1}^{+1}\frac{\tilde{Q}(\xi)}{\xi - Z}\,\mathrm{d}\xi, \tag{A1.5.2}$$

where

$$Q\left(\frac{b-a}{2}\,\xi+\frac{b+a}{2}\right)\equiv\tilde{Q}(\xi)=\frac{\hat{Q}(\xi)}{\sqrt{1-\xi^2}}\,. \tag{A1.5.3}$$

Assuming the form (A1.5.3) for the function $\tilde{Q}(\xi)$ implies that the derivatives of the displacements have singularities at the crack tips. The type of the singularity follows from the analysis of the integral equation of the problem (Muskhelishvili (1953)). For the problem under consideration here it will be shown that expression (A1.5.3) of the function $\tilde{Q}(\xi)$ allows us to satisfy all the boundary conditions of the problem. The function $\hat{Q}(\xi)$ will be developed in a series of Chebychev polynomials

$$\hat{Q}(\xi)=\sum_{n=0}^{\infty}{}' \,Q_n T_n(\xi),\quad |\xi|<1, \tag{A1.5.4}$$

where $T_n(\xi)=\cos(n\arccos\xi)$, while the prime denotes that in the sum the coefficient Q_0 is divided by 2. Introducing (A1.5.3) and (A1.5.4) into (A1.5.2) we have

$$\Phi_{\overline{\mathrm{AB}}}(z)=-\frac{1}{2\sqrt{Z^2-1}}\sum_{n=0}^{\infty}{}'\frac{Q_n}{(Z+\sqrt{Z^2-1})^n}\,. \tag{A1.5.5}$$

Here the formula (Homentcovschi (1986))

$$\frac{1}{\pi}\int_{-1}^{+1}\frac{T_n(\xi)}{\sqrt{1-\xi^2}}\,\frac{\mathrm{d}\xi}{\xi-Z}=-\frac{1}{\sqrt{Z^2-1}}\,\frac{1}{(Z+\sqrt{Z^2-1})^n}$$

has also been used; the complex radical is defined on the plane Z with the cut $[-1, 1]$ on the real axis and we consider the square-root determination which is real and positive for $Z=X>1$. Similarly the relations (A1.3.8) and (A1.3.9) yield the expressions

$$\Psi_{\overline{\mathrm{AB}}}(z)=\frac{\bar{b}-\bar{a}}{b-a}\,\frac{1}{\sqrt{Z^2-1}}\left\{\sum_{n=0}^{\infty}{}'\left(\frac{Q_n-\overline{Q_n}}{2}+i\overline{q_n}\right)\frac{1}{(Z+\sqrt{Z^2-1})^n}-\right.$$

$$\left.-\frac{Z+(\bar{b}+\bar{a})/(\bar{b}-\bar{a})}{2(Z^2-1)}\sum_{n=0}^{\infty}{}'\,Q_n\,\frac{Z+n\sqrt{Z^2-1}}{(Z+\sqrt{Z^2-1})^n}\right\} \tag{A1.5.6}$$

$$S_{\overline{\mathrm{AB}}}(z,\bar{z})=\frac{\bar{b}-\bar{a}}{b-a}\,\frac{i}{\sqrt{Z^2-1}}\left\{\sum_{n=0}^{\infty}{}'\left(\frac{Q_n-\overline{Q_n}}{2i}+\overline{q_n}\right)\frac{1}{(Z+\sqrt{Z^2-1})^n}+\right.$$

$$\left.+\frac{i}{2}\,\frac{Z-\overline{Z}}{Z^2-1}\sum_{n=0}^{\infty}{}'\,Q_n\,\frac{Z+n\sqrt{Z^2-1}}{(Z+\sqrt{Z^2-1})^n}\right\}. \tag{A1.5.7}$$

Similar as in (A1.5.3) and (A1.5.4) we have used

$$q\left(\frac{b-a}{2}\,\xi+\frac{b+a}{2}\right)\equiv\tilde{q}(\xi)=\frac{\sum\limits_{n=0}^{\infty}{}' \, q_n T_n(\xi)}{\sqrt{1-\xi^2}}\,. \tag{A1.5.8}$$

The influence of the elementary crack $\overline{\mathrm{AB}}$ on the stress and displacement state in any point of the elastic domain can be obtained directly from the formulae (A1.5.5), (A1.5.7) (via the Kolosow—Muskhelishvili formula). Further, the relations (A1.2.7) and (A1.2.9) determine the stress vector and displacement on any curvilinear element. Let us observe also that the functions $\Phi_{\overline{\mathrm{AB}}}(z)$, $S_{\overline{\mathrm{AB}}}(z, \bar{z})$ are continuous outside the cut $\overline{\mathrm{AB}}$ so that the influence of the elementary crack $\overline{\mathrm{AB}}$ on any other element is obtained directly by replacing the variable z by its corresponding value.

On the crack $\overline{\mathrm{AB}}$ the functions $\Phi_{\overline{\mathrm{AB}}}(z)$ and $S_{\overline{\mathrm{AB}}}(z, \bar{z})$ are discontinuous. We will estimate the influence of the elementary crack $\overline{\mathrm{AB}}$ on itself by considering the two sides separately. Putting $Z = \cos\theta$ we have

$$\begin{aligned}(N+iT)^{\pm}_{\overline{\mathrm{AB}}} &= \sum_{n=1}^{\infty}(Q_n+iq_n)\frac{\sin n\theta}{\sin\theta} \pm \sum_{n=0}^{\infty}{}' \, q_n\frac{\cos n\theta}{\sin\theta}\\ 2G\left[\frac{\mathrm{d}}{\mathrm{d}t}(u+iv)\right]^{\pm}_{\overline{\mathrm{AB}}} &= \sum_{n=1}^{\infty}\left(\frac{K-1}{2}Q_n-iq_n\right)\frac{\sin n\theta}{\sin\theta}\mp\\ &\quad\mp\sum_{n=0}^{\infty}{}'\left(\frac{K+1}{2i}Q_n+q_n\right)\frac{\cos n\theta}{\sin\theta}\,.\end{aligned} \tag{A1.5.9}$$

Making an identification of these formulae with (A1.3.4) we also obtain

$$\begin{aligned}&\sum_{n=0}^{\infty}{}' \, q_n\frac{\cos n\theta}{\sin\theta}=q(t)\\ &\sum_{n=0}^{\infty}{}'\left(\frac{K+1}{2i}Q_n+q_n\right)\frac{\cos n\theta}{\sin\theta}=\frac{K+1}{2i}g'(t)\\ &\sum_{n=1}^{\infty}(Q_n+iq_n)\frac{\sin n\theta}{\sin\theta}+(N+iT)_0=p(t)\\ &\sum_{n=1}^{\infty}\left(\frac{K-1}{2}Q_n-iq_n\right)\frac{\sin n\theta}{\sin\theta}+2G\left[\frac{\mathrm{d}}{\mathrm{d}t}(u+iv)\right]_0=f'(t).\end{aligned} \tag{A1.5.10}$$

In these relations

$$t = \frac{b-a}{2}\cos\theta + \frac{b+a}{2}.$$

The first two formulae (A1.5.10) are identities if we take into account (A1.5.3), (A1.5.4), and (A1.5.8). Finally, the formulae (A1.5.10) provide, via the boundary condition, the equations for the determination of the coefficients Q_n or q_n.

It must be observed that in the last two relations (A1.5.10) the terms containing Q_0 and q_0 are missing. In this case the system of equations are supplemented with the compatibility conditions (A1.4.4) and (A1.4.5). Using in these formulae the representations given above for the functions $Q(t)$ and $q(t)$ we obtain the relations

$$\begin{aligned} Q_0 &= \frac{2}{(b-a)\pi}\,\frac{1}{1+K}\,(\hat{X}_{\mathrm{R}} + i\hat{Y}_{\mathrm{R}}) \\ q_0 &= \frac{i}{(b-a)\pi}\,(\hat{X}_{\mathrm{R}} + i\hat{Y}_{\mathrm{R}}) \end{aligned} \tag{A1.5.11}$$

which are used in the first and second main problems, respectively.

A1.6. The Griffith crack in the plane problem

In order to point out how the formulae established above are used, the solution of the problem of an elastic plane containing a stress-free rectilinear crack $\overline{\mathrm{AB}}$ will be considered. The elastic state at infinity is characterized by the principal stresses p_∞, q_∞ (A1.2.10). Let 2ℓ be the length of the crack, and let the system of coordinates be chosen so that the origin is placed is the middle of the crack

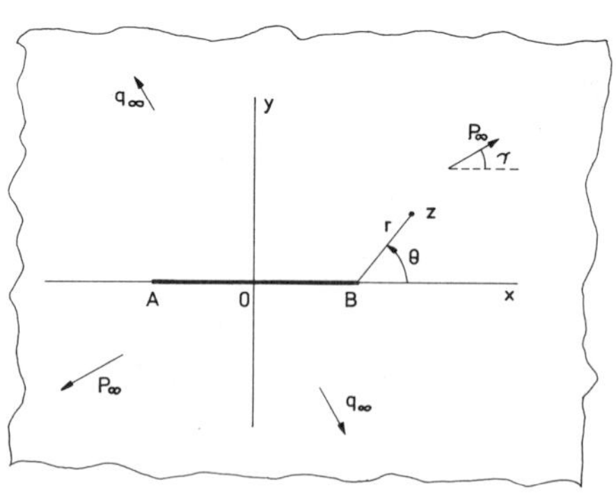

Figure A1.2. The Griffith crack.

and the direction of the $0x$-axis is directed along $\overline{\mathrm{AB}}$ (Fig. A1.2). In this case we have

$$\Phi_0 = \frac{p_\infty + q_\infty}{2}, \qquad \Psi_0 = -\frac{p_\infty - q_\infty}{2}\exp(-2i\gamma)$$

and taking into account the relations (A1.2.10) in (A1.5.10) we obtain

$$\begin{aligned}
&\sum_{n=0}^{\infty}{}' q_n \frac{\cos n\theta}{\sin\theta} = 0 \\
&\sum_{n=1}^{\infty} (Q_n + iq_n)\frac{\sin n\theta}{\sin\theta} + \frac{p_\infty + q_\infty}{2} - \frac{p_\infty - q_\infty}{2}\exp(2i\gamma) = 0.
\end{aligned} \tag{A1.6.1}$$

From the first relation (A1.6.1) we get $q_n = 0$ $(n = 0, 1, \ldots)$, and from the second

$$\begin{aligned}
&Q_1 = -\frac{p_\infty + q_\infty}{2} + \frac{p_\infty - q_\infty}{2}\exp(2i\gamma) \\
&Q_j = 0 \quad (j = 2,3, \ldots).
\end{aligned} \tag{A1.6.2}$$

The coefficient Q_0 is determined by the first relation (A1.5.11) as $Q_0 = 0$.

Since the origin is placed in the middle of the crack it follows $Z = z/\ell$, $a = -\ell$, $b = \ell$ and therefore

$$\begin{aligned}
&\Phi_{\overline{\mathrm{AB}}}(z) = -\frac{Q_1}{2}\left(\frac{z}{\sqrt{z^2 - \ell^2}} - 1\right) \\
&S_{\overline{\mathrm{AB}}}(z, \bar{z}) = i\left\{\frac{Q_1 - \overline{Q_1}}{2i}\left(\frac{z}{\sqrt{z^2 - \ell^2}} - 1\right) + \right. \\
&\qquad \left. + i\frac{Q_1}{2}\ell^2\frac{z - \bar{z}}{(z^2 - \ell^2)^{3/2}}\right\}.
\end{aligned} \tag{A1.6.3}$$

Let $z - \ell = r\exp(i\theta')$ (Fig. A1.2). In the neighbourhood of the tip $z = b$ of the crack the functions $\Phi_{\overline{\mathrm{AB}}}(z)$ and $S_{\overline{\mathrm{AB}}}(z, \bar{z})$ have a development of the form

$$\begin{aligned}
&\Phi_{\overline{\mathrm{AB}}}(z) = -\frac{Q_1\sqrt{\ell}}{2\sqrt{2}}\frac{\exp(-i(\theta'/2))}{\sqrt{r}} + O(r^0) \\
&S_{\overline{\mathrm{AB}}}(z, \bar{z}) = i\frac{Q_1 - \overline{Q_1}}{2i}\frac{\sqrt{\ell}}{\sqrt{2r}}\exp(-i\theta'/2) - \\
&\qquad - \frac{Q_1\sqrt{\ell}}{4\sqrt{2}}(1 - \exp(-2i\theta'))\frac{\exp(-i\theta'/2)}{\sqrt{r}} + O(r^0).
\end{aligned}$$

$O(r^0)$ denotes the terms which are bounded for $r \to 0$. The Kolosow–Muskhelishvili relations yield the following expressions for the stress and displacement fields

$$\begin{bmatrix} \sigma_{xx} \\ \sigma_{yy} \\ \sigma_{xy} \end{bmatrix} = \frac{k_1}{4\sqrt{2\pi r}} \begin{bmatrix} 3\cos\frac{\theta'}{2} + \cos\frac{5\theta'}{2} \\ 5\cos\frac{\theta'}{2} - \cos\frac{5\theta'}{2} \\ -\sin\frac{\theta'}{2} + \sin\frac{5\theta'}{2} \end{bmatrix} + $$

$$+ \frac{k_2}{4\sqrt{2\pi r}} \begin{bmatrix} -7\sin\frac{\theta'}{2} - \sin\frac{5\theta'}{2} \\ -\sin\frac{\theta'}{2} + \sin\frac{5\theta'}{2} \\ 3\cos\frac{\theta'}{2} + \cos\frac{5\theta'}{2} \end{bmatrix} + O(r^0) \tag{A1.6.4}$$

$$4G(u + iv) = (k_1 - ik_2)\sqrt{\frac{r}{2\pi}}\,(2K\exp(i\theta'/2) - \exp(-i\theta'/2)) -$$

$$-(k_1 + ik_2)\sqrt{\frac{r}{2\pi}}\exp(3i\theta'/2) +$$

$$+ 2ik_2\sqrt{\frac{r}{2\pi}}\exp(-i\theta'/2)$$

with the notation

$$k_1 - ik_2 = -\sqrt{\pi\ell}\,Q_1 = \sqrt{\pi\ell}\left(\frac{p_\infty + q_\infty}{2} - \frac{p_\infty - q_\infty}{2}\exp(2i\gamma)\right). \tag{A1.6.5}$$

From the relations (A1.6.4) written on the two sides of the crack (for $\theta' = \pi$ and $\theta' = -\pi$, respectively) it follows also that

$$k_1 - ik_2 = -\lim_{\substack{t \to \ell \\ t < \ell}} [\sqrt{2\pi|t - \ell|}\,g'(t)] \tag{A1.6.6}$$

and also for $\theta' = 0$

$$k_1 + ik_2 = \lim_{\substack{z \to \ell \\ z = \ell + r}} \sqrt{2\pi|z - \ell|}(\sigma_{yy} + i\sigma_{xy}). \tag{A1.6.7}$$

The relation (A1.6.6) expresses the coefficient $k_1 - ik_2$ in terms of parameters which are independent of the particular coordinate system used.

The above considerations can be extended to the more general case of an elementary crack situated in a bounded domain subjected to an arbitrary stress and displacement field (on the boundary and at infinity). In this case the contribution of the elementary crack $\overline{AB}$ is expressed by the relations (A1.5.5), (A1.5.7) while the behaviour of the stresses and displacements in the neighbourhood of the tip $z = b$ (in a system of coordinates attached to the crack as in Fig. A.1.2) is also of the form (A1.6.4), where in this case

$$k_1 - ik_2 = -\sqrt{\pi \frac{|b - a|}{2}} \sum_{n=0}^{\infty} Q_n. \tag{A1.6.8}$$

A1.7. Stress intensity factors and criteria for the propagation of the crack

The coefficients k_1 and k_2 defined by the relation (A1.6.6) are referred to as stress intensity factors corresponding to a symmetric distribution and an antisymmetric distribution, respectively, of the loading with respect to the crack. The mechanical meaning follows from the analysis of the two fundamental modes of plane displacement. In the first mode (Fig. A1.3*a*) $\gamma = 0$, $p_\infty = 0$, $q_\infty \neq 0$, and the forces to which the body is subjected are shown in Fig. A1.3*b*. It follows that

$$k_1 = \sqrt{\pi \ell}\, q_\infty, \qquad k_2 = 0.$$

The coefficient k_2 corresponds to the second mode of plane displacement of the body with a crack (Fig. A1.4*a*). The loading is characterized in this case by the relations $\gamma = \frac{1}{4}\pi$, $p_\infty \neq 0$, $q_\infty = -p_\infty$, and is shown in Fig. A1.4*b*. In this case the relation (A1.6.5) yields

$$k_2 = \sqrt{\pi \ell}\, p_\infty, \qquad k_1 = 0.$$

The relations (A1.6.4) and (A1.6.6)–(A1.6.7) established for the case of elementary crack hold also for a stress-free curvilinear crack (Cherepanov (1979)) in the interior of an arbitrary elastic body.

The examination of the solution obtained for the elementary crack shows that, in the neighbourhood of the tip, the stresses become infinite, which is unacceptable from the mechanical point of view. In fact, for most real materials, high stresses lead to the appearance of some enclaves around the tips of the

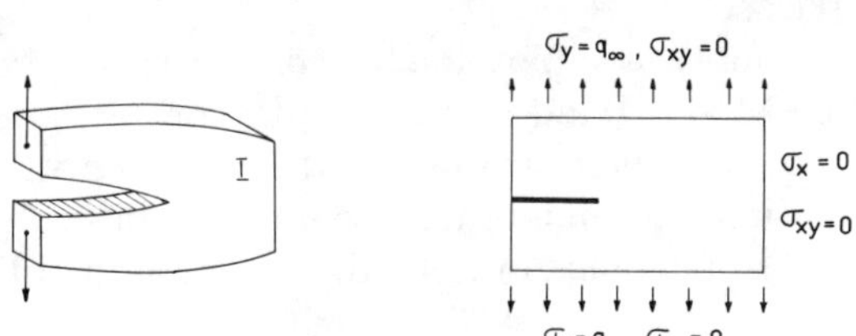

Figure A1.3. The first basic mode of deformation of the crack surface (opening).

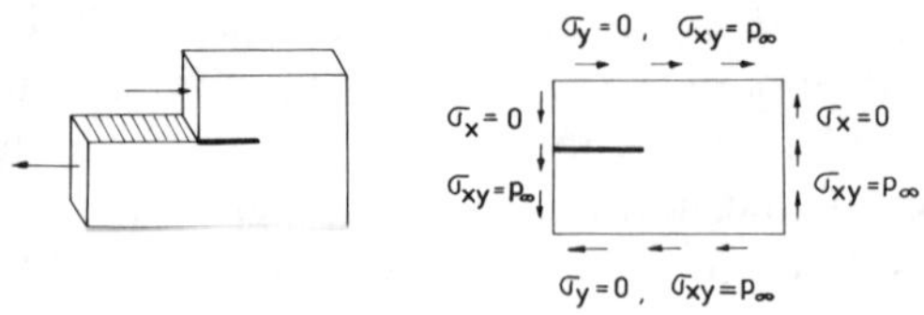

Figure A1.4. The second basic mode of deformation of the crack surface (sliding).

crack (for instance some plastic domains) where the material response becomes non-linear, and where the stress and displacement distributions are not the same as those obtained for linear elasticity. For quasi-brittle materials, the domains of non-linear effects are assumed to be quite small with respect to the dimensions of the elastic body. As a consequence, the shape of these 'enclaves' of non-linear behaviour, as well as the intensity of plastic strains in them, are completely determined by the properties of the material and by the stress state from the neighbouring domain characterized by the relations (A1.6.4). This shows the importance of the coefficients k_1 and k_2 in the problems concerning the mechanics of cracks in quasi-brittle elastic media; i.e., the relations (A1.6.4) describe completely the stress state in the neighbourhood of the tips of crack. In other words, the influence of the geometrical shape of the elastic body (and the geometrical shape of the crack) as well as the external loadings, are expressed through the coefficients k_1 and k_2 only; they can be considered to be a measure of the loading of the body as 'felt' by the material in the neighbourhood of the crack tip.

Since for quasi-brittle materials the stress intensity factors characterize in a unique way the loading state at the tips of the crack, the criteria which describe crack propagation are formulated in terms of coefficients k_1 and k_2. These criteria are related to the response of the material to various loadings and therefore they cannot be formulated only in terms of the parameters characterizing the elastic state of the corresponding body.

The appearance and the development of the crack are accompanied by plastic deformation of the material in the neighbourhood of the crack tips, which results in a decrease in the elastic energy of the medium considered. In the case of a symmetric loading (Mode I), a rectilinear crack grows in its initial plane. The initiation of the propagation phenomenon is reached when the Mode I stress intensity factor k_1 takes on the value k_1^* which is experimentally determined. These results are based on energetic considerations according to which the limit equilibrium state is characterized by a maximal value of the rate of elastic energy release.

In the case of a more complex loading of the elastic body (for instance a combination of the Modes I and II loadings) the problem becomes much more involved: it requires the determination of the direction of propagation of the crack and also its limiting equilibrium length. The difficulties of the problem have resulted in the appearance of a multitude of criteria of crack propagation. We mention some of them:

(a) The criterion of maximal rate of elastic energy release: the crack propagates in the direction which ensures the maximum of the rate of elastic energy release.

(b) The condition of local symmetry according to which the direction of propagation of the crack is characterized by the vanishing of the coefficient k_2 at the tip of the extended crack.

(c) A maximum condition for the hoop component of the stress on the extension of the crack.

Further, the limit equilibrium state is characterized by making precise a critical value (determined experimentally) of the rate of the elastic energy release or of the coefficient k_1 of the stress intensity factor at the tip of the extended crack. The length and orientation of the crack is determined with the help of these criteria.

In order to use the criteria presented above we must determine the stress state on the virtually extended cracks in various directions.

The first criterion is physically more appealing, since it may be related to the second law of thermodynamics (Nemat-Nasser (1979)). In the case of cracks which propagate in directions close to the initial one, it has been shown that criteria (*a*) and (*b*) are equivalent, while the numerical results by Nemat-Nasser and Horii (1982) show that in the problems concerning the out-of-plane extension of cracks under overall compression the results obtained with the criteria (*b*) and (*c*) are essentially the same. Though significant efforts have been devoted, the problem of the best criterion for crack propagation (and of the direction of propagation) is still open. The interested reader may find additional information in Husain *et al.* (1974), and Palaniswamy and Knauss (1978).

Taking into account the above discussion, in what follows we apply the criterion of local symmetry in a slightly modified version: the direction of propagation of the crack is the one which ensures the maximum value for the Mode I stress intensity factor k_1 at the tip of the extended crack.

A1.8. Systems of rectilinear cracks

Let us consider now a pattern of cracks which form a system of J elementary cracks. The contribution of the elementary crack $[a_j, b_j]$ to the solution is expressed by

$$\Phi_j(Z_j) = -\frac{1}{2}\,\frac{1}{\sqrt{Z_j^2-1}}\,\sum_{n=0}^{\infty}{}'\,\frac{Q_{jn}}{(Z_j+\sqrt{Z_j^2-1})^n}$$

$$S_j(Z_j) = \frac{i\exp(-2i\alpha_j)}{\sqrt{Z_j^2-1}}\,\sum_{n=0}^{\infty}{}'\left\{\frac{Q_{jn}-\overline{Q_{jn}}}{2i}+\overline{q_{jn}}+\right.$$

$$\left.+\frac{i}{2}\,(Z_j+n\sqrt{Z_j^2-1})\,\frac{Z_j-\overline{Z_j}}{Z_j^2-1}\,Q_{jn}\right\}\frac{1}{(Z_j+\sqrt{Z_j^2-1})^n} \qquad \text{(A1.8.1)}$$

where

$$\exp\{-2i\alpha_j\} = (\bar{b}_j - \bar{a}_j)/(b_j - a_j),$$

and

$$Z_j = \frac{2}{b_j - a_j}\left(z - \frac{a_j + b_j}{2}\right) \tag{A1.8.2}$$

is the current point of the complex plane with respect to a system of coordinates attached to the crack. In relations (A1.8.1), q_{jn} and Q_{jn} are coefficients which have to be determined from the boundary conditions.

The influence of the elementary crack $[a_j, b_j]$ on the elementary crack $[a_m, b_m]$ $(m \neq j)$ is obtained from (A1.8.1) by replacing Z_j by Z_{jm}:

$$Z_{jm} = \frac{2}{b_j - a_j}\left(\frac{b_m - a_m}{2}\cos\theta + \frac{b_m + a_m}{2} - \frac{b_j + a_j}{2}\right). \tag{A1.8.3}$$

The expressions

$$\Phi_{jm} = \Phi_j(Z_{jm}), \qquad \hat{S}_{jm} = \exp(2i\alpha_m)S_j(Z_{jm}) \quad (j \neq m) \tag{A1.8.4}$$

are continuous across the crack $[a_m, b_m]$.

The loading $N + iT$ and the displacement $v_n + iv_t$ on the two sides of the crack $[a_m, b_m]$ follow from

$$(N - iT)_m^{\pm} = \sum_{\substack{j=1\\ j\neq m}}^{J} \{\Phi_{jm} + \overline{\Phi_{jm}} + \hat{S}_{jm}\} + (N_0 - iT_0)_m + \\ + \sum_{n=1}^{\infty}{}' (\overline{Q_{mn}} - i\overline{q_{mn}})\frac{\sin n\theta}{\sin\theta} \pm \sum_{n=0}^{\infty}{}' q_{mn}\frac{\cos n\theta}{\sin\theta} \tag{A1.8.5}$$

$$2Gi\frac{\mathrm{d}}{\mathrm{d}s}(v_n - iv_t)_m^{\pm} = \sum_{\substack{j=1\\ j\neq m}}^{J} \{K\overline{\Phi_{jm}} - \Phi_{jm} - \hat{S}_{jm}\} + \\ + 2Gi\frac{\mathrm{d}}{\mathrm{d}s}(v_{n0} - iv_{t0})_m + \\ + \sum_{n=1}^{\infty}\left(\frac{K-1}{2}\overline{Q_{mn}} + i\overline{q_{mn}}\right)\frac{\sin n\theta}{\sin\theta} \pm \\ \pm \sum_{n=0}^{\infty}{}'\left(\frac{K+1}{2i}\overline{Q_{mn}} - \overline{q_{mn}}\right)\frac{\cos n\theta}{\sin\theta}. \tag{A1.8.6}$$

Let us now consider the consequences of various types of boundary conditions on the representations (A1.8.5), (A1.8.6).

(a) If at the crossing of the elementary crack $[a_m, b_m]$ the stresses are continuous, relation (A1.8.5) yields

$$q_{mn} = 0 \qquad (n = 0, 1, 2, \dots). \tag{A1.8.7}$$

(b) For stress-free cracks, besides (A1.8.7), the condition

$$\sum_{n=1}^{\infty} \overline{Q_{mn}} \frac{\sin n\theta}{\sin \theta} + \sum_{\substack{j=1 \\ j \neq m}}^{J} \{\Phi_{jm} + \overline{\Phi_{jm}} + \hat{S}_{jm}\} + (N_0 - iT_0)_m = 0 \tag{A1.8.8}$$

also holds.

(c) The continuity condition of the displacement normal to the (closed) crack together with condition (a) yields

$$Q_{mn} = iQ''_{mn} \quad \text{(pure imaginary)} \quad \operatorname{Re} Q_{mn} = 0, \quad (n = 0, 1, 2, \dots).$$

(d) Finally, the condition of slip with friction along $[a_m, b_m]$ associated to the conditions (a) and (c) yields

$$- \sum_{n=1}^{\infty} Q_{mn} \frac{\sin n\theta}{\sin \theta} + \mu^* \sum_{\substack{j=1 \\ j \neq m}}^{J} \{\Phi_{jm} + \overline{\Phi_{jm}}\} +$$

$$+ \operatorname{Re} \left\{ (\mu^* - i) \left[\sum_{\substack{j=1 \\ j \neq m}}^{J} \hat{S}_{jm} + (N_0 - iT_0)_m \right] \right\} - \tau_c = 0. \tag{A1.8.9}$$

A1.9. Application to the crack kinking problem

We shall apply the above method to study the problem of kinked crack extension from an initially closed rectilinear crack which is under an overall far field compression and which may undergo simple frictional slip. The pre-existing rectilinear flaw $[a_1, b_1]$ is subjected at infinity to a compression characterized by the principal stresses p_∞, q_∞ (both negative, Fig. A1.5). The out-of-plane growth of the pre-existing straight flaw takes place along the symmetric cracks $[a_2, b_2]$ and $[a_3, b_3]$ making the angle α with the initial direction of the flaw. The problem is solved for various values of the angle α and lengths ℓ of the extended crack (modelled by rectilinear cracks); the direction of the crack growth follows from the application of the criteria established at the end of §A1.7. Normalizing all the lengths by the length of the pre-existing flaw, we obtain three elementary cracks $[a_1, b_1]$, $[a_2, b_2]$, $[a_3, b_3]$ characterized by the geometrical parameters

$$\begin{aligned} a_1 &= -1, & b_1 &= 1, & \alpha_1 &= 0 \\ a_2 &= 1, & b_2 &= 1 + \ell \exp(i\alpha), & \alpha_2 &= \alpha \\ a_3 &= -1, & b_3 &= -1 - \ell \exp(i\alpha), & \alpha_3 &= \alpha + \pi. \end{aligned} \tag{A1.9.1}$$

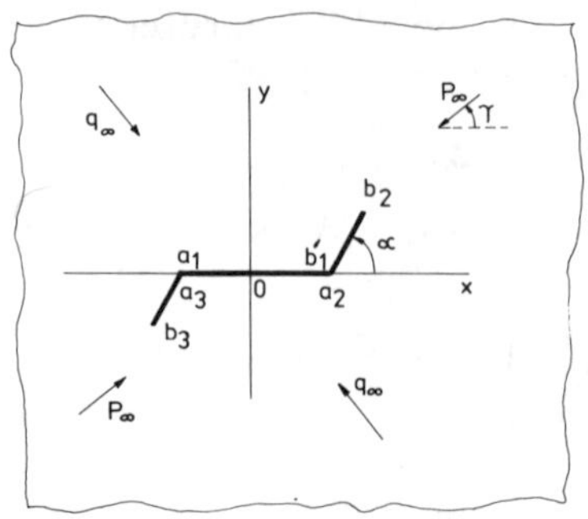

Figure A1.5. Pre-existing closed crack $a_1\,b_1$ and its compression-induced kinked extensions $a_2\,b_2$ and $a_3\,b_3$.

From the stress continuity along the whole crack, it follows that

$$q_{1n} = q_{2n} = q_{3n} = 0, \quad (n = 0, 1, 2, \ldots). \tag{A1.9.2}$$

The symmetry of the problem with respect to the point 0 yields also

$$Q_{2n} = Q_{3n} \qquad (n = 0, 1, 2, \ldots), \quad Q_{1,\,2k+1} = 0 \quad (k = 0, 1, 2, \ldots) \tag{A1.9.3}$$

Let us now examine the other boundary conditions. On the pre-existing flaw $[a_1,\, b_1]$ we assume the continuity of the component v_y of the displacement as well as a friction law of the form (A1.8.9). Thus

$$Q_{1n} = iQ''_{1n} \quad \text{(pure imaginary)}, \quad \mathrm{Re}\, Q_{1n} = 0 \tag{A1.9.4}$$

$$- \sum_{n=1}^{\infty} Q''_{1n} \sin n\theta + \mu^* \sin\theta \left\{ \Phi_{21} + \overline{\Phi_{21}} + \Phi_{31} + \overline{\Phi_{31}} \right\} +$$

$$+ \mathrm{Re}\left\{ (\mu^* - i)[\hat{S}_{21} + \hat{S}_{31} + N_{01} - iT_{01}] \sin\theta \right\} - \tau_c \sin\theta = 0. \tag{A1.9.5}$$

The kinked extensions are free of tensions because the kinks are tension cracks that do not transmit force. Hence we have

$$\sum_{n=1}^{\infty} \overline{Q_{2n}} \sin n\theta + \sin\theta \left\{ \Phi_{12} + \overline{\Phi_{12}} + \hat{S}_{12} + \Phi_{32} + \overline{\Phi_{32}} + \right.$$

$$\left. + \hat{S}_{32} + N_{02} - iT_{02} \right\} = 0. \tag{A1.9.6}$$

In the above relations $N_{01} - iT_{01}$, $N_{02} - iT_{02}$ characterize the influence of the far field stress-state on the pre-existing flaw $[a_1,\, b_1]$ and crack $[a_2,\, b_2]$. Thus according to the relation (A1.2.10) we get

$$\begin{aligned} N_{01} - iT_{01} &= \frac{p_\infty + q_\infty}{2} - \frac{p_\infty - q_\infty}{2} \exp(-2i\gamma) \\ N_{02} - iT_{02} &= \frac{p_\infty + q_\infty}{2} - \frac{p_\infty - q_\infty}{2} \exp(-2i(\gamma - \alpha)). \end{aligned} \tag{A1.9.7}$$

Let us now discuss the additional conditions. The condition of single-valuedness of the displacements (A1.4.4) becomes

$$\int_{[b_3,\, a_3]} Q(t)\,\mathrm{d}t + \int_{[a_1,\, b_1]} Q(t)\,\mathrm{d}t + \int_{[a_2,\, b_2]} Q(t)\,\mathrm{d}t = 0.$$

Due to symmetry, the two integrals of the tension cracks cancel and we have

$$\int_{[a_1,\, b_1]} Q(t)\,\mathrm{d}t = 0.$$

Therefore,

$$Q_{10} = 0. \tag{A1.9.8}$$

On the other hand, since the points b_1 (or a_2) and a_1 (or b_2) are not the tips of the cracks, the singularities of the functions $f'(t)$ and $g'(t)$ at these points are weaker than $\frac{1}{2}$. In this case, from (A1.5.3) and (A1.5.4) it follows that

$$\sum_{n=0}^{\infty}{}' \; Q_{1n} = 0, \qquad \sum_{n=0}^{\infty}{}' \; (-1)^n Q_{1n} = 0 \tag{A1.9.9}$$

or

$$\sum_{n=0}^{\infty}{}' \; (-1)^n Q_{2n} = 0. \tag{A1.9.10}$$

The parameter of interest in this problem is the coefficient $k_1^{(2)} - ik_2^{(2)}$ corresponding to the kinked extension $[a_2,\, b_2]$. It can be estimated from (A1.6.8) as

$$k_1^{(2)} - ik_2^{(2)} = -\sqrt{\pi\ell/2}\, \sum_{n=0}^{\infty}{}' \; Q_{2n}.$$

The relations (A1.9.5), (A1.9.6) are the system for the determination of Q''_{1n}, Q'_{2n}.

A1.10. Some numerical and experimental results

The numerical computations carried out according to the above algorithm are not as yet complete. Hence, in what follows, some of the numerical results obtained by Nemat-Nasser and Horii (1982) and Horii and Nemat-Nasser (1986) will be presented. They have studied analytically the problem of kinked crack extension by using singular integral equations (instead of our infinite system of linear equations). The solution of the integral equations was obtained numerically.

Under the axial compression ($p_\infty < 0$) the initial flaw $(a_1,\, b_1)$ extends in the directions (a_2, b_2) and (a_3, b_3) making the angle α (of about 72°) with the direction of initial flaw (the exact value of this angle depends on the friction factor and the direction of the pre-existing flaw). Subsequently the crack grows with the increasing axial load. In Fig. A1.6 the dependence of the kink angle α on the initial flaw orientation γ is shown for various values of the kink length ℓ, $\mu = 0.3$, and $q_\infty = 0$.

The crack grows toward the direction of maximum far field compressive stress, and eventually α attains the value γ. An experimental check of this result is due to Nemat-Nasser and Horii (1982).

The variation of the maximum value of the stress intensity factor $K = k_{1\max}/(|p_\infty|\sqrt{\pi})$ with respect to the kink length ℓ for various values of p_∞ and q_∞ is shown in Fig. A1.7. If k_c is the critical value of k_1 (the material's fracture toughness for the opening mode) then the kink length can be evaluated by taking $k_{1\max} = k_c$ in Fig. A1.7. When $q_\infty = 0$, this coefficient decreases monotonically, approaching zero when ℓ becomes very large. Therefore, in order to increase the kink length, the axial compression must be increased and thus the crack growth process is stable.

If a lateral compression ($q_\infty \neq 0$) accompanies the axial one, then k_1

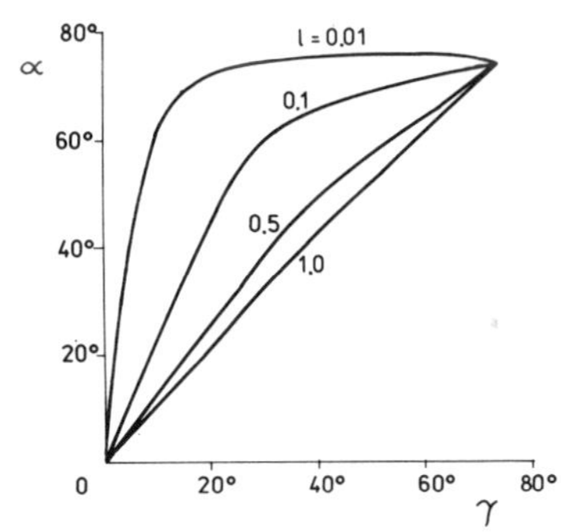

Figure A1.6. Optimal crack orientation angle as function of initial flaw orientation for $\mu = 0.3$, $q_\infty = 0$ and $q_\infty/p_\infty = 0$ (Horii and Nemat-Nasser (1986)).

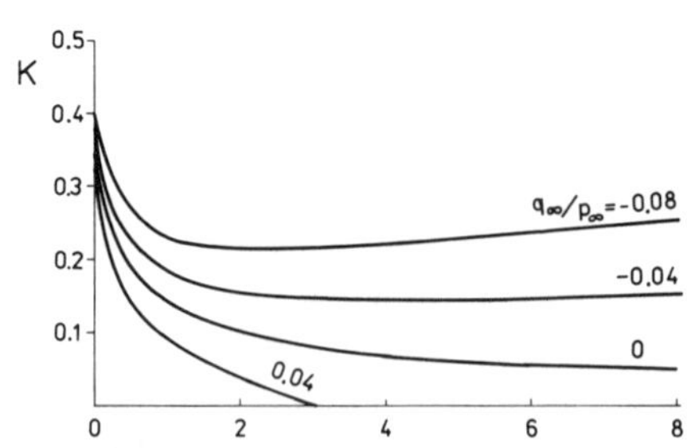

Figure A1.7. Normalized maximum opening mode stress intensity factor as a function of crack length for indicated rate q_∞/p_∞; the curves are the exact (numerical) calculations (Horii and Nemat-Nasser (1986)).

decreases, becoming zero for a finite value of ℓ. The process of crack extension then is stable and stops at a certain length of the extended crack.

If a lateral tension accompanies the axial compression ($q_\infty > 0$), the coefficient k_1 first decreases with increasing ℓ up to a certain critical value (the minimum point on the corresponding curves) and afterwards increases with ℓ. In other words, in order to increase the crack up to the critical value the axial compression must increase, but afterwards the kink crack continues to increase even if the axial compression decreases. Therefore the kink growth is stable up to the point of minimum on the curves shown in Fig. A1.7 and afterwards becomes unstable, leading to axial splitting. For further experimental and theoretical study of this subject, the reader may consult Horii and Nemat-Nasser (1985, 1986).

Appendix 2

Creep and stress variation around a well or a tunnel. A numerical approach

In order to obtain a general solution for the creep and stress variation around a well or a tunnel (see §11.4.2 and 12.4) a numerical method has been used. For this purpose the following system of partial differential equations must be integrated: the balance of momentum (11.1.6)

$$\frac{\partial \sigma_r}{\partial r} + \frac{\sigma_r - \sigma_\theta}{r} = 0, \tag{A2.1}$$

the compatibility condition (11.1.7)

$$\frac{\partial \varepsilon_\theta}{\partial r} + \frac{\varepsilon_\theta - \varepsilon_r}{r} = 0, \tag{A2.2}$$

and the constitutive equation (11.4.11)

$$\begin{aligned} \dot{\varepsilon}_r &= \left(\frac{1}{3K} - \frac{1}{2G}\right)\dot{\sigma} + \frac{1}{2G}\dot{\sigma}_r + k\left\langle 1 - \frac{W^{\mathrm{I}}}{H}\right\rangle \frac{\partial H}{\partial \sigma_r} \\ \dot{\varepsilon}_\theta &= \left(\frac{1}{3K} - \frac{1}{2G}\right)\dot{\sigma} + \frac{1}{2G}\dot{\sigma}_\theta + k\left\langle 1 - \frac{W^{\mathrm{I}}}{H}\right\rangle \frac{\partial H}{\partial \sigma_\theta} \\ 0 &= \left(\frac{1}{3K} - \frac{1}{2G}\right)\dot{\sigma} + \frac{1}{2G}\dot{\sigma}_z + k\left\langle 1 - \frac{W^{\mathrm{I}}}{H}\right\rangle \frac{\partial H}{\partial \sigma_z}. \end{aligned} \tag{A2.3}$$

In these equations the stresses are secondary (actual stresses after excavation) and the strains are relative.

As already mentioned in §11.4.2, this system is a parabolic–hyperbolic system with five characteristic lines: r = const. (three times) and t = const. (twice).

The initial conditions at $t = 0$ are obtained from the initial secondary stress and relative strain states

$$
\begin{aligned}
\sigma_r &= \sigma_r^{S_0} = (p - \sigma_h)\frac{a^2}{r^2} + \sigma_h \\
\sigma_\theta &= \sigma_\theta^{S_0} = -(p - \sigma_h)\frac{a^2}{r^2} + \sigma_h \\
\sigma_z &= \sigma_z^{S_0} = \sigma_v \\
\varepsilon_r &= \varepsilon_r^{R_0} = \frac{p - \sigma_h}{2G}\frac{a^2}{r^2} \\
\varepsilon_\theta &= \varepsilon_\theta^{R_0} = -\frac{p - \sigma_h}{2G}\frac{a^2}{r^2} \\
\varepsilon_z &= \varepsilon_z^{R_0} = 0.
\end{aligned}
\tag{A2.4}
$$

The boundary conditions formulated at the orifice $r = a$ and at infinity are

$$r = a, \qquad t \geqslant 0 : \sigma_r = p \tag{A2.5}$$

$$\left.\begin{matrix} r \to \infty \\ t \geqslant 0 \end{matrix}\right\} : \begin{matrix} \sigma_r = \sigma_\theta = \sigma_h, \qquad \sigma_z = \sigma_v \\ \varepsilon_r = \varepsilon_\theta = \varepsilon_z = 0. \end{matrix} \tag{A2.6}$$

For practical purposes the boundary conditions at infinity will be formulated at a certain great (conventionally chosen) distance from the opening $r = r_\infty$ ($= 10a$, say), where it will be assumed that the stresses and strains coincide with the initial values (A2.4) computed for $r = r_\infty$.

In the system of five equations (A2.1)—(A2.3), five unknown quantities are involved: the secondary stresses σ_r, σ_θ, and σ_z, and the relative strains ε_r and ε_θ. Solving $(A2.3)_3$ with respect to $\dot{\sigma}_z$ and replacing in $(A2.3)_{1,2}$ we get

$$
\begin{aligned}
\dot{\varepsilon}_r &= \frac{3K + 4G}{4G(3K + G)}\dot{\sigma}_r + \frac{2G - 3K}{4G(3K + G)}\dot{\sigma}_\theta + \\
&+ k\left\langle 1 - \frac{W^{I}}{H}\right\rangle\left[\frac{3K - 2G}{2(3K + G)}\frac{\partial H}{\partial \sigma_z} + \frac{\partial H}{\partial \sigma_r}\right] \\
\dot{\varepsilon}_\theta &= \frac{2G - 3K}{4G(3K + G)}\dot{\sigma}_r + \frac{3K + 4G}{4G(3K + G)}\dot{\sigma}_\theta + \\
&+ k\left\langle 1 - \frac{W^{I}}{H}\right\rangle\left[\frac{3K - 2G}{2(3K + G)}\frac{\partial H}{\partial \sigma_z} + \frac{\partial H}{\partial \sigma_\theta}\right].
\end{aligned}
\tag{A2.7}
$$

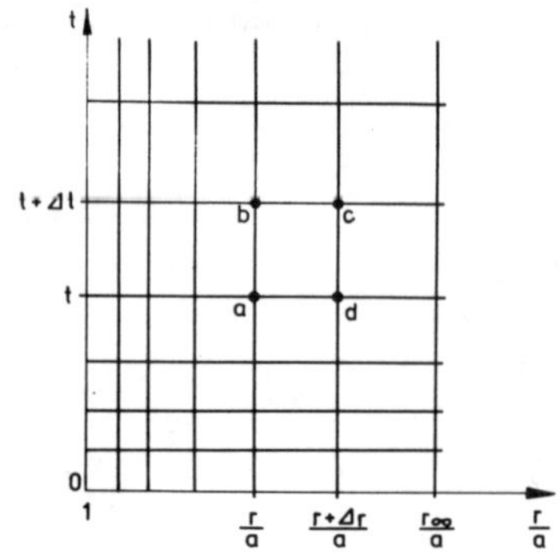

Figure A2.1. A grid used in numerical integration.

Thus the system which is written in finite differences is (A2.1), (A2.2), and (A2.7), and subsequently $(A2.3)_3$ too. For a space and time discretization in the plane r, t, a net with variable mesh sizes Δr and Δt is considered (Fig. A2.1). The integration is carried out in the strip $a \leqslant r \leqslant r_\infty$, $t \geqslant 0$. Let us consider a generic loop a, b, c, d, and assume that at the moment t the solution is already known from previous integration. For $t + \Delta t$ the integration starts from $r = a$, taking into account (A2.5), and assuming to a first approximation that σ_r, σ_θ and σ_z at $t + \Delta t$ are the same as at moment t. The strains are obtained from (11.4.7) as if a creep under constant stress occurs, i.e.

$$\varepsilon_r(t+\Delta t) = \varepsilon_r(t) + \left[\frac{\left\langle 1 - \dfrac{W^{\mathrm{I}}}{H} \right\rangle \dfrac{\partial H}{\partial \sigma_r}}{\dfrac{1}{H}\dfrac{\partial H}{\partial \boldsymbol{\sigma}} \cdot \boldsymbol{\sigma}} \times \left\{ 1 - \exp\left[-\frac{k}{H}\frac{\partial H}{\partial \boldsymbol{\sigma}} \cdot \boldsymbol{\sigma}\,\Delta t \right] \right\} \right]\Bigg|_t, \tag{A2.8}$$

where all the functions are estimated at time t, including W^{I}. In particular, at the first point $t = \Delta t$ we have $W^{\mathrm{I}}(\Delta t) = W^{\mathrm{IP}}$. Thus the solution along $r = a$ is obtained to a first approximation.

For the following vertices along $t =$ const. the formulae (A2.1), (A2.2) and (A2.7) are written in finite differences as

$$(\sigma_r)_{c(i+1)} = (\sigma_r)_b - \left[\left(\frac{\sigma_r - \sigma_\theta}{r}\right)_b (1-q) + \left(\frac{\sigma_r - \sigma_\theta}{r}\right)_{c(i)} q\right] \Delta r$$

$$(\varepsilon_\theta)_{c(i+1)} = (\varepsilon_\theta)_b - \left[\left(\frac{\varepsilon_\theta - \varepsilon_r}{r}\right)_b (1-q) + \left(\frac{\varepsilon_\theta - \varepsilon_r}{r}\right)_{c(i)} q\right] \Delta r$$

$$(\sigma_\theta)_{c(i+1)} = (\sigma_\theta)_d + \frac{4G(3K+G)}{3K+4G} \left\{ (\varepsilon_\theta)_{c(i+1)} - (\varepsilon_\theta)_d - \right.$$

$$- \frac{2G-3K}{4G(3K+G)} ((\sigma_r)_{c(i+1)} - (\sigma_r)_d) -$$

$$\left. - k[(A)_d(1-q) + (A)_{c(i)}\, q]\, \Delta t \right\} \qquad \text{(A2.9)}$$

$$(\varepsilon_r)_{c(i+1)} = (\varepsilon_r)_d + \frac{3K+4G}{4G(3K+G)} ((\sigma_r)_{c(i+1)} - (\sigma_r)_d) +$$

$$+ \frac{2G-3K}{4G(3K+G)} ((\sigma_\theta)_{c(i+1)} - (\sigma_\theta)_d) + k[(B)_d(1-q) + (B)_{c(i)}\, q]\, \Delta t$$

$$(\sigma_z)_{c(i+1)} = (\sigma_z)_d - \frac{9KG}{3K+G} \left\{ \frac{1}{3} \left(\frac{1}{3K} - \frac{1}{2G} \right) ((\sigma_r)_{c(i+1)} + (\sigma_\theta)_{c(i+1)} - \right.$$

$$\left. - (\sigma_r)_d - (\sigma_\theta)_d) + k[(C)_d (1-q) + (C)_{c(i)}\, q]\, \Delta t \right\},$$

where $(\sigma_r)_{c(i+1)}$ means the $(i+1)$th iteration of σ_r at point c, etc., and

$$A = \left\langle 1 - \frac{W^{\mathrm{I}}}{H} \right\rangle \left[\frac{3K-2G}{2(3K+G)} \frac{\partial H}{\partial \sigma_z} + \frac{\partial H}{\partial \sigma_\theta} \right]$$

$$B = \left\langle 1 - \frac{W^{\mathrm{I}}}{H} \right\rangle \left[\frac{3K-2G}{2(3K+G)} \frac{\partial H}{\partial \sigma_z} + \frac{\partial H}{\partial \sigma_r} \right] \qquad \text{(A.2.10)}$$

$$C = \left\langle 1 - \frac{W^{\mathrm{I}}}{H} \right\rangle \frac{\partial H}{\partial \sigma_z}.$$

In each vertex, W^{I} is computed according to the formula (11.4.3) which is written in finite differences as

$$W^{\mathrm{I}}(t) = W^{\mathrm{IP}} + \sum^{t} \Delta W^{\mathrm{I}}(t) \quad (\text{fixed } r), \tag{A2.11}$$

with

$$(\Delta W^{\mathrm{I}}(t))_{c(i+1)} = k\{(\sigma_r)_{(i+1)}[(B)_d(1-q) + (B)_{c(i)}\, q] + \\ + (\sigma_\theta)_{(i+1)}[(A)_d(1-q) + (A)_{c(i)}\, q]\}\, \Delta t. \tag{A2.12}$$

The summation in (A2.11) is carried out with respect to t, i.e. for $r = \text{const}$. The first iteration corresponds to $i = 0$, when the terms with the subscript $c(0)$ are zero.

A variable mesh size Δr was considered in the space discretization, with smaller values near $r = a$ and increasing values as r increases. No more than three or four iterations are necessary at each mesh point.

In the above formulae, $0 \leqslant q \leqslant 1$ is a parameter generally introduced to accelerate the convergence (Zienkiewicz (1977) Chapter 18), but here it is used to match the boundary conditions from $r = a$ with those from $r = r_\infty$. In other words q was varied between 0 and 1 in order to obtain at $r = r_\infty$ a numerical solution as close as possible to the 'elastic' solution (A2.4) since it is considered that the irreversible components are at $r = r_\infty$ already negligible. Thus the most appropriate value of q for increasing r was obtained.

After reaching $r = r_\infty$, the computations are repeated backward for decreasing r (and constant t), starting from the exact values of the variables according to the 'elastic' solution (A2.4). For decreasing r, the following formulae are used:

$$(\sigma_r)_{b(i+1)} = (\sigma_r)_c + \left\{\left(\frac{\sigma_r - \sigma_\theta}{r}\right)_c (1-q) + \left[\left(\frac{\sigma_r - \sigma_\theta}{r}\right)_{b(-1)} X + \right.\right. \\ \left.\left. + \left(\frac{\sigma_r - \sigma_\theta}{r}\right)_{b(i)} (1-X)\right] q\right\} \Delta r$$

$$(\varepsilon_\theta)_{b(i+1)} = (\varepsilon_\theta)_c + \left\{\left(\frac{\varepsilon_\theta - \varepsilon_r}{r}\right)_c (1-q) + \left[\left(\frac{\varepsilon_\theta - \varepsilon_r}{r}\right)_{b(-1)} X + \right.\right. \\ \left.\left. + \left(\frac{\varepsilon_\theta - \varepsilon_r}{r}\right)_{b(i)} (1-X)\right] q\right\} \Delta r$$

$$(\sigma_\theta)_{b(i+1)} = (\sigma_\theta)_a + \frac{2G\left(\dfrac{1}{G} + \dfrac{1}{3K}\right)}{\dfrac{1}{2G} + \dfrac{2}{3K}} \left\{ (\varepsilon_\theta)_{b(i+1)} - (\varepsilon_\theta)_a + \right.$$

$$+\frac{\frac{1}{2G}-\frac{1}{3K}}{2G\left(\frac{1}{G}+\frac{1}{3K}\right)}((\sigma_r)_{b(i+1)}-(\sigma_r)_a)-k[(A)_a(1-q)+$$

$$\left.+((A)_{b(-1)}X+(A)_{b(i)}(1-X))q]\,\Delta t\right\} \qquad \text{(A2.13)}$$

$$(\varepsilon_r)_{b(i+1)}=(\varepsilon_r)_a+\frac{3K+4G}{4G(3K+G)}((\sigma_r)_{b(i+1)}-(\sigma_r)_a)-$$

$$-\frac{3K-2G}{4G(3K+G)}((\sigma_\theta)_{b(i+1)}-(\sigma_\theta)_a)+k[(B)_a(1-q)+$$

$$+((B)_{b(-1)}X+(B)_{b(i)}(1-X))q]\,\Delta t$$

$$(\sigma_z)_{b(i+1)}=(\sigma_z)_a+\frac{9KG}{3K+G}\left\{\frac{1}{3}\left(\frac{1}{3K}-\frac{1}{2G}\right)((\sigma_r)_{b(i+1)}+\right.$$

$$+(\sigma_\theta)_{b(i+1)}-(\sigma_r)_a-(\sigma_\theta)_a)+k[(C)_a(1-q)+$$

$$\left.+((C)_{b(-1)}X+(C)_{b(i)}(1-X))q]\,\Delta t\right\},$$

where

$$X=\begin{cases}0 & \text{if } r=a\\ 1 & \text{if } r\neq a.\end{cases} \qquad \text{(A2.14)}$$

$W^{\mathrm{I}}(t)$ is again obtained from the approximate formula (A2.11) with

$$(\Delta W^{\mathrm{I}}(t))_{b(i+1)}=k\{(\sigma_r)_{b(i+1)}[(B)_a(1-q)+((B)_{b(-1)}X+$$
$$+(B)_{b(i)}(1-X))q]+(\sigma_\theta)_{b(i+1)}[(A)_a(1-q)+$$
$$+((A)_{b(-1)}X+(A)_{b(i)}(1-X))q]\}\,\Delta t. \qquad \text{(A2.15)}$$

In the above formulae the index (-1) corresponds to the solution at the same mesh point obtained when r was increased. According to (A2.13) the first iteration corresponds to $i = 0$, when all the terms labelled with the index $b(0)$ are zero. Again, a variable space discretization is used with decreasing Δr, when r decreases. The parameter X defined by (A2.14) is introduced in order to avoid useless computations for $r \neq a$ but to get an accurate solution at $r = a$, by iterations.

The parameter q was varied again so that when reaching for $t =$ const. the last vertex $r = a$, the value of the component σ_r be as close as desired to the prescribed data (A2.5). It was found that for increasing r the optimal value of q

varied mainly in the interval $0.6 \leqslant q \leqslant 1$, while during decreasing r mainly in the interval $0.4 \leqslant q \leqslant 1$. By varying the parameter q it was possible to match very well the boundary conditions prescribed at $r = a$ with those at $r = r_\infty$.

The time is further increased by another increment Δt, etc., until stabilization of the creep and stress variation occur. In the time discretization, a variable mesh size Δt was used, Δt increasing with t. Since it is important to determine the time of stabilization as accurately as possible by the numerical procedure, the interval Δt was again decreased in the neighbourhood of the time of stabilization.

Generally, only two or three iterations were necessary at a vertex with the exception of those vertices, where by increasing r by a step Δr, one passes from a space interval where the bracket $\langle\ \ \rangle \neq 0$, to another one where $\langle\ \ \rangle = 0$ or vice versa. Along a row of vertices with $t =$ const. this happens only once.

References

Abou-Sayed, A. S., Brechtel, C. E., and Clifton, R. J. 1978. *In situ* stress determination by hydrofracturing: a fracture mechanics approach. *J. Geophys. Res.* **83**, No. B6, 2851—2862.

Adachi, T., Tamura, T., and Shinkawa, M. 1982. Analytical and experimental study on tunnel support system. In: *Numerical Methods in Geomechanics*, Edmonton 1982, Proc. Fourth Int. Conf. Edited by Z. Eisenstein, Balkema, Rotterdam, 513—522.

Aitaliev, Sh. M., Banichuk, N. V., and Kayupov, M. A. 1986. *Optimal Design of Extended Underground Structures*, Nauka, Alma-Ata. In Russian.

Akagi, T. 1985. Application of rheological models to nonlinear creep properties of rock. In: *Numerical Methods in Geomechanics*, Nagoya 1985, Proc. Fifth Int. Conf. Edited by T. Kawamoto, Y. Ichikawa, Balkema, Rotterdam, 415—420.

Albrecht, H. and Langer, M. 1974. The rheological behaviour of rock salt and related stability problems of storage caverns. *Proc. 3rd Congr. Int. Soc. Rock Mech.*, Denver, vol. 2, Part B, 967—974.

Al-Chalabi, M. and Huang, C. L. 1974. Stress distribution within circular cylinders in compression. *Int. J. Rock Mech. Min. Sci. &Geomech. Abstr.* **11**, 45—56.

Allirot, D. and Boehler, J. P. 1979. Evolution des propriétés mécaniques d'une roche stratifiée sous pression de confinement. *Fourth Int. Congr. on Rock Mechanics*, Montreux, Balkema, Rotterdam, **I**, 15—22.

Amadei, B. 1983. *Rock Anisotropy and the Theory of Stress Measurements.* Springer-Verlag, Berlin.

Ardashev, K. A., Amusin, B. Z., and Koshelev, V. F. 1985. Numerical methods of analysis of underground constructions. In: *Numerical Methods in Geomechanics*, Nagoya 1985. Proc. Fifth Int. Conf. Edited by T. Kawamoto, Y. Ichikawa, Balkema, Rotterdam, 1077—1084.

Asszonyi, C. and Richter, R. 1979. *The Continuum Theory of Rock Mechanics.* Trans Tech Publ., Clausthal-Zellerfeld.

Baar, C. A. 1973. *Applied Salt-Rock Mechanics.* Vol. I, Elsevier, Amsterdam.

Baidyuk, B. V. 1963. *Mechanical Properties of Rocks at Pressures and Temperatures.* Gostotehizdat, Moscow. In Russian.

Baklashov, I. V. and Kartoziya, B. A. 1975. *Mechanics of Rocks.* Nedra, Moscow. In Russian.

Balashov, D. B. and Volarovich, M. P. 1971. Study of isothermal compressibility of eruptive rocks at pressures up to 32000 kg/cm^2. *Izv. Akad. Nauk SSSR, Ser. Fiz. Zemli*, No. 2, 24—32. In Russian.

Baoshen, L. 1979. A study of the mechanism of support. *Fourth Int. Congr. on Rock Mechanics*, Montreux, Balkema, Rotterdam, **I**, 157—160.

Barnabe, Y. 1986. The effective pressure law for permeability in Chelmford granite and Barre granite. *Int. J. Rock Mech. Min. Sci. &Geomech. Abstr.* **23**, 3, 267—275.

Barnabe, Y. 1987. The effective pressure law for permeability during pore pressure and confining pressure cycling of several crystalline rocks. *J. Geophys. Res.* **92**, No. B1, 649—657.

Bayuk, E. I. 1966. Velocities of elastic waves in specimens of eruptive and metamorphic rocks at pressures up to 4000 kg/cm^2. *Trudy Inst. Fiz. Zemli.* No. 37 (204), 16—36. In Russian.

Bažant, Z. P. 1978. Endochronic inelasticity and incremental plasticity. *Int. J. Solids Structures* **14**, 691—714.

Bell, J. F. 1973. *The Experimental Foundations of Solid Mechanics.* Handbuch des Physik, vol. VI a/1, Springer-Verlag, Berlin.

Berest, P., Habib, P., and Nguyen, M. D. 1980, Tentative d'interprétation des déformations observées aux tunnels du Fréjus et du Gran Sasso. *Revue Française de Geotechnique*, No. 12, 44—55.

Berest, P., Ledoux, E., Legait, B., and Marsily, G. de, 1979. Effects thermiques dans les cavités en couche salifère. *Fourth Int. Congr. on Rock Mechanics*, Montreux, Balkema, Rotterdam, **I**, 31—35.

Berest, P. and Nguyen, M. D. 1979. Etude en grandes déformations de la contraction et de l'expansion de cylindres creux de révolution élastoplastiques avec radoucissement. *J. de Mécanique Appliquée* **3**, 4, 463—487.

Berest, P. and Nguyen, M. D. 1983*a*. Comportement mécanique des cavités profondes de stockage d'hydrocarbures dans le sel. *5th Int. Congr. on Rock Mechanics*, Melbourne, Preprints, Section D, D227—D231.

Berest, P. and Nguyen, M. D. 1983*b*. Time-dependent behaviour of lined tunnels in soft rocks. *Eurotunnel '83 Conference*, Basle, Switzerland, 57—62.

Bergman, M., Editor, 1978. *Storage in Excavated Rock Caverns*, Proc. First. Int. Symp. Pergamon Press, New York.

Bieniawski, Z. T. 1984. *Rock Mechanics Design in Mining and Tunneling.* Balkema, Rotterdam.

Biot, M. A. 1973. Nonlinear and semilinear rheology of porous solids. *J. Geophys. Res.* **78**, 23, 4924—4937.

Birger, I. A. 1977. On a failure and plasticity criterion. *Izv. Akad. Nauk SSSR, Meh. Tverd. Tela* **4**, 143—150. In Russian.

Blackwood, R. L. 1979. An inference of crustal rheology from stress observations. *Fourth Int. Congr. on Rock Mechanics*, Montreux, Balkema, Rotterdam, **I**, 37—44.

Boley, B. A. and Weiner, J. H. 1960. *Theory of Thermal Stresses.* John Wiley and Sons, New York.

Borchert, K. M., Klapperich, H., Richter, T., and Walz, B. 1979. Berechnungen von Verformungen und Spannungen in Salzkavernen unter Berücksichtigung eines Visko-Plastischen Materialverhaltens. *Fourth Int. Congr. on Rock Mechanics*, Montreux, Balkema, Rotterdam, **I**, 51—60.

Bordia, S. K. 1971. The effects of size and stress concentration on the dilatancy and fracture of rock. *Int. J. Rock Mech. Min. Sci. & Geomech. Abstr.* **8**, 629—640.

Bordia, S. K. 1972. Complete stress-volumetric strain equation for brittle rock up to strength failure. *Int. J. Rock Mech. Min. Sci. & Geomech. Abstr.* **9**, 17—24.

Borm, G. 1983. Analysis of creep and relaxation around an underground opening in rock salt. *5th Int. Congr. on Rock Mechanics*, Melbourne, Preprints, Section D, D23—D27.

Boucly, P. 1984. *In situ* experience and mathematical representation of the behavior of rock salt used in storage of gas. In: *The mechanical behavior of salt.* Proc. First Conf. Edited by H. R. Hardy, Jr. and M. Langer, Trans Tech Publ., Clausthal-Zellerfeld, 453—471.

Brace, W. F. 1964. Brittle fracture of rocks. In: *State of Stress in the Earth's Crust*, Proc. Int. Conf. 1963, The Rand Corporation, Edited by W. R. Judd, Elsevier, New York, 111—178.

Brace, W. F. 1965. Some new measurements of linear compressibility of rocks. *J. Geophys. Res.* **70**, 2, 391—398.

Brace, W. F., 1978. Volume change during fracture and frictional sliding, a review. *Pageoph* **116**, 603—614.

Brace, W. F. 1981. The effect of size on mechanical properties of rock. *Geophys. Res. Letters* **8**, 7, 651—652.

Brace, W. F. and Bombolakis, E. G. 1963. A note on brittle crack growth in compression. *J. Geophys. Res.* **68**, 12, 3709—3713.

Brace, W. F., Paulding, B. W. Jr., and Scholz, C. 1966. Dilatancy in the fracture of crystalline rocks. *J. Geophys. Res.* **71**, 16, 3939—3953.

Brady, B. H. G. 1979. A direct formulation of the boundary element method of stress analysis for complete plane strain. *Int. J. Rock Mech. Min. Sci. &Geomech. Abstr.* **16**, 235—244.

Brady, B. H. G. and Bray, J. W. 1978a, The boundary element method for determining stresses and displacements around long openings in a triaxial stress field. *Int. J. Rock Mech. Min. Sci. &Geomech. Abstr.* **15**, 21—28.

Brady, B. H. G. and Bray, J. W. 1978b. The boundary element method for elastic analysis of tubular orebody extraction, assuming complete plane strain. *Int. J. Rock. Mech. Min. Sci. & Geomech. Abstr.* **15**, 29—37.

Brady, B. H. G. and Wassyng, A. 1981. A coupled finite element-boundary element method of stress analysis. *Int. J. Rock Mech. Min. Sci. &Geomech. Abstr.* **18**, 475—485.

Brady, B. T. 1970. A mechanical equation of state for brittle rock, I. *Int. J. Rock Mech. Min. Sci. &Geomech. Abstr.* **7**, 385—421.

Brady, B. T. 1971. An exact solution to the radially end-constrained circular cylinder under triaxial loading. *Int. J. Rock Mech. Min. Sci. &Geomech. Abstr.* **8**, 165—178.

Bridgman, P. 1952. *Studies in Large Plastic Flow and Fracture.* McGraw-Hill, New York.

Brown, E. T. and Hoek, E. 1978. Trends in relationships between measured in-situ stresses and depth. *Int. J. Rock Mech. Min. Sci. &Geomech. Abstr.* **15**, 211—215.

Burgert, W. and Lippmann, H. 1981 . Models of translatory rock bursting in coal. *Int. J. Rock Mech. Min. Sci. &Geomech. Abstr.* **18**, 285—294.

Byerlee, J. D. 1975. The fracture strength and frictional strength of Weber sandstone. *Int. J. Rock Mech. Min. Sci. &Geomech. Abstr.* **12**, 1—4.

Byerlee, J. D. 1978. Friction of rocks. *Pageoph* **116**, 615—626.

Carini, A. and Gioda, G. 1986. A boundary integral equation technique for visco-elastic stress analysis. *Int. J. Numer. Analyt. Meth. in Geomech.* (preprint).

Carneiro, F. 1947. Une nouvelle methode d'essai pour déterminer la resistance à la traction du beton. *Reunion des Laboratoires d'Essai de Materiaux*, Paris.

Carter, J. P. and Booker, J. R. 1982. Elastic consolidation around a deep circular tunnel. *Int. J. Solids Structures* **18**, 12, 1059—1074.

Carter, J. P., Booker, J. R. 1985. Thermomechanical analysis of some proposed schemes for radioactive waste disposal. In: *Numerical Methods in Geomechanics*, Nagoya, 1985, Proc. Fifth Int. Conf. Edited by T. Kawamoto, Y. Ichikawa, Balkema, Rotterdam-Boston, 1249—1256.

Carter, N. L., Anderson, D. A., Hansen, F. D., and Kranz, R. L. 1981. Creep and creep rupture of granitic rock. In: *Mechanical Behaviour of Crustal Rocks.* Geophysical Monograph **24**, Amer. Geophys. Union, 61—82.

Carter, N. L. and Hansen, F. D. 1983. Creep of rocksalt. *Tectonophysics* **92**, 275—333.

Carter, N. L. and Kirby, S. H. 1978. Transient creep and semibrittle behaviour of crystalline rock. *Pageoph* **116**, 807—839.

Celle, C. C., Cheatham, J. B. 1981. Anisotropic hardening of an initially isotropic porous limestone. *Rock Mechanics* **13**, 221—233.

Cheatham, J. B., Jr., Allen, M. B., Celle, C. C. 1984. Kinematic hardening of a porous limestone. *Rock Mechanics* **17**, 233—242.

Chen, R. 1979. Acoustic emission of rocks under triaxial compression with various stress paths. *Int. J. Rock Mech. Min. Sci. &Geomech. Abstr.* **16**, 401—405.

Chen, R.,, Xin, Y. X., and Sen, X. H. 1979. Studies of the fracture of Gabbro. *Int. J. Rock Mech. Min. Sci. &Geomech. Abstr.* **16**, 187—193.

Chen, W. F. and Han, D. J. 1985. A five-parameter mixed hardening model for concrete materials. In: *Plasticity Today. Modelling, Methods and Applications.* Edited by A. Sawczuk, G. Bianchi, Elsevier, London, 587—602.

Cherepanov, G. P. 1979. *Mechanics of Brittle Fracture.* McGraw-Hill, New York.

Cherry, J. T., Schock, R. N., Sweet, J. 1975. A theoretical model of the dilatant behavior of a brittle rock. *Pure Appl. Geophys.* **113**, 183—196.

Chong, K. P., Uenishi, K., Smith, J. W., and Munari, A. C. 1980. Nonlinear three dimensional mechanical characterization of Colorado oil shale. *Int. J. Rock Mech. Min. Sci. & Geomech. Abstr.* **17**, 339—347.

Cividini, A., Gioda, G., and Barla, G. 1985. Calibration of a rheological model on the basis of field measurements. In: *Numerical Methods in Geomechanics*, Nagoya, 1985. Proc. Fifth Int. Conf. Edited by T. Kawamoto, Y. Ichikawa, Balkema, Rotterdam, 1621—1628.

Coates, D. F. 1981. *Rock Mechanics Principles.* CANMET, Canada.

Constantinescu, M. 1981. Experimental formulation of constitutive equations for solid porous materials. Ph.D. Dissertation, Bucharest.

Constantinescu, M. and Cristescu, N. 1982. Anelastic properties of cement-concrete. *Rev. Roumaine Sci. Tech. Sér. Méc. Appl.* **27**, 667—680.

Constantinescu, M. and Cristescu, N. 1983. Creep of rock-like materials. *Int. J. Engng. Sci.* **21**, 1. 45—49.

Constantino, M. C. 1978. Statistical variation in stress volumetric strain behaviour of Westerly granite. *Int. J. Rock Mech. Min. Sci. &Geomech. Abstr.* **15**, 105—111.

Costin, L. S. 1983. A microcrack damage model for brittle rock. Sandia Report SAND83-1590, Sandia National Laboratories, Albuquerque, N. M.

Costin, L. S. 1985. Damage mechanics in the past-failure regime. *Mechanics of Materials* (preprint)

Costin, L. S. and Holcomb, D. J. 1981. Time-dependent failure of rocks under cyclic loading. *Tectonophysics* **79**, 279—296.

Costin, L. S. and Holcomb, D. J. 1983. A continuum model of inelastically deformed brittle rock based on the mechanics of microcracks. In: *Proc. Int. Conf. On Constitutive Laws for Engineering Materials; Theory and Applications.* Edited by C. S. Desai, R. H. Gallagher, University of Arizona, Tucson, 185—195.

Coulomb, C. A. 1773. Essai sur une application des règles de maximis et minimis à quelsues problèmes de statique rèlatifs à l'architecture. *Mémoires de Mathématique et de Physique* **7**, 343—382.

Crawford, A. M. and Curran, J. H. 1981. The influence of shear velocity on the frictional resistance of rock discontinuities. *Int. J. Rock Mech. Min. Sci. &Geomech. Abstr.* **18**, 505—515.

Cristescu, N. 1963. On the propagation of elastic-plastic waves in metallic rods. *Bull. Acad. Pol. Sci. Sér. Sci. Tech.* **11**, 129—133.

Cristescu, N. 1967. *Dynamic Plasticity*, North-Holland, Amsterdam.

Cristescu, N. 1979. A viscoplastic constitutive equation for rocks. *INCREST Preprint in Mathematics*, Nr. 49/1979, Bucharest.

Cristescu, N. 1981. Compressibility and/or dilatancy of rocks. *Anal. Sti. Univ. "Al. I. Cuza" Iaşi, Secţ. I a Mat.* **27**, 125—130.

Cristescu, N. 1982. Rock dilatancy in uniaxial tests. *Rock Mechanics* **15**, 133—144.

Cristescu, N. 1983*a*. *Mechanics of Composite Materials.* University of Bucharest. In Romanian.

Cristescu, N. 1983*b*. *Mechanics of Rocks.* University of Bucharest. In Romanian.

Cristescu, N. 1984. *Rock Mechanics.* University of Bucharest. In Romanian.

Cristescu, N. 1985*a*. Irreversible dilatancy or compressibility of viscoplastic rock-like materials and some applications. *Int. J. Plasticity* **1**, 3, 189—204.

Cristescu, N. 1985*b*. Fluage, dilatance et/ou compressibilité des roches autour des puits verticaux et des forages pétroliers. *Rev. Française Géotechnique* No. 31, 11—22.

Cristescu, N. 1985*c*. Rock plasticity. In: *Plasticity Today. Modelling, Methods and Applications.* Edited by A. Sawczuk, G. Bianchi, Elsevier, London, 643—655.

Cristescu, N. 1985*d*. Plasticity of compressible/dilatant rock-like materials. *Int. J. Engng. Sci.* **23**, 10, 1091—1100.

Cristescu, N. 1985*e*. Viscoplastic creep of rocks around horizontal tunnels. *Int. J. Rock Mech. Min. Sci. &Geomech. Abstr.* **22**, 6, 453—459.

Cristescu, N. 1985*f*. Mathematical modelling of the mining support and of the rock-support interaction. Report, University of Bucharest.

Cristescu, N. 1986. Damage and failure of viscoplastic rock-like materials. *Int. J. Plasticity* **2**, 2, 189—204.

Cristescu, N. 1987. Elastic-viscoplastic constitutive equations for rock. *Int. J. Rock Mech. Min. Sci. &Geomech. Abstr.* **24**, 5, 271—282.

Cristescu, N. 1988*a*. An elastic/viscoplastic constitutive equation for rocks and the associated mechanical – AE/MA analogy. *Fourth Conference on Acoustic Emission/Microseismic Activity in Geologic Structures and Materials*, Penn. State Univ., Oct. 1985 (in press).

Cristescu, N. 1988b. Viscoplastic creep of rocks around a lined tunnel. *Int. J. Plasticity* **4**, (in press).

Cristescu, N., Fotă, D, and Medveş, E. 1987. Tunnel support analysis incorporating rock creep. *Int. J. Rock Mech. Min. Sci. &Geomech. Abstr.* **24**, 6, 321—330.

Cristescu, N., Fotă, D., Medveş, E. 1988. Rock-support interaction in lined tunnels. *Euromech Colloq. No. 194*, Sept. 1985, Bucharest, Lecture Notes in Earth Science, **14**, Springer-Verlag, Berlin,245— 272.

Cristescu, N., Massier, D., Dobre-Pall, M., Nicolae, M., Cristescu, C. and Teodorescu, S. 1985. Slow deformation in time of horizontal circular mining works. *Mine, Petrol, Gaze.* **36**, 10, 509—517. In Romanian.

Cristescu, N. and Suliciu, I. 1982. *Viscoplasticity*, Martinus Nijhoff, The Hague — Ed. Tehnică, Bucureşti. (First edition *Vîscoplasticitate*. 1976, Ed. Tehnică, Bucureşti. In Romanian).

Crotogino, F. R. 1984. Salt cavern *in situ* testing from the constructor's and from the operator's viewpoint. In: *The Mechanical Behavior of Salt.* Proc. First Conf. Edited by H. R. Hardy, Jr. and M. Langer. Trans Tech Publ., Clausthal-Zellerfeld, 613—628.

Dawson, P. R. and Munson, D. E. 1983. Numerical simulation of creep deformations around a room in a deep potash mine. *Int. J. Rock Mech. Min. Sci. &Geomech. Abstr.* **20**, 1, 33—42.

Demiris, C. A. 1985. Surfaces de rupture et de limite de dilatance du marble de Thasses soumis à des efforts de compression multiaxiaux. *C.R. Acad. Sc. Paris* **301**, Série II, No. 17, 1225—1230.

Desai, C. S., Frantziskonis, G. and Somasundaram, S. 1985. Constitutive modelling for geological materials. In: *Numerical Methods in Geomechanics*, Nagoya, 1985. Proc. Fifth Int. Conf. Edited by T. Kawamoto, Y. Ichikawa, Balkema, Rotterdam, 19—34.

Desai, C. S., Somasundaram, S. and Frantziskonis, G. 1986. A hierarchical approach for constitutive modeling of geologic materials. *Int. J. Numer. Analyt. Meth. in Geomech.* **10**, 3.

Detournay, E. 1985. Solution approximative de la zone plastique autour d'une galerie souterraine soumise à un champ de contrainte non hydrostatique. *C.R. Acad. Sci. Paris* **301**, Série II, No. 12, 857—860.

Detournay, E., Cheng, A. H.-D. and Roegiers, J.-C. 1987. Some new examples of poroelastic effects in rock mechanics. *28th U.S. Rock Mech. Symp.*, Tucson, AZ, (preprint).

Detournay, E. and Fairhurst, C. 1982a. Generalization of the ground reaction curve concept. *23rd U.S. Symp. on Rock. Mech.*, Berkeley, 926—936.

Detournay, E. and Fairhurst, C. 1982*b*. Approximate statical solution of the elastoplastic interface in the problem of an infinite medium with a circular hole. *Numerical Methods in Geomechanics*, Edmonton, 1982. Proc. Fourth Int. Conf. Edited by Z. Eisenstein, Balkema, Rotterdam, 27—35.

Detournay, E. and Fairhurst, C. 1987. Two-dimensional elasticoplastic analysis of a long, cylindrical cavity under non-hydrostatic loading. *Int. J. Rock Mech. Min. Sci. & Geomech. Abstr.* **24**, 4, 197—211.

Detournay, E., McLennan, J. D., Roegiers, J.-C. 1986. Poroelastic concepts explain some of the hydraulic fracturing mechanisms. *Unconventional Gas Technology Symp.* SPE 15262, 629—637.

Detournay, E. and Vardoulakis, I. 1985. Determination of the ground reaction curve using the hodograph method. *Int. J. Rock Mech. Min. Sci. &Geomech. Abstr.* **22**, 3, 173—176.

Dorris, J. F. and Nemat-Nasser, S. 1982. A plasticity model for flow of granular materials under triaxial stress states. *Int. J. Solids Structures* **18**, 497—531.

Dortman, N. B., Editor. 1976. *Physical Properties of Rocks and Minerals*, Nedra, Moscow. In Russian.

Dragon, A., Mróz, Z. 1979. A model for plastic creep of rock-like materials accounting for the kinetics of fracture. *Int. J. Rock Mech. Min. Sci. &Geomech. Abstr.* **16**, 253—259.

Dreyer, W. 1973. *The Science of Rock Mechanics*, Part 1. Trans Tech Publ., Clausthal-Zellerfeld.

Dreyer, W. E. 1984. Crude oil storage in a system of salt caverns. In: *The Mechanical Behavior of Salt*. Proc. First Conf. Edited by H. R. Hardy, Jr. and M. Langer, Trans Tech. Publ., Clausthal-Zellerfeld, 629—660.

Duffaut, P. and Feuga, B. 1985. Towards a theory of tunnels in heterogeneous formations. Numerical analysis of some cases of heterogeneity around tunnels. In: *Numerical Methods in Geomechanics*, Nagoya, 1985. Proc. Fifth Int. Conf. Edited by T. Kawamoto, Y. Ichikawa, Balkema, Rotterdam-Boston, 1039—1047.

Duncan Fama, M. E. and Pender, M. J. 1980. Analysis of the hollow inclusion technique for measuring *in situ* rock stress. *Int. J. Rock Mech. Min. Sci. &Geomech. Abstr.* **17**, 137—146.

Eekelen, H. A., Hulsebos, T., and Urai, J. L. 1984. Creep of Bischofite. In: *The Mechanical Behavior of Salt*. Proc. First Conf. Edited by H. R. Hardy, Jr. and M. Langer, Trans Tech Publ., Clausthal-Zellerfeld, 389—405.

Egger, P. and Gindroz, C. 1979. Tunnels ancrés à faible profondeur. Etude comparative sur modèles physique et mathèmatique. *Fourth Int. Congr. on Rock Mechanics*, Montreux, Balkema, Rotterdam, **II**, 121—130.

Eisenstein, Z. 1986. The contribution of numerical analysis to the design of shallow tunnels. In: *Geomechanical Modelling in Engineering Practice*. Edited by R. Dungar and J. A. Studer, Balkema, Rotterdam, 135—165.

Eringen, A. C. 1967. *Mechanics of Continua*. John Wiley, New York.

Eringen A. C. 1981. On nonlocal plasticity. *Int. J. Engng. Sci.* **19**, 12, 1461—1467.

Eringen, A. C. and Şuhubi, E. S. 1974. *Elastodynamics*, vol. 1. Academic Press, New York.

Erzhanov, J. S. and Bergman, E. I. 1977. *Creep of Rock Salt*. Nauka, Alma Ata. In Russian.

Farmer, J. W. and Gilbert, M. J. 1984. Time dependent strength reduction of rock salt. In: *The Mechanical Behavior of Salt*. Proc. First. Conf. Edited by H. R. Hardy, Jr. and M. Langer. Trans Tech Publ. Clausthal-Zellerfeld, 3—18.

Filatov, N. A., Beliakov, V. D., Dokukin, A. V., Trumbachev, V. F., and Slavin, O. K. 1983. Photomechanical practice in studying conditions of solid rock near underground working driven at great depth. *5th Int. Congr. on Rock Mechanics*, Melbourne. Preprints. Section D, D121—D128.

Galperin, A. M. and Shafarenko, E. M. 1977. *Rheological Computation of Structures in Rocks*. Nedra, Moscow, In Russian.

Gangi, A. F. 1983. Transient and steady-state deformation of synthetic rocksalt. *Tectonophysics,* **91**, 137—156.

Gangi, A. F., Parrish, D. K. and Handin, J. 1984. Transient and steady-state deformation of synthetic rock salt. In: *The Mechanical Behavior of Salt*. Proc. First Conf. Edited by H. R. Hardy, Jr. and M. Langer. Trans Tech Publ, Clausthal-Zellerfeld, 37—51.

Genevois, R. and Prestininzi, A. 1979. Time-dependent behaviour of granitic rocks related to their alteration grade. *Fourth Int. Congr. on Rock Mechanics*, Montreux, Balkema, Rotterdam, **I**, 153—158.

Germain, P. 1973. *Cours de Mécanique des Milieux Continues*, I. Masson, Paris.

Gioda, G. 1981. A finite element solution of non-linear creep problems in rocks. *Int. J. Rock Mech. Min. Sci. &Geomech. Abstr.* **18**, 35—46.

Gioda, G. 1984. A simple boundary equation technique for elastic stress analysis of underground cavities. *Rock Mechanics and Rock Engineering* **17**, 147—165.

Gioda, G. and Cividini, A. 1981. Viscous behaviour around an underground opening in a two-phase medium. *Int. J. Rock Mech. Min. Sci. &Geomech. Abstr.* **18**, 437—443.

Gioda, G. and Maier, G. 1980. Direct search solution of an inverse problem in elastoplasticity: identification of cohesion, friction angle and *in situ* stress by pressure tunnel tests. *Int. J. Numer. Methods Engrg.* **15**, 1823—1848.

Goodman, R. E. 1980. *Introduction to rock mechanics*. John Wiley, New York.

Gowd, T. N. and Rummel, F. 1977. Effect of fluid injection on the fracture behavior of porous rock. *Int. J. Rock Mech. Min. Sci. &Geomech. Abstr.* **14**, 203—208.

Gowd, T. N. and Rummel, F. 1980. Effect of confining pressure on the fracture behaviour of porous rocks. *Int. J. Rock Mech. Min. Sci. &Geomech. Abstr.* **17**, 225—229.

Grady, D. E. and Kipp, M. E. 1980. Continuum modelling of explosive fracture in oil shale. *Int. J. Rock Mech. Min. Sci. &Geomech. Abstr.* **17**, 147—157.

Granryd, L., Getting, I. C. and Spetzler, H. 1983. Path dependence of acoustic velocity and attenuation in experimentally deformed Westerly granite. *Geophys. Res. Letters* **10**, 1, 71—74.

Green, A. E. and Naghdi, P. M. 1965. A general theory of an elastic-plastic continuum. *Arch. Rational Mech. Anal.* **18**, 4, 251—281.

Griggs, D. 1939. Creep of rocks. *J. Geol.* **47**, 3, 225—251.

Gurtin, M. E., Williams, W. O. and Suliciu, I. 1980. On rate type constitutive equations and the energy of viscoelastic and viscoplastic materials. *Int. J. Solids Structures* **16**, 607—617.

Gustkiewicz, J. 1975. Uniaxial compression testing of brittle rock specimens with special consideration given to bending moment effects. *Int. J. Rock Mech. Min. Sci. & Geomech. Abstr.* **12**, 13—25.

Habib, P. 1973. *Précis de Géotechnique*. Dunod, Paris.

Haimson, B. C. 1978. The hydrofracturing stress measuring method and recent field results. *Int. J. Rock Mech. Min. Sci. &Geomech. Abstr.* **15**, 167—178.

Handin, J. and Carter, N. 1979. Rheological properties of rocks at high temperatures. *Fourth Int. Congr. on Rock Mechanics*, Montreux, Balkema, Rotterdam, 97—106.

Hansen, F. D. 1977. Evaluation of an inelastic law for salt creep. *Proc. 18th U.S. Symp. on Rock Mechanics*, June 22—24, 1977, Keystone, Colorado, Colorado School of Mines Press. Preprint.

Hansen, F. D. and Carter, N. L. 1980. Creep of rock salt at elevated temperature. *21st U.S. Symp. on Rock Mechanics*, May 28—30, 1980, University of Missouri — Rolla. Preprint.

Hansen, F. D. and Carter, N. L. 1984. Creep of Avery Island rock salt. In: *The Mechanical Behavior of Salt*. Proc. First Conf. Edited by H. R. Hardy, Jr. and M. Langer, Trans Tech Publ., Clausthal-Zellerfeld, 53—69.

Hansen, F. D. and Melegard, K. D. 1980. Creep of 50-mm diameter specimens of dome salt from Avery Island, Louisiana. *ONWI-104, Technical Report RE/SPEC Inc.* Rapid City, SD.

Hansen, F. D., Mellegard, K. D., and Senseny, P. E. 1984. Elasticity and strength of ten natural rock salts. In: *The Mechanical Behavior of Salt*. Proc. First Conf. Edited by H. R. Hardy, Jr. and M. Langer, Trans Tech Publ., Clausthal-Zellerfeld, 71—83.

Hardy, H. R., Jr. 1972. Application of acoustic emission technique to rock mechanics research. In: *Acoustic Emission*. Edited by R. G. Liptai, D. O. Harris, C. A. Tatro. ASTM STP 505, Amer. Soc. for Testing and Materials, Philadelphia, 41—83.

Hardy, H. R., Jr. 1984. Geotechnical applications of acoustic emission technique: present status and future goals. *7th Int. Acoustic Emission Symp.*, Sendai, Japan. Preprint.

Hardy, H. R. Jr., Bakhtar, K., Mrugala, M. and Kimble, E. J., Jr. 1984. Development of laboratory facilities for evaluating the creep behavior of salt. In: *The Mechanical Behavior of Salt*. Proc. First Conf. Edited by H. R. Hardy, Jr. and M. Langer. Trans Tech Publ., Clausthal-Zellerfeld, 85—117.

Hardy, H. R., Jr. and Leighton, F. W. Editors. 1977. *First Conf. on Acoustic Emission/Microseismic Activity in Geologic Structures and Materials*, Trans Tech Publ., Clausthal-Zellerfeld.

Hardy, H. R., Jr. and Leighton, F. W. Editors. 1980. *Second Conf. on Acoustic Emission/ Microseismic Activity in Geologic Structures and Materials*. Trans Tech Publ., Clausthal-Zellerfeld.

Hardy, H. R., Jr. and Leighton, F. W. Editors. 1984. *Acoustic Emission/Microseismic Activity in Geologic Structures and Materials*. Proc. of the Third Conf., Trans Tech Publ., Clausthal-Zellerfeld.

Hardy, H. R., Jr. and Leighton, F. W. Editors, 1988. *Fourth Conf. on Acoustic Emission/ Microseismic Activity in Geologic Structures and Materials*. Trans Tech Publ., Clausthal-Zellerfeld. (in press).

Heard, H. C. and Duba, A. 1978. Capabilities for measuring physicochemical properties at high pressure. *Lawrance Livermore Laboratory Rept.* UCRL-52420.

Heard, H. C. and Page, L. 1982. Elastic moduli, thermal expansion, and inferred permeability of two granites to 350°C and 55 Megapascals. *J. Geophys. Res.* **87**, No. B11, 9340—9348.

Heerden, W. L., van, 1985. *In Situ* determination of the dynamic moduli of elasticity of coal. *Int. J. Rock Mech. Min. Sci. &Geomech. Abstr.* **22**, 5, 339—343.

Hendron, A. J. 1968. Mechanical properties of rocks. In: *Rock Mechanics in Engineering Practice*. Edited by K. G. Stagg, O. C. Zienkiewicz, John Wiley, New York.

Heuze, F. E., 1983. High-temperature mechanical, physical and thermal properties of granitic rocks — a review. *Int. J. Rock Mech. Min. Sci. &Geomech. Abstr.* **20**, 1, 3—10.

Hill, R. 1950. *The Mathematical Theory of Plasticity*. Clarendon Press, Oxford.

Hoek, E. 1983. Strength of jointed rock masses. *Geotechnique* **23**, 3, 187—223.

Hoek, E. and Brown, E. T. 1980. *Underground Excavations in Rock*. The Institution of Mining and Metallurgy, London.

Holcomb, D. J. and Costin, L. S. 1986. Detecting damage surfaces in brittle materials using acoustic emission. *Trans. ASME, J. Appl. Mech.* **53**, 536—544.

Holcomb, D. J. and Stevens, J. L. 1980. The reversible Griffith crack: a viable model for dilatancy. *J. Geophys. Res.* **85**, No. B12, 7101—7104.

Homentcovschi, D. 1986. *Complex Functions with Applications in Science and Technology*. Ed. Tehnică, Bucureşti. In Romanian.

Horii, H. and Nemat-Nasser, S. 1985. Compression-induced microcrack growth in brittle solids: axial splitting and shear failure. *J. Geophys. Res.* **90**, 3105—3125.

Horii, H. and Nemat-Nasser, S. 1986. Brittle failure in compression: splitting, faulting and brittle-ductile transition. *Phil. Trans. Royal Soc. London* **319**, 1549, 337—374.

Horseman, S., Passaris, E. 1984. Creep tests for storage cavity closure prediction. In: *The Mechanical Behavior of Salt*. Proc. First Conf. Edited by H. R. Hardy, Jr. and M. Langer. Trans Tech Publ., Clausthal-Zellerfeld, 119—157.

Houpert, R. and Homand-Etienne, F. 1979. Influence de la température sur le comportement mécanique des roches. *Fourth Int. Congr. on Rock Mechanics*, Montreux, Balkema, Rotterdam, **I**, 177—180.

Hueckel, T. 1982. A model for dilatant rock behaviour under variable loading. In: *Numerical Models in Geomechanics*. Int. Symp. Edited by R. Dungar, G. N. Pande and J. A. Studer, Balkema, Rotterdam, 198—207.

Hunsche, U. 1984*a*. Fracture experiments on cubic rock salt samples. In: *The Mechanical Behavior of Salt*. Proc. First Conf. Edited by H. R. Hardy, Jr. and M. Langer, Trans Tech Publ., Clausthal-Zellerfeld, 169—179.

Hunsche, U. 1984*b*. Results and interpretation of creep experiments on rock salt. In: *The Mechanical Behavior of Salt*. Proc. First Conf. Edited by H. R. Hardy, Jr. and M. Langer, Trans Tech Publ., Clausthal-Zellerfeld, 159—168.

Hunsche, U., 1984*c*. Measurement of creep in rock salt at small strain rates. *Second Conf. on the Mechanical Behavior of Salt*, Sept. 1984, Hannover. Preprint.

Hunsche, U. and Plischke, I. 1984. *In situ* creep experiments under controlled stress in rock salt pillars — design, instrumentation and evaluation. *CEC/NEA Workshop on Design and Instrumentation of In-situ Experiments in Underground Laboratories for Radioactive Waste Disposal*, May 1984, Brussels. Preprint.

Husain, M. A., Pu, S. L. and Underwood, I. 1974. Strain energy release rate for a crack under combined mode I and mode II. *Proc. of the 1973 National Symp. on Fracture Mechanics*, part II, ASTM STP 560, Amer. Soc. for Testing and Materials, 2—28.

Ichikawa, Y., Kyoya, T., and Kawamoto, T. 1985. Incremental theory of plasticity for rock. In: *Numerical Methods in Geomechanics*, Nagoya, 1985. Proc. Fifth Int. Conf. Edited by T. Kawamoto, Y. Ichikawa, Balkema, Rotterdam, 451—462.

Inada, Y. 1985. Stability of underground openings for storage of L.N.G. In: *Numerical Methods in Geomechanics*, Nagoya, 1985. Proc. Fifth Int. Conf. Edited by T. Kawamoto, Y. Ichikawa, Balkema, Rotterdam, 1225—1232.

ISRM. 1978, 1979. International Society for Rock Mechanics. *Int. J. Rock Mech. Min. Sci. & Geomech. Abstr.* **15**, 47—51; **16**, 135—140, 141—156.

Ishizuka, Y., Kinoshita, N., and Okuno, T. 1985. Stability analysis of rock cavern with L.P.G. storage under thermal stresses. In: *Numerical Methods in Geomechanics*, Nagoya, 1985. Proc. Fifth Int. Conf. Edited by T. Kawamoto, Y. Ichikawa, Balkema, Rotterdam, 1233—1246.

Isobe, T. and Tominga, Y. 1979. Effects formerly acted stress on stress-strain behaviour of rock and its anisotropical property. *Fourth Int. Congr. on Rock Mechanics*, Montreux, Balkema, Rotterdam, **I**, 181—187.

Itô, H. 1979. Rheology of the crust based on long-term tests of rocks. *Tectonophysics* **52**, 629—641.

Itô, H. 1983. Creep of rock based on long-term experiments. *5th Int. Congr. on Rock Mechanics*, Melbourne, Preprints, Section A, A117—A120.

Jaeger, J. C. and Cook, N. G. W. 1979. *Fundamentals of Rock Mechanics*. Chapman and Hall, London.

Jamison, D. B. and Cook, N. G. W. 1980. Note on measured values for the state of stress in the Earth's crust. *J. Geophys. Res.* **85**, No. B4, 1833—1838.

Jaunzemis, W. 1967. *Continuum Mechanics*. Macmillan, New York.

Johnson, J. N. and Green, S. J. 1976. The mechanical response of porous media subject to static loads. In: *Effects of Voids on Material Deformation*. Edited by S. C. Cowan, M. M. Carroll, ASME, 93—123.

Jones, R. M. 1975. *Mechanics of Composite Materials*. Scripta Book Company, Washington, D.C.

Jumikis, A. R. 1979. *Rock Mechanics*. Trans Tech Publ., Clausthal-Zellerfeld.

Kachanov, L. M. 1960. *Theory of Creep*. Fizmatgiz, Moscow, In Russian.

Kaiser, P. K. 1979. Tunnel design — conclusions from long-term model tests. *Fourth Int. Congr. on Rock Mechanics*, Montreux, Balkema, Rotterdam, **I**, 236—238.

Kaiser, P. K. 1981. A new concept to evaluate tunnel performance — influence of excavation procedure. *22nd U.S. Rock Mechanics Symp.*, Boston, 264—271.

Kaiser, P. K., Guenot, A., and Morgenstern, N. R. 1985. Deformation of small tunnels — IV. Behaviour during failure. *Int. J. Rock Mech. Min. Sci. &Geomech. Abstr.* **22**, 3, 141—152.

Kaiser, P. K. and Maloney, S. M. 1982. Deformation properties of a sub-bituminous coal mass. *Int. J. Rock Mech. Min. Sci. &Geomech. Abstr.* **19**, 247—252.

Kaiser, P. K., Maloney, S., and Morgenstern, N. R. 1983. Time-dependent behaviour of tunnels in highly stressed rock. *5th Int. Congr. on Rock Mechanics*, Melbourne, Preprints Section D, D329—D336.

Kaiser, P. K. and Morgenstern, N. R. 1982. Time-independent and time-dependent deformation of small tunnels -III- Pre-failure behaviour. *Int. J. Rock Mech. Min. Sci. & Geomech. Abstr.* **19**, 307—342.

Kaneko, N., Shinokawa, T., Yoshida, N., and Kawahara, M. 1985. Numerical analysis of viscoelasticity using combined finite and boundary element methods. In: *Numerical Methods in Geomechanics*, Nagoya, 1985. Proc. Fifth Int. Conf. Edited by T. Kawamoto, Y. Ichikawa, Balkema, Rotterdam, 1049—1054.

Kapustyanskii, S. M. and Nikolaevskii, V. N. 1985. Parameters of elastic-plastic dilatant model for geomaterials. *Z. Prikl. Meh. i Tech. Fiz.* **6**, 145—150. In Russian.

Kármán, Th. von. 1911. Festigkeitsversuche unter allseitigen Druck. *Z. Ver. dt. Ing.* **55**, 1749—1757.

Kartashov, I. M. 1973. *Speeded Methods for the Determination of Rheological Properties of Rocks*. Nedra, Moscow. In Russian.

Kern, H. and Franke, J. H. 1984. The effect of temperature on the chemical and mechanical behavior of Carnallite-Halite rocks. In: *The Mechanical Behavior of Salt*. Proc. First Conf. Edited by H. R. Hardy, Jr. and M. Langer, Trans Tech Publ., Clausthal-Zellerfeld, 181—191.

Kidybinski, A. 1981. Bursting liability indices of coal. *Int. J. Rock Mech. Min. Sci. & Geomech. Abstr.* **18**, 295—304.

Kim, M. K. and Lade, P. V. 1984. Modelling rock strength in three dimensions. *Int. J. Rock Mech. Min. Sci. &Geomech. Abstr.* **21**, 21—33.

Kimura, F., Okabayashi, N., Ono, K., and Kawamoto, T. 1985. Rock-mechanical discussion for the mechanism of supporting system in severe swelling rock tunnel. In: *Numerical Methods in Geomechanics*, Nagoya, 1985. Proc. Fifth Int. Conf. Edited by T. Kawamoto, Y. Ichikawa, Balkema, Rotterdam, 1265—1271.

King, M. S. 1966. Wave velocities in rocks as function of changes in overburden pressure and pore fluid saturants. *Geophysics* **31**, 1, 50—73.

King, M. S. 1970. Static and dynamic elastic moduli of rocks under pressure. In: *Rock Mechanics — Theory and Practice*. Proc. 11th Symp. on Rock Mechanics. Edited by W. H. Somerton. Amer. Inst. Mining Metallurgical and Petroleum Engrs, New York, 329—351.

King, M. S. 1984. Elastic-wave velocities in quartz monzonite at different levels of water saturation. *Int. J. Rock Mech. Min. Sci. &Geomech. Abstr.* **21**, 1, 35—38.

King, M. S. and Paulson, B. N. P. 1981. Acoustic velocities in heated block of granite subjected to uniaxial stress. *Geophys. Res. Letters* **8**, 7, 699—702.

Kobayashi, S. 1985. Applications of boundary integral equation method to geomechanics. In: *Numerical Methods in Geomechanics*, Nagoya, 1985, Proc. Fifth Int. Conf. Edited by T. Kawamoto, Y. Ichikawa, Balkema, Rotterdam-Boston, 83—92.

Kolsky, H. 1963. *Stress Waves in Solids*. Dover, New York.

Kovári, K. 1977. The elasto-plastic analysis in the design practice of underground openings. In: *Finite Elements in Geomechanics*, Edited by G. Gudehus, John Wiley, New York. 377—412.

Kranz, R. L. 1979. Crack growth and development during creep of Barre granite. *Int. J. Rock Mech. Min. Sci. &Geomech. Abstr.* **16**, 23—35.

Kranz, R. L. 1980. The effects of confining pressure and stress difference on static fatigue of granite. *J. Geophys. Res.* **85**, No. B4, 1854—1866.

Kranz, R. L., Harris, W. J., and Carter, N. L. 1982. Static fatigue of granite at 200°C. *Geophys. Res. Letters* **9**, 1, 1—4.

Kranz, R. L. and Scholz, C. H. 1977. Critical dilatant volume of rocks at the onset of tertiary creep. *J. Geophys. Res.* **82**, 4893—4898.

Krenk, S. 1978. Internally pressurized spherical and cylindrical cavities in rock salt. *Int. J. Rock Mech. Min. Sci. &Geomech. Abstr.* **15**, 219—224.

Krieg, R. D. 1982. A unified creep-plasticity model for halite, In: *Mechanical Testing for Deformation Model Development*. ASTM STP 765, Edited by R. W. Rohde and J. C. Swearengen, ASTM, 139—147.

Kumagai, N. and Itô, H. 1970. Creep of granite observed in a laboratory for 10 years. *Proc. Fifth Int. Congr. on Rheology*, University of Tokyo Press and University Park Press, **2**, 579—590.

Kurita, K., Swanson, P. L., Getting, I. C., and Spetzler, H. 1983. Surface deformation of Westerly granite during creep. *Geophys. Res. Letters* **10**, 1, 75—78.

Ladanyi, B. and Gill, D. E. 1983. *In-situ* determination of creep properties of rock salt. *5th Int. Congr. on Rock Mechanics*, Melbourne, Preprints Section A, A219—A225.

Lajtai, E. Z. and Schmidtke, R. H. 1986, Delayed failure in rock loaded in uniaxial compression. *Rock Mechanics and Rock Engineering* **19**, 11—25.

Lama, R. D. and Vutukuri, V. S. 1978. *Handbook on Mechanical Properties of Rocks,* **II** and **III**, Trans Tech Publ., Clausthal-Zellerfeld.

Lamé, G. 1852. *Laçons sur la Théorie de l'Elasticité*. Paris.

Langer, M. 1979. Rheological behaviour of rock masses. *Fourth Int. Congr. on Rock Mechanics*, Montreux, Balkema, Rotterdam, **I**, 29—62.

Langer, M. 1982. Geotechnical investigation methods for rock salt. *Bull. Int. Assoc. Engng. Geology*. No. 25, 155—164.

Langer, M. 1984. The rheological behavior of rock salt. In: *The Mechanical Behavior of Salt*. Proc. First Conf. Edited by H. R. Hardy, Jr. and M. Langer, Trans Tech Publ., Clausthal-Zellerfeld, 201—240.

Lee, E. H. 1969. Elastic-plastic deformation at finite strains. *Trans. ASME, J. Appl. Mech.* **36**, 1, 1—6.

Lee, E. H. 1970. Constitutive relations for dynamic loading and plastic waves. In: *Inelastic Behaviour of Solids*. Battelle Inst. Mat. Sci. Colloq., 1969, Edited by M. F. Kanninen, W. F. Adler, A. R. Rosenfields, R. I. Jaffee, McGraw-Hill, New York, 423—439.

Lee, I. K. and Ingles, O. G. 1968. Strength and deformation of soils and rocks. In: *Soil Mechanics, Selected Topics*. Edited by I. K. Lee, American Elsevier, New York, 195—294.

Leighton, F. 1984. Microseismic activity associated with outburst in coal mines. In: *Acoustic Emission/Microseismic Activity in Geologic Structures and Materials*, Proc. of the Third Conf. Edited by H. R. Hardy, Jr. and F. W. Leighton, Trans Tech Publ., Clausthal-Zellerfeld, 467—477.

Lekhnitskii, S. G. 1977. *Theory of Elasticity of an Anisotropic Body.* Mir, Moscow.

Lemaitre, J. 1986. Local approach of fracture. *Engineering Fracture Mechanics* **25**, 5/a, 523—537.

Lemaitre, J. and Chaboche, J. L. 1985. *Mécanique de Materiaux Solides.* Dunod, Paris.

Levykin, A. I. and Vavakin, V. V. 1978. The study of elastic wave velocities and densities of rocks and minerals at pressures up to 20 kbars and temperatures up to 500°C. *Izv. Akad. Nauk SSSR, Ser. Fiz. Zemli* **5**, 42—51. In Russian.

Lindholm, U. S., Yeakley, L. M., and Nagy, A. 1974. The dynamic strength and fracture properties of Dresser basalt. *Int. J. Rock Mech. Min. Sci. &Geomech. Abstr.* **11**, 181—191.

Lindner, E. N. and Brady, B. H. G. 1984. Memory aspects of salt creep. In: *The Mechanical Behavior of Salt.* Proc. First Conf. Edited by H. R. Hardy, Jr. and M. Langer, Trans Tech Publ., Clausthal-Zellerfeld, 241—273.

Lindner, E. N. and Halpern, J. A. 1978. *In-situ* stress in North America. A compilation. *Int. J. Rock Mech. Min. Sci. &Geomech. Abstr.* **15**, 183—203.

Lippmann, H. 1984. Constitutive relations for rock-like materials, based on a shear crack model. *Mech. Res. Comm.* **11**, 3, 177—184.

Litwiniszyn, J. 1985. A model for the initiation of coal gas outburst. *Int. J. Rock Mech. Min. Sci. &Geomech. Abstr.* **22**, 1, 39—46.

Lo. T.-W., Coyner, K. B., and Toksöz, M. N. 1986. Experimental determination of elastic anisotropy of Berea sandstone, Chicopee shale, and Chelmsford granite. *Geophysics* **51**, 1, 164—171.

Lockner, D. and Byerlee, J. 1977. Acoustic emission and creep in rock at high confining pressure and differential stress. *Bull. Seismol. Soc. of America* **67**, 2, 247—258.

Lockner, D. and Byerlee, J. 1980. Development of fracture planes during creep in granite. In: *Second Conf. on Acoustic Emission/Microseismic Activity in Geologic Structures and Materials*, Edited by H. R. Hardy, Jr. and F. W. Leighton, Trans Tech Publ., Clausthal-Zellerfeld, 11—25.

Loureiro-Pinto, I. 1979. Determination of the elastic constants of anisotropic bodies by diametral compression tests. *Fourth Int. Congr. on Rock Mechanics*, Montreux, Balkema, Rotterdam, **II**, 359—363.

Love, A. E. H. 1944. *A Treatise on the Mathematical Theory of Elasticity.* Dover, New York.

Lux, K. H. and Rokahr, R. 1984. Laboratory investigations and theoretical statements as a basis for the design of caverns in rock salt formation. In: *The Mechanical Behavior of Salt.* Proc. First Conf. Edited by H. R. Hardy, Jr. and M. Langer, Trans Tech Publ., Clausthal-Zellerfeld, 275—310.

MacDonald, P. and Brown, K. M. 1984. A microseismic study of an outburstprone coal seam. In: *Acoustic Emission/Microseismic Activity in Geologic Structures and Materials*, Proc. of the Third Conf. Edited by H. R. Hardy, Jr. and F. W. Leighton, Trans Tech Publ., Clausthal-Zellerfeld, 479—497.

Maier, G. and Hueckel, T. 1979. Nonassociated and coupled flow rules of elastoplasticity for rock-like materials. *Int. J. Rock Mech. Min. Sci. &Geomech. Abstr.* **16**, 77—92.

Malvern, L. E. 1969. *Introduction to the Mechanics of a Continuous Medium.* Prentice-Hall, Englewood Cliffs, N.J.

Margolin, L. G. 1984. Microphysical models for inelastic material response. *Int. J. Engng. Sci.* **22**, 8—10, 1171—1179.

Massier, D. and Cristescu, N. 1981. *In situ* creep of rocks. *Rev. Roumaine Sci. Techn. Sér. Méc. Appl.* **26**, 5, 687—702.

Maxwell, D. E. 1984. Variable power-law creep model. In: *The Mechanical Behavior of Salt.* Proc. First Conf. Edited by H. R. Hardy, Jr. and M. Langer, Trans Tech Publ., Clausthal-Zellerfeld, 697—707.

McElroy, J. J., Koerner, R. M., and Lord, A. E. 1985. An acoustic jack to assess *in situ* rock behaviour. *Int. J. Rock Mech. Min. Sci. &Geomech. Abstr.* **22**, 1, 21—29.

Mellegard, K. D., Senseny, P. E., and Hansen, F. D. 1981. Quasi-static strength under creep characteristics of 100-mm-diameter specimens of salt from Avery Island, Louisiana. ONWI-250, Topical Report RSI-0140, RE/SPEC Inc. Rapid City, SD.

Michelis, P. N. 1981. Work-softening and hardening behaviour of granular rocks. *Rock Mechanics* **14**, 187—200.

Michell, J. H. 1899. On the determination of stress in an elastic solid with application to the theory of plates. *Proc. London Math. Soc.* **31**, 100.

Mier, J. G. M., van. 1986. Multiaxial strain-softening of concrete. Part. I: Fracture. *Matériaux et Constructions* **19**, No. 111, 179—200.

Mier, J. G. M., van, and Borst, R. de. 1986. 2-D representation of 3-D limit surfaces. *J. Engrg. Mech. Div. ASCE* **112**, 11, 1255—1258.

Mihăilescu-Suliciu, M. and Suliciu, I. 1979. Energy for hypoelastic constitutive equations. *Arch. Rational Mech. Anal.* **71**, 327—344.

Mihăilescu-Suliciu, M. and Suliciu, I. 1981. Hypoelasticity and rate-independent plasticity theory. INCREST Preprint Series in Mathematics Nr. 88/1981.

Mirza, U. A. 1984. Prediction of creep deformations in rock salt pillars. In: *The Mechanical Behavior of Salt.* Proc. First. Conf. Edited by H. R. Hardy, Jr. and M. Langer, Trans Tech Publ., Clausthal-Zellerfeld, 311—337.

Mises, R. von, 1913. Mechanik der festen Körper in plastisch deformable Zustand. *Nachr. Akad. Wiss. Göttingen Math.-Phys.* K1. H. 4, 582—592.

Mogi, K. 1968. *Bulletin of the Earthquake Research Institute, TDJKA* **46**, 1103—1125.

Mogi, K. 1971. Fracture and flow of rocks under high triaxial compression. *J. Geophys. Res.* **76**, 5, 1255—1269.

Mogi, K. 1977. Dilatancy of rocks under general triaxial stress states with special references to earthquake precursors. *J. Phys. Earth* **25**, Suppl. S., 203—217.

Mogi, K. 1979. Flow and fracture of rocks under general triaxial compression. *Fourth Int. Congr. on Rock Mechanics*, Montreux, Balkema, Rotterdam, **I**, 125—130.

Moore, I. D. and Booker, J. R. 1982. A circular boundary element for the analysis of deep underground openings. In: *Numerical Methods in Geomechanics*, Edmonton, 1982. Proc. Fourth Int. Conf. Edited by Z. Eisenstein, Balkema, Rotterdam, 53—60.

Morgenstern, N. R. and Tamuly Phukan, A. L. 1969. Non-linear stress-strain relation for a homogeneous sandstone. *Int. J. Rock Mech. Min. Sci.* **6**, 127—142.

Mróz, Z. and Angelillo, M. 1982. Rate-dependent degradation model for concrete and rock. In: *Numerical Models in Geomechanics*, Proc. Int. Symp. Edited by R. Dungar, G. N. Pande and J. A. Studer, Balkema, Rotterdam, 208—217.

Munson, D. E. and Dawson, P. R. 1984. Salt constitutive modeling using mechanism maps. In: *The Mechanical Behavior of Salt.* Proc. First Conf. Edited by H. R. Hardy, Jr. and M. Langer, Trans Tech Publ., Clausthal-Zellerfeld, 717—737.

Murakami, S. 1983. Notion of continuum damage mechanics and its application to anisotropic creep damage theory. *Trans ASME, J. Engrs. Materials and Technology* **105**, 99—105.

Murakami, S. 1987. Anisotropic damage theory and its application to creep crack growth analysis. In: *Constitutive Laws for Engineering Materials. Theory and Applications.* Proc. Int. Conf. Edited by C. S. Desai, R. H. Gallagher. Elsevier, Amsterdam, 187—194.

Muskhelishvili, N. I. 1953. *Some Basic Problems in the Mathematical Theory of Elasticity*, Noordhoff, Groningen.

Mühlhaus, H.-B. 1985. Lower bound solutions for circular tunnels in two and three dimensions. *Rock Mechanics and Rock Engineering* **18**, 37—52.

Nadai, A. 1950. *Theory of Flow and Fracture of Solids.* McGraw-Hill, New York. Vol. 1.

Natau, O. P., Frölich, B. O., and Mutschler, Th. O. 1983. Recent developments of the large-scale triaxial test. *5th Int. Congr. on Rock Mechanics*, Melbourne, Preprints Section A, A65—A74.

Nemat-Nasser, S. 1979. The second law of thermodynamics and non-colinear crack growth. *Proc. Third Congr. Mech. Div. Spec. Conf.*, Amer. Soc. Civil Eng., Austin, Texas, 441.

Nemat-Nasser, S. 1982. On dynamic and static behaviour of granular materials. In: *Soil Mechanics — Transient and Cyclic Loads.* Edited by G. N. Pande, O. C. Zienkiewicz, John Wiley, New York, 439—458.

Nemat-Nasser, S. and Horii, H. 1982. Compression-induced nonplanar crack extension with application to splitting, exfolation and rockburst. *J. Geophys. Res.* **87**, 6805—6825.

Nemat-Nasser, S. and Shokooh, A. 1980. On finite plastic flow of compressible materials with internal friction. *Int. J. Solids Structures* **16**, 495—514.

Nguyen, M. D. and Berest, P. 1979. Etude de la stabilité des cavités souterraines avec un modèle de comportement élastoplastique radoucisant. *Fourth Int. Congr. on Rock Mechanics*, Montreux, Balkema, Rotterdam, **I**, 249—256.

Nguyen, M. D., Berest, P., and Bergues, J. 1983. Analyse du comportement differs des ouvrages souterrains. *5th Int. Congr. on Rock Mechanics*, Melbourne, Preprints Section D, D233—D239.

Nguyen, M. D., Habib, P., and Guerpillon, Y. 1984. Time dependent behaviour of a pilot tunnel driven in hard marls. In: *Design and performance of underground excavations*. ISRM/BGS, Cambridge, 453—459.

Niwa, Y., Kobayashi, S., and Fukui, T. 1979. Stresses and displacements around an advancing face of a tunnel. *Fourth Int. Congr. on Rock Mechanics*, Montreux, Balkema, Rotterdam, **I**, 703—710.

Nova, R. 1980. The failure of transversely isotropic rocks in triaxial compression. *Int. J. Rock Mech. Min. Sci. &Geomech. Abstr.* **17**, 325—332.

Nur, A. 1972. Dilatancy, pore fluids, and premonitory variations of t_s/t_p travel times. *Bull. Seism. Soc. Amer.* **62**, 5, 1217—1222.

Nur, A. 1975. A note on the constitutive law for dilatancy. *Pure Appl. Geophys.* **113**, 197—206.

Nur, A., Bell, L., and Talwani, P. 1973. Fluid flow and faulting. 1: a detailed study of the dilatancy mechanism and premonitory velocity changes. *Proc. San Andreas Conf.*, Stanford, 391—404.

Nur, A. and Schultz, P. 1973. Fluid flow and faulting. 2: a stiffness model for seismicity. *Proc. San Andreas Conf.*, Stanford, 405—416.

Obert, L. 1972. Brittle fracture in rocks. In: *Fracture. An Advanced Treatise*. Edited by H. Liebowitz, vol. 1, part. I. Ch. 2, Academic Press. New York.

Obert, L. and Duvall, W. I. 1967. *Rock Mechanics and the Design of Structures in Rock*. John Wiley, New York.

Oka, F. and Adachi, T. 1985. An elastic-plastic constitutive equation of geologic materials with memory. In: *Numerical Methods in Geomechanics*, Nagoya, 1985. Proc. Fifth Int. Conf. Edited by T. Kawamoto, Y. Ichikawa, Balkema, Rotterdam-Boston, 293—300.

Okamoto, T. 1985. Application of deformation model with no yield surface to various over-consolidated geologic materials. In: *Numerical Methods in Geomechanics*, Nagoya, 1985, Proc. Fifth Int. Conf. Edited by T. Kawamoto, Y. Ichikawa, Balkema, Rotterdam, 301—308.

Ottosen, N. S. 1986. Viscoelastic-viscoplastic formulas for analysis of cavities in rock salt. *Int. J. Rock Mech. Min. Sci. &Geomech. Abstr.* **23**, 3, 201—212.

Ottosen, N. S. and Krenk, S. 1979. Nonlinear analysis of cavities in rock salt. *Int. J. Rock Mech. Min. Sci. &Geomech. Abstr.* **16**, 245—252.

Palaniswamy, K. and Knauss, W. G. 1978. On the problem of crack extension in brittle solids under general loading. In: *Mechanics Today*, Edited by S. Nemat-Nasser, vol. 4, p. 87, Pergamon Press, New York.

Pandit, B. I. and King, M. S. 1979. The variation of elastic wave velocities and quality factor Q of a sandstone with moisture content. *Canadian J. Earth Sciences.* **16**, 12, 2187—2193.

Panet, M. 1969. Quelques problèmes de mécanique des roches posés par le tunnel du Mont Blanc. *Bull. Liaison Labo. Routiers. P. et Ch.* no. 42, 115—145.

Panet, M. 1979*a*. Time-dependent deformations in underground works. *Fourth Int. Congr. on Rock Mechanics*, Montreux, Balkema, Rotterdam, **I**, 279—289.

Panet, M. 1979*b*. Les déformations différées dans les ouvrages souterrains. *Fourth Int. Congr. on Rock Mechanics*, Montreux, Balkema, Rotterdam, **I**, 291—301.

Passaris, E. K. S. 1979. The rheological behaviour of rocksalt as determined in an *in situ* pressurized test cavity. *Fourth Int. Congr. on Rock Mechanics*, Montreux, Balkema, Rotterdam, **I**, 257—264.

Paterson, L. 1986. A model for outburst in coal. *Int. J. Rock Mech. Min. Sci. & Geomech. Abstr.* **23**, 4, 327—332.

Paul, B. 1968. Macroscopical criteria of plastic flow and brittle fracture. In: *Fracture. An Advanced Treatise*. Edited by H. Liebowitz, vol. 2, Ch. 4, Academic Press, New York.

Perkins, R. D., Green, S. J., and Friedman, M. 1970. Uniaxial stress behaviour of porphyritic tonalite at strain rates up to 10^3/sec. *Int. J. Rock Mech. Min. Sci. & Geomech. Abstr.* **7**, 527—535.

Petukhov, I. M. and Liukov, A. M. 1983. Theoretical principles and fundamentals of rock burst prediction and control. *5th Int. Congr. on Rock Mechanics*, Melbourne, Preprints Section D, D113—D120.

Pipkin, A. C. and Rivlin, R. S. 1965. Mechanics of rate-independent materials. *Z. Angew, Math. Phys.* **16**, 3, 313—327.

Podio, A. L., Gregory, A. R., and Gray, K. E. 1968. Dynamic properties of dry and water-saturated Green River shale under stress. *Soc. Petrol Engrs. J.* **389**, 404.

Podio-Guidugli, P. and Suliciu, I. 1984. On rate-type viscoelasticity and second law of thermodynamics. *Int. J. Non-Linear Mech.* **19**, 545—564.

Popović, B., Marković, O., and Monojlović, M. 1979. Stresses and strains at the contact of rigid tunnel lining and soft rock. *Fourth Int. Congr. on Rock Mechanics*, Montreux, Balkema, Rotterdam, **I**, 525—531.

Pratt, H. R., Black, A. D., Brown, W. S., and Brace, W. F. 1972. The effect of specimen size on the mechanical properties of unjointed diorite. *Int. J. Rock Mech. Min. Sci. & Geomech. Abstr.* **9**, 513—529.

Preece, D. S. and Wawersik, W. R. 1984. Leached salt cavern design using a fracture criterion for rock salt. In: *Rock Mechanics in Productivity and Protection.* Proc. Twenty-Fifth Symp. on Rock Mechanics, Northwestern University, Evanston, Illinois, 25—27 June 1984, Edited by C. H. Dowding and M. M. Singh. Soc. Mining Engrs. of Amer. Inst. Mining Metal. Petroleum Engrs, New York, 556—565.

Proskuryakov, N. M., Permyakov, R. S., and Chernikov, A. K. 1973. *Physico-Mechanical Properties of Rock Salt.* Nedra, Moscow. In Russian.

Rezowalli, J. J., King, M. S., and Myer, L. R. 1984. Cross-hole acoustic surveying in basalt. *Int. J. Rock Mech. Min. Sci. & Geomech. Abstr.* **21**, 213—216.

Rice, J. R. 1975. On the stability of dilatant hardening for saturated rock masses. *J. Geophys. Res.* **80**, 11, 1531—1537.

Rindorf, H. J. 1984. Location of microseismic activity. In: *Acoustic Emission/Microseismic Activity in Geologic Structures and Materials*, Proc. of the Third Conf., Edited by H. R. Hardy, Jr. and F. W. Leighton, Trans Tech Publ., Clausthal-Zellerfeld, 695—706.

Rinehart, J. S. 1965. Dynamic fracture strength of rocks. *Proc. 7th Symp. on Rock Mechanics*, Pennsylvania State University.

Rodrigues, F. P. 1979. The anisotropy of the moduli of elasticity and of the ultimate stresses in rocks and rock masses. *Fourth Int. Congr. on Rock Mechanics*, Montreux, Balkema, Rotterdam, **II**, 517—523.

Rodriguez-Roa, F. 1985. Lining ground interaction in circular tunnels. In: *Numerical Methods in Geomechanics*, Nagoya, 1985, Proc. Fifth Int. Conf. Edited by T. Kawamoto, Y. Ichikawa, Balkema, Rotterdam, 1257—1264.

Rowe, R. K., Lo, K. Y., and Tham, L. G. 1982. The analysis of tunnels and shafts in dense (oil) sand. In: *Numerical Methods in Geomechanics*, Edmonton, 1982, Proc. Fourth Int. Conf. Edited by Z. Eisenstein, Balkema, Rotterdam, 587—596.

Rudnicki, J. W. 1982. Discussion of 'On finite plastic flows of compressible materials with internal friction', *Int. J. Solids Structures* **18**, 4, 357—360.

Rudnicki, J. W. 1984. A class of elastic-plastic constitutive laws for brittle rock. *J. Rheology* **28**, 6, 759—778.

Rudnicki, J. W. and Rice, J. R. 1975. Conditions for the localization of deformation in pressure-sensitive dilatant materials. *J. Mech. Phys. Solids* **23**, 371—394.

Rummel, F., 1974. Changes in the P-wave velocity with increasing inelastic deformation in rock specimens under compression. In: *Advances in Rock Mechanics*, Proc. Int. Congr. on Rock Mechanics, Washington D.C., 517—523.

Rummel, F., Alheid, H. J., and Frohn, C. 1978. Dilatancy and fracture induced velocity change in rock and their relation to frictional sliding. *Pageoph* **116**, 743—764.

Rummel, F. and Fairhurst, C. 1970. Determination of the post-failure behaviour of brittle rock using a servo-controlled testing machine. *Rock Mechanics* **2**, 189—204.

Rummel, F. and Frohn, C. 1982. Variation of ultrasonic velocity in granite and serpentinita during dilatant fracture under general triaxial compression. In: *High-Pressure Researches in*

Geosciences. Edited by W. Schreyer. E. Schweizerbart'sche Verlagsbuchhandlung, Stuttgart, 103—111.

Rusch, H. 1960. Research toward a general flexural theory for structural concrete. *Proc. Amer. Conc. Inst.* **57**, 1—28.

Russell, J. E. and Lomenick, T. H. 1984. Analysis of long-term creep tests on model pillars. In: *The Mechanical Behavior of Salt.* Proc. First Conf. Edited by H. R. Hardy, Jr. and M. Langer, Trans Tech Publ., Clausthal-Zellerfeld, 355—368.

Sakurai, S., Shimizu, N., and Matsumuro, K. 1985. Evaluation of plastic zone around underground openings by means of displacement measurements. In: *Numerical Methods in Geomechanics*, Nagoya, 1985, Proc. Fifth Int. Conf. Edited by T. Kawamoto, Y. Ichikawa, Balkema, Rotterdam, 111—118.

Sano, O. 1981. A note on the sources of acoustic emissions associated with subcritical crack growth. *Int. J. Rock Mech. Min. Sci. &Geomechs. Abstr.* **18**, 259—263.

Sano, O., Itô, I., and Terada, M. 1981. Influence of strain rate on dilatancy and strength of Oshima granite under uniaxial compression. *J. Geophys. Res.* **86**, No. B10, 9299—9311.

Sano, O., Terada, M., and Ehara, S. 1982. A study on the time-dependent microfracturing and strength of Oshima granite. *Tectonophysics* **84**, 343—362.

Sasajima, S. and Itô, H. 1980. Long-term creep experiment of rock with small deviator of stress under high confining pressure and temperature. *Tectonophysics* **68**, 183—198.

Savage, W. Z. and Swolfs, H. S. 1980. *In situ* geomechanics of crystalline and sedimentary rocks. Part I: The long-term deformation and time-temperature correspondence of viscoelastic rock — an alternative theoretical approach. *U.S. Geol. Surv. Open File Rept.* 80—708, 1—21.

Savarensky, E. 1975. *Seismic Waves.* Mir Publishers, Moscow.

Savruk, M. P. 1981. *Two-dimensional Problems of elasticity for Bodies with Cracks.* Naukova Dumka, Kiev. In Russian.

Saxena, P. C., Mokhashi, S. L., and Rame Gowda, B. M. 1979. Rock stress measurements at Nagjhari tunnels, Kalinadi Hydro-electric project, India. *Fourth Int. Congr. on Rock Mechanics*, Montreux, Balkema, Rotterdam, **II**, 589—594.

Schleicher, F. 1926. Der Spannungszustand an der Fliessgrenze. *Z. Angew. Math. Mech.* **6**.

Schmidtke, R. H. and Lajtai, E. Z. 1985. The long-term strength of Lac du Bonnet granite. *Int. J. Rock Mech. Min. Sci. &Geomech. Abstr.* **22**, 6, 461—465.

Schock, R. N. 1976. A constitutive relation describing dilatant behavior of Climax Stock granodiorite. *Int. J. Rock Mech. Min. Sci. &Geomech. Abstr.* **13**, 221—223.

Schock, R. N. 1977. The response of rocks to large stresses. In: *Impact and Explosion Cratering.* Edited by D. J. Roddy, R. O. Pepin, R. B. Merrill, Pergamon Press, New York, 657—668.

Schock, R. N. and Duba, A. G. 1972. Quasistatic deformation of solids with pressure. *J. Appl. Phys.* **43**, 5, 2204—2210.

Schock, R. N. and Heard, H. C. 1974. Static mechanical properties and shock loading response of granite. *J. Geophys. Res.* **79**, 11, 1662—1666.

Schock, R. N., Heard, H. C., and Stephens, D. R. 1973. Stress-strain behavior of a granodiorite and two graywackes on compression to 20 kilobars. *J. Geophys. Res.* **78**, 26, 5922—5941.

Schock, R. N. and Louis, H. 1982. Strain behavior of a granite and a graywacke sandstone in tension. *J. Geophys. Res.* **87**, B9, 7817—7823.

Scholz, C. H. 1968*a*. Experimental study of the fracturing process in brittle rock. *J. Geophys. Res.* **73**, 1447—1454.

Scholz, C. H. 1968*b*. Mechanism of creep in brittle rock. *J. Geophys. Res.* **73**, 3295—3302.

Scholz, C. H. and Kranz, R. L. 1974. Notes on dilatancy recovery. *J. Geophys. Res.* **79**, 2132—2135.

Schreiber, E., Anderson, O. L., and Soga, N. 1973. *Elastic Constants and Their Measurement*, McGraw-Hill, New York.

Sekiguchi, H., Nishida, Y., Matsumoto, T., and Uesawa, M. 1985. Characterization of a diatomaceous mudstone by elasto-viscoplasticity. In: *Numerical Methods in Geomechanics*, Nagoya, 1985, Proc. Fifth Int. Conf. Edited by T. Kawamoto, Y. Ichikawa, Balkema, Rotterdam, 437—444.

Senseny, P. E. 1984. Specimen and history effects on creep of salt. In: *The Mechanical Behavior of Salt*. Proc. First Conf. Edited by H. R. Hardy, Jr. and M. Langer, Trans Tech. Publ., Clausthal-Zellerfeld, 369—379.

Serafim, J. L. 1968. Influence of interstitial water on the behaviour of rock masses. In: *Rock Mechanics in Engineering Practice*. Edited by K. G. Stagg, O. C. Zienkiewicz, John Wiley, London, 55—97.

Serata, S., Sakurai, S., and Adachi, T. 1972. Theory of aggregate rock behavior based on absolute three-dimensional testing (ATT) of rock salt. In: *Basic and Applied Rock Mechanics*. Edited by K. E. Gray. The Amer. Inst. Mining, Metallurgical and Petroleum Engineers, Inc. New York, 431—473.

Sharma, K. G., Varadarajan, A., and Srivastava, R. K. 1985. Elasto-visco-plastic finite element analysis of tunnels. In: *Numerical Methods in Geomechanics*, Nagoya, 1985, Proc. Fifth Int. Conf. Edited by T. Kawamoto, Y. Ichikawa, Balkema , Rotterdam, 1141—1148.

Shimamoto, T. 1986. Strengthening of phyllosilicate and gypsum Ganges with increasing temperature: effect of temperature or moisture elimination? *Int. J. Rock Mech. Min. Sci. & Geomech. Abstr.* **23**, 6, 439—443.

Smith, M. B. and Cheatham, J. B., Jr. 1980. An anisotropic compacting yield condition applied to porous limestone. *Int. J. Rock Mech. Min. Sci. &Geomech. Abstr.* **17**, 159—165.

Soga, H., Mizutani, H., Spetzler, H., and Martin, R. J. III. 1978. The effect of dilatancy on velocity anisotropy in Westerly granite. *J. Geophys. Res.* **83**, No. B9, 4451—4458.

Sondergeld, C. H. and Estey, L. H. 1981. Acoustic emission study of micro-fracturing during the cyclic loading of Westerly granite. *J. Geophys. Res.* **86**, No. B4, 2915—2924.

Stagg, K. G. and Zienkiewicz, O. C. 1968. Pore pressure and stress distribution in porous materials. In: *Rock Mechanics in Engineering Practice*. Edited by K. G. Stagg, O. C. Zienkiewicz. John Wiley, London, 86—89.

Stamatiu, M. 1962. *Rock Mechanics*. Editura Didactică şi Pedagogică, Bucureşti. In Romanian.

Staupendahl, G. and Schmidt, M. W. 1984. Field investigation of the long-term deformational behavior of a 10,000 m^3 cavity at the ASSE salt mine. In: *The Mechanical Behavior of Salt*. Proc. First Conf. Edited by H. R. Hardy, Jr. and M. Langer, Trans Tech Publ., Clausthal-Zellerfeld, 511—525.

Stavrogin, A. N. and Protosenya, A. G. 1979. *Plasticity of Rocks*. Nedra, Moscow. In Russian.

Stavrogin, A. N. and Protosenya, A. G. 1985. *Strength of Rocks and Stability of Deep Underground Excavation*. Nedra, Moscow. In Russian.

Stephens, D. R., Lilley, E. M., and Louis, H. 1970. Pressure-volume equation of state of consolidated and fractured rocks to 40 kb. *Int. J. Rock Mech. Min. Sci.* **7**, 257—296.

Stevens, J. L. and Holcomb, D. J. 1980. A theoretical investigation of the sliding crack model of dilatancy. *J. Geophys. Res.* **85**, No. B12, 7091—7100.

Sugawara, K. and Obara, Y. 1986. Measurement of *in-situ* rock stress by hemispherical-ended borehole technique. *Mining Science and Technology* **3**, 287—300.

Sugawara, K., Obara, Y., Kaneko, K., and Aoki, T. 1986. Hemispherical-ended borehole technique for measurement of absolute rock stress. In: *Rock Stress and Rock Stress Measurements*. Proc. Int. Symp. Edited by O. Stephansson, CENTEK Publ. 207—216.

Suliciu, I. 1984. On the disappearance of shear waves and longitudinal waves in a solid body subject to high stresses and large strains. *Meccanica* **19**, 38—42.

Suliciu, I. 1988. Some remarks on the instantaneous response in rate-type viscoplasticity. *Int. J. Plasticity* **4** (in press).

Sun, J. and Lee, Y. S. 1985. A viscous elasto-plastic numerical analysis of the underground structure interacted with families of multilaminate rock mass using FEM. In: *Numerical Methods in Geomechanics*, Nagoya, 1985, Proc. Fifth Int. Conf. Edited by T. Kawamoto, Y. Ichikawa, Balkema, Rotterdam, 1127—1134.

Sundaram P. N. and Corrales, J. M. 1980. Brazilian tensile strength of rocks with different elastic properties in tension and compression. *Int. J. Rock Mech. Min. Sci. & Geomech. Abstr.* **17**, 131—133.

Swenson, C. A. 1960. *Physics at High Pressure*, Academic Press, New York.

Swoboda, G. 1985. Interpretation of field measurements under consideration of the three-dimensional state of stress, the viscoelasticity of shotcrete, and the viscoplastic behaviour of rock. In: *Numerical Methods in Geomechanics*, Nagoya, 1985, Proc. Fifth Int. Conf. Edited by T. Kawamoto, Y. Ichikawa, Balkema, Rotterdam, 1651—1658.

Tanimoto, K. and Nakamura, J. 1984. Use of AE Technique in field investigation of soil. In: *Acoustic Emission/Microseismic Activity in Geologic Structures and Materials*. Proc. of the Third Conf. Edited by H. R. Hardy, Jr. and F. W. Leighton, Trans Tech Publ., Clausthal-Zellerfeld, 601—612.

Tanimoto, K and Tanaka, Y. 1985. A method of determining yield locus of sandy soil. *Proc. Eleventh Int. Conf. Soil Mech. Fund. Eng.*, San Francisco, Balkema, Rotterdam, 1069—1072.

Tapponier, P and Brace, W. F. 1976. Development of stress-induced microcracks in Westerly granite. *Int. J. Rock Mech. Min. Sci. &Geomech. Abstr.* **13**, 103—112.

Taylor, L. N., Chen, E. P., and Kuszmaul, J. S. 1986. Microcrack-induced damage accumulation in brittle rock under dynamic loading. *Computer Methods in Applied Mechanics and Engineering* **55**, 301—320.

Telles, J. C. F. 1983. *The Boundary Element Method Applied to Inelastic Problems*. Springer-Verlag, Berlin.

Teodorescu, P. 1977. *Tuneluri*, Ed. Tehnică, Bucureşti. In Romanian.

Teodorescu, P. P. 1966. *Plane Problems in the Theory of Elasticity*. II. Editura Academiei R. S. Romania, Bucureşti. In Romanian.

Tijani, S. M., Vouille, G., and Hugot, B. 1983. Le sel gemme en tant que liquide visqueux. *5th Int. Congr. on Rock Mechanics*, Melbourne, Preprints Section A, A241—A246.

Timoshenko, S. and Goodier, J. M. 1961.*Théorie de l'Elasticité*. Libr. Polytechnique, Paris.

Truesdell, C. 1972. *A First Course In Rational Continuum Mechanics*. The Johns Hopkins University Press, Baltimore, Md.

Varadarajan, A., Sharma, K. G., and Singh, R. B. 1985. Analysis of underground openings by FEBEM. In: *Numerical Methods in Geomechanics*, Balkema, 1985. Proc. Fifth Int. Conf. Edited by T. Kawamoto, Y. Ichikawa, Balkema, Rotterdam, 1063—1069.

Vardoulakis, I. 1984. Rock bursting as a surface instability phenomenon. *Int. J. Rock Mech. Min. Sci. &Geomech. Abstr.* **21**, 3, 135—144.

Vardoulakis, I. and Detournay, E. 1982. Determination of the ground reaction curve in deep tunnels using Biot's hodograph method. In: *Numerical Methods in Geomechanics*, Edmonton, 1982. Proc. Fourth Int. Conf. Edited by Z. Eisenstein, Balkema, Rotterdam, 619—624.

Volarovich, M. P. 1978. Study of physical properties of rocks and minerals at high pressures and temperatures. *Izv. Akad. Nauk SSSR, Ser, Fiz, Zemli* **10**, 21—33. In Russian.

Volarovich, M. P. and Bayuk, E. I. 1960. The influence of hydrostatic pressure up to 400 kg/cm^2 on the elastic properties of rock specimens. *Dokl. Akad. Nauk SSSR* **135**, 1, 65—68.

Volarovich, M. P., Bayuk, E. I., Levykin, A. I., and Tomashevskaya, I. S. 1974. *Physico-mechanical Properties of Rocks and Minerals at High Pressures and Temperatures*, Nauka, Moscow. In Russian.

Volarovich, M. P., Stiller, H., Lebedev, T. S., Vollstädt, H., Pros, Z., Bayuk, E. I., and Savenko, B. Yu. 1977. *High Pressure and Temperature Studies of Physical Properties of Rocks and Minerals*. Naukova Dumka, Kiev. In Russian.

Vouille, G., Tijani, S. M., and Grenier, F. de, 1984. Experimental determination of the rheological behavior of Tersanne rock salt. In: *The Mechanical Behavior of Salt*. Proc. First Conf. Edited by H. R. Hardy, Jr. and M. Langer, Trans Tech Publ., Clausthal-Zellerfeld, 407—420.

Vutukuri, V. S., Lama, R. D., and Saluja, S. S. 1974. *Handbook on Mechanical Properties of Rocks*, vol. 1, Trans Tech Publ., Clausthal-Zellerfeld.

Wai, R. S. C., Lo, K. Y. 1982. Temperature effects on strength and deformation behaviour of rocks in Southern Ontario, Can. *Geotech. J.* **19**, 307—319.

Wallner, M. 1984. Analysis of thermomechanical problems related to the storage of heat producing radioactive waste in rock salt. In: *The Mechanical Behavior of Salt*. Proc. First Conf. Edited by H. R. Hardy, Jr. and M. Langer, Trans Tech Publ., Clausthal-Zellerfeld, 739—763.

Wallner, M. and Wulf, A. 1982. Thermomechanical calculations concerning the design of a radioactive waste repository in rock salt. *ISRM Symposium*, Aachen, Balkema, Rotterdam, 1003—1012.

Walrath, D. E. and Adams, D. F. 1983. The Iosipescu shear test as applied to composite materials. *Exper. Mech.* **23**, 1, 105—110.

Walsh, J. B. 1965. The effect of cracks on the compressibility of rock. *J. Geophys. Res.* **70**, 2, 381—389.

Walsh, J. B. 1973. Theoretical bounds for thermal expansion, specific heat, and strain energy due to internal stress. *J. Geophys. Res.* **78**, 7637—7646.

Walsh, J. B. 1980. Static deformation of rock. *J. Engng. Mech. Div., Proc. ASCE,* **106** (EMS) 1005—1019.

Walsh, J. B. and Brace, W. F. 1966. Elasticity of rock: A review of some recent theoretical studies. In: *Rock Mechanics and Engineering Geology*, vol. IV/4, Springer-Verlag, Wien, 283—297.

Wawersik, W. R. 1975. Technique and apparatus for strain measurements on rock in constant confining pressure experiments. *Rock Mechanics,* **7**, 231—241.

Wawersik, W. R. 1985. Determination of steady state creep rates and activation parameters for rock salt. In: *Measurement of Rock Properties at Elevated Pressures and Temperatures*, ASTM STP869, Edited by J. J. Pincus and E. R. Hoskins, American Society for Testing and Materials, Philadelphia, 72—92.

Wawersik, W. R. 1984. Alternatives to a power-law creep model for rock salt at temperatures below 160°C. *Proc. 2nd Conf. Mech. Behavior of Salt.* Edited by H. R. Hardy, Jr. and M. Langer. Hannover, Sept. 1984. Preprint.

Wawersik, W. R. and Hannum, D. W. 1980. Mechanical behavior of New Mexico rock salt in triaxial compression up to 200°C. *J. Geophys. Res.* **85**, No. B2, 891—900.

Wawersik, W. R., Herrmann, W., Montgomery, S. T., and Lauson, H. S. 1982. Excavation design in rock salt-laboratory experiments. Material modelling and validation. *Proc. Int. Symp. Rock Mechanics Related to Caverns and Pressure Shafts*, Aachen, Balkema, Rotterdam.

Wawersik, W. R. and Preece, D. S. 1984. Creep testing of salt — Procedures, problems and suggestions. In: *The Mechanical Behavior of Salt*, Proc. First Conf. Edited by H. R. Hardy, Jr. and M. Langer, Trans Tech Publ., Clausthal-Zellerfeld, 421—449.

Wawersik, W. R. and Stone, C. M. 1986. Experience with hydraulic fracturing tests for stress measurements in the WIPP. *Proc. 27th U.S. Rock Mech. Symp.*, Univ. of Alabama, 23—25 June, 1986.

Wawersik, W. R. and Zeuch, D. H. 1986. Modeling and mechanistic interpretation of creep of rock salt below 200°C. *Tectonophysics* **121**, 125—152.

Wijk, G. 1978. Some new theoretical aspects of indirect measurements of the tensile strength of rocks. *Int. J. Rock Mech. Min. Sci. &Geomech. Abstr.* **15**, 149—160.

Wittke, W. 1977. New design concept for underground openings in rock. In: *Finite Elements in Geomechanics.* Edited by G. Gudehus, John Wiley and Sons, New York, 413—478.

Wittke, W., Pierau, B. and Schetelig, K. 1978. Planning of a compressed-air pumped-storage scheme at Vianden/Luxembourg. *Proc. 1st Int. Symp. on Storage in Excavated Rock Caverns*, Stockholm, 1977, Edited by M. Bergman, Pergamon Press, New York, 367—376.

Wong, T.-F. 1982. Shear fracture energy in Westerly granite from post-failure behavior. *J. Geophys. Res.* **87**, No. B2, 990—1000.

Wong, T.-F., Brace, W. F. 1979. Thermal expansion of rocks: some measurements at high pressures. *Tectonophysics,* **57**, 95—117.

Woodruff, S. D. 1966. *Methods of Working Coal and Metal Mines*, vol. 2, Pergamon Press, Oxford.

Worotnicki, G. and Denham, D. 1976. The state of stress in the upper part of the Earth's crust in Australia according to measurements in tunnels and mines and from seismic observations. In: *Investigation of Stress in Rock. Advances in Stress Measurements*, Pergamon Press, Oxford, 71—82.

Wunderlich, W., Cramer, H., Kutter, H. K., and Rahn, W. 1981. *Finite-Element-Modelle für die Beschreibung des Materialverhaltens von Fels.* Institut für Konstruktiven Ingenieurbau, Ruhr-Universität Bochum, Mitteilung Nr. 81—10.

Xiong, W. 1985. Symmetric formulation of tangential stiffnesses for non-associated plasticity. In: *Numerical Methods in Geomechanics* Nagoya, 1985, Proc. Fifth Int. Conf. Edited by T. Kawamoto, Y. Ichikawa, Balkema, Rotterdam, 341—347.

Xu, W.-H., Chen, Q., and Tu, S. 1985. The stress analysis of tunnel surrounding rock by the method of weighted residuals. In: *Numerical Methods in Geomechanics*, Nagoya, 1985, Proc. Fifth Int. Conf. Edited by T. Kawamoto, Y. Ichikawa, Balkema, Rotterdam, 239—243.

Yamada, S. E., Schatz, J. F., Abou-Sayed, A., and Jones, A. H. 1981. Elasto plastic behaviour of porous rocks under undrained conditions. *Int. J. Rock Mech. Min. Sci. & Geomech. Abstr.* **18** 177—179.

Yanagidani, T., Ehara, S., Nishizawa, O., Kusunose, K., and Terada, M. 1985. Localization of dilatancy in Oshima granite under constant uniaxial stress. *J. Geophys. Res.* **90**, No. B8, 6840—6858.

Yang, M. and Toksöz, M. N. 1981. Time-dependent deformation and stress relaxation after strike slip Earthquakes. *J. Geophys. Res.* **86**, No. B4, 2889—2901.

Yuen, D. A., Sabadini, R., and Boschi, E. V. 1982. Viscosity of the lower mantle as inferred from rotational data. *J. Geophys. Res.* **87**, No. B13, 10745—10762.

Yufin, S. A., Postolskaya, O. K., Shavarchko, I. R., and Titkov, V. I. 1985. Some aspects of underground structure mechanism in the finite element method analysis. In: *Numerical Methods in Geomechanics*, Nagoya, 1985, Proc. Fifth Int. Conf. Edited by T. Kawamoto, Y. Ichikawa, Balkema, Rotterdam, 1093—1100.

Yuritzinn, T., Panet, M., and Guenot, A. 1982. Analysis of tunnels in strain-softening grounds. In: *Numerical Methods in Geomechanics*, Edmonton, 1982, Proc. Fourth Int. Conf. Edited by Z. Eisenstein, Balkema, Rotterdam, 635—644.

Zhang, M. J. and Brinson, H. F. 1986. Cumulative creep damage for polycarbonate and polysulfone. *Experimental Mechanics* **26**, 2, 155—162.

Zhu, K., Zhu, J., Pan, W., and Fan, Z. 1985. Elastic-plastic analysis of underground openings by BEM and FEM. In: *Numerical Methods in Geomechanics*, Nagoya, 1985, Proc. Fifth Int. Conf. Edited by T. Kawamoto, Y. Ichikawa, Balkema, Rotterdam, 1055—1061.

Zienkiewicz, O. C. 1977. *The Finite Element Method*. McGraw-Hill, New York.

Zienkiewicz, O. C. and Mróz, Z. 1984. Generalized plasticity formulation and applications to geomechanics. In: *Mechanics of Engineering Materials*. Edited by C. S. Desai and R. H. Gallagher, John Wiley and Sons, New York, 655—679.

Zimmerman, R. W. and King, M. S. 1985. Propagation of acoustic waves through cracked rock. *Proc. 26th U.S. Symp. on Rock Mechanics*, June 1985, Rapid City, S.D. Preprint.

Zimmerman, R. W., King, M. S., Monteiro, P. J. M. 1986. The elastic moduli of mortar as a porous-granular material. *Cement and Concrete Research* **16**, 239—245.

Zoback, M. D. and Byerlee, J. D. 1975. The effect of cyclic differential stress on dilatancy in Westerly granite under uniaxial and triaxial conditions. *J. Geophys. Res.* **80**, No. 11, 1526—1530.

Zubelewicz, A. and Mróz, Z. 1983. Numerical simulation of rock burst processes treated as problems of dynamic instability. *Rock Mech. and Rock Engineering* **16**, 253—274.

Author index

Subject index